IONIC TRANSPORT PROCESSES

Ionic Transport Processes

In Electrochemistry and Membrane Science

KYÖSTI KONTTURI and LASSE MURTOMÄKI
Aalto University, Finland

JOSÉ A. MANZANARES
University of Valencia, Spain

OXFORD
UNIVERSITY PRESS

OXFORD
UNIVERSITY PRESS

Great Clarendon Street, Oxford, OX2 6DP,
United Kingdom

Oxford University Press is a department of the University of Oxford.
It furthers the University's objective of excellence in research, scholarship,
and education by publishing worldwide. Oxford is a registered trade mark of
Oxford University Press in the UK and in certain other countries

First published 2008
First published in paperback 2015

Impression: 1

Published in the United States of America by Oxford University Press
198 Madison Avenue, New York, NY 10016, United States of America

British Library Cataloguing in Publication Data
Data available

ISBN 978–0–19–953381–7 (Hbk.)

ISBN 978–0–19–871999–1 (Pbk.)

Printed and bound in Great Britain by
Clays Ltd, St Ives plc

Preface

Transport phenomena constitute an integral part of electrode and membrane processes. Electrode reactions are heterogeneous and take place on the electrode surface, thus creating concentration differences in the electrode vicinity, and these differences cause mass transport. In membrane processes, analogous surface phenomena occur in many cases, but in addition, transport processes inside the membrane phase can also be decisive.

The characteristic feature of the information obtained from a transport phenomenon is that measurable quantities represent integral values over the entire surface under study. This means that in, e.g., electrode processes, only the response function of the chosen perturbation function that has been fed into the system can be measured. In some cases, however, it is possible to get complementary spectroscopic information of the surface. When modelling a heterogeneous process, transport phenomena are of great importance but they alone do not describe the entire process sufficiently. Various surface phenomena and reactions must be included in the model. And it has to be realized that the study of these surface phenomena, e.g. adsorption, is possible only after the solution of the inherent transport problem and after the subtraction of its effect on the entire process. Thus, a comprehensive model of a heterogeneous process is mathematically rather demanding.

This book has originated from the lecture notes of a course held since 1987 at Helsinki University of Technology, Laboratory of Physical Chemistry and Electrochemistry. The course is principally directed to post-graduates who already have an electrochemistry background or who are simultaneously attending an electrochemistry course. Therefore, the current presentation does not discuss electrochemical methods in depth, rather it concentrates on topics of transport processes that are usually not encountered in the electrochemical literature. Also, hydrodynamics is only briefly introduced because of the abundance of textbooks in this area.

In order to keep the contents within bounds, this text concentrates on passive transport processes in isothermal and incompressible liquid systems. Passive processes involve the evolution of the system towards the state of thermodynamic equilibrium, and biological active-transport processes are thus excluded. The liquid solutions are described as multicomponent continua, where the solutes are often electrolytes that dissociate into ions and homogeneous chemical reactions may take place in its interior. The transport equations are presented within the theoretical background of macroscopic thermodynamics. The paradigm adopted here is based on the work by L. Onsager in the early 1930s, which has achieved an undisputed status in the description of transport phenomena. This theoretical construction is known as the thermodynamics of irreversible processes and is based on the study of entropy production and the properties of phenomenological equations derived thereof.

Chapter 1 presents an introduction to the thermodynamics of irreversible processes. The fundamental thermodynamic concepts required to study the irreversible processes taking place in a moving fluid are described in Section 1.1. A local approach is followed in which the state variables are functions of time and the spatial co-ordinates. The local equilibrium hypothesis and its key role in irreversible thermodynamics are also explained. The transport processes involve the exchange of matter, electric charge, linear momentum, energy, entropy, etc., between the neighbouring volume elements in the fluid. Section 1.2 aims at establishing the balance equations that rule these exchanges. Among them, the entropy balance equation receives special attention because it is the starting point for the statement of the phenomenological transport equations in Chapter 2.

However, Chapter 1 is something more than just groundwork for Chapter 2. The concept of electric potential in thermodynamics and electrochemistry is controversial and, without entering into subtle details, we explain here the rationale under the treatment of this quantity in thermodynamics of irreversible processes. Moreover, a sound understanding of the differences between the transport mechanisms of convection, chemical diffusion, electrodiffusion, ionic diffusion, ionic migration, and electric conduction requires Chapters 1 and 2, particularly Sections 1.2.2, 2.1.4, 2.1.5, and 2.3.3.

Convection refers to those processes in which there is motion of the fluid mass with respect to an inertial laboratory reference frame. The maintenance of the fluid motion requires mechanical forces. In electroneutral solutions and in the absence of external forces such as gravitational and centrifugal forces, the only mechanical force that can induce the fluid motion is an applied pressure gradient. Thus, it is commonly accepted to talk of convection when there is a pressure gradient.

Chemical diffusion refers to the motion of a neutral component (e.g. a non-electrolyte) or an electroneutral combination of at least two charged particles (e.g. a dissociated electrolyte) driven by its concentration gradient. It must be stressed that whenever we talk about motion, we must specify the reference frame used to describe such motion. From a theoretical point of view, the preferred reference frame is the one bound to the local centre of mass of the solution, which is known as the barycentric reference frame, because only then chemical diffusion and convection are separate mechanisms. From an experimental point of view, however, diffusion measurements are carried out in the Fick's or volume-average reference frame. The relation between different reference frames is described in Sections 1.2.2 and 2.1.2. Yet, this is done rather superficially because this text concentrates on dilute solutions and this issue then becomes of secondary importance.

Electrodiffusion is the transport mechanism for charged species, such as the ions that result from the dissociation of electrolytes. The motion of an ionic species (in the barycentric reference frame) is driven by the gradient of its electrochemical potential (at least, within the Nernst–Planck approximation). When this gradient is considered as a single force, the transport should be described as electrodiffusion. However, it is customary to decompose the electrochemical potential as the sum of the chemical potential and a term proportional to the electric potential, and hence the gradient of the electrochemical potential of a charged species can be expressed as a term proportional to the concentration gradient and another one proportional to the electric field. When these gradients are considered as two different driving forces for transport, their associated mechanisms are denoted as ionic diffusion and ionic migration. That is, migration refers to the motion of charged species under the influence of an electric field regardless of whether there is passage of electric current through the solution.

Electric conduction is the transport mechanism for electric charge in ionic conducting solutions. Thus, we can only talk of conduction when a (conduction) electric current passes through the solution. Since ions carry the electric charge in solution, every ion contributes to this current in an amount that is proportional to the current. On the contrary, in the ionic migration mechanism every ion contributes in an amount that is proportional to the electric field. In the absence of current, there can still be migration but not conduction.

The difference between conduction and migration mentioned above is closely related to the coupling of driving forces, and hence of the associated fluxes. Even though we can think of ionic diffusion and ionic migration as additive transport mechanisms, they cannot be considered independent because there exists an electrical coupling between the ionic concentration gradient and fluxes. Thus, when we evaluate the contribution of the different transport mechanism to the entropy production rate (or to the dissipation function) it is found that electric conduction and chemical conduction make separate contributions. On the contrary, the contributions from ionic diffusion and ionic migration are not independent and must be grouped in a single electrodiffusion term. Similarly, the contributions of the chemical diffusion of different electroneutral electrolytes to the entropy production rate cannot be separated in additive terms because these processes are also electrically coupled. That is, the chemical diffusion of a neutral electrolyte is driven not only by its own concentration gradient but also by the concentration gradient of other electrolytes present in solution.

These transport processes, their description and the coupling phenomena are thoroughly studied in Chapter 2. It is established there that their rate is proportional to the extent of the deviation from the equilibrium, that is, to the gradients of electrochemical potential and of mechanical pressure. In Chapter 2 we show the most common theoretical approaches to describe transport processes: the phenomenological, the Fickian, the Stefan–Maxwell, and the Nernst–Planck approaches. Chapters 3 to 5 consider the description of transport processes in electrochemical and membrane systems making use of the Nernst–Planck formalism and, therefore, special attention is paid to it in Chapter 2. We describe the assumptions made in the derivation of the Nernst–Planck transport equations, and outline the main ideas of the alternative formulations that could be useful when the Nernst–Planck equations are no longer valid.

For those readers who are not so much interested in the foundations of the description of transport phenomena as in the practical solution of the transport equations in electrochemical and membrane systems, the core of the text is certainly formed by Chapters 3 to 5. These readers might skip Chapters 1 and 2, except for Section 2.3, in a first reading.

Chapter 3 is concerned with transport in the vicinity of electrodes, and hence on the coupling between Faradaic electrode processes and mass transport. This chapter covers, at an introductory level, transport in stationary and transient conditions, planar and spherical geometries, the presence and absence of supporting electrolytes, as well as convective transport in hydrodynamic electrodes. Some common electrochemical techniques are also discussed and the solutions of the corresponding transient transport problems are worked out in detail.

Chapter 4 describes transport processes in membrane systems. The emphasis is placed on stationary processes, although some examples of the solution of the transport equations in transient conditions are also worked out. This chapter covers homogeneous and porous membranes, both neutral and charged. Section 4.1 deals with transport through neutral porous membranes under applied concentration gradients, electric current, and pressure gradient. The use of mass balances to analyse the changes in the bathing solution concentration receives particular attention. Donnan equilibria and the description of the electrical double layer at the membrane/external solution interfaces are then presented in Section 4.2. Section 4.3 describes transport through homogeneous charged membranes. The solution of the transport equations in multi-ionic systems is worked out in detail and applied to the study of classical topics such as the bi-ionic potential and uphill transport. The influence of the diffusion boundary layers is also analysed. Finally, Section 4.4 describes transport through charged porous membranes. This is done from a very practical point of view and the space-charge model is only briefly referred to, although relevant references are given to the interested reader.

Chapter 5 describes transport in liquid membranes. This chapter aims at introducing the concepts of carrier-mediated transport and coupled transport. The topics of facilitated, competitive, co-transport and countertransport are covered, making use of examples of practical interest. Although we concentrate on relatively simple transport problems involving neutral solutes and heterogeneous complexation reactions, this chapter provides a good introduction to the solution of reaction–diffusion problems.

Our co-operation with the discipline of pharmaceutical technology has certainly pressed its footprint in the problem set-ups in Chapter 4, and Chapter 5 approaches the problems of liquid/liquid interfacial electrochemistry also carried out in our laboratories.

Each chapter contains a few exercises. Some of them are rather demanding. Solutions are available to lecturers at the web site of this book.

K. K., L. M., and J. A. M.

Acknowledgments

Any progress in science is hardly possible without previous generations and tutoring. KK and LM wish to acknowledge their teachers, lecturer Aarne Ekman, Professors Simo Liukkonen, Jussi Rastas and Göran Sundholm for valuable lessons in irreversible thermodynamics and electrochemistry. JAM acknowledges his teachers and colleagues, Professors Julio Pellicer and Salvador Mafé for twenty years of interesting discussions on membrane electrochemistry. All of them have certainly had an influence on the shape and contents of the present book.

We also would like to thank Professors Javier Garrido and Salvador Mafé for reading and commenting on the manuscript, and the students for their continued comments and questions.

KK also acknowledges the Academy of Finland for a sabbatical year, which facilitated finishing the book.

Finally, our thanks to the copyright holders, Elsevier, CRC Press, and VSSD (Vereniging voor Studie- en Studentenbelangen te Delft), for granting permission to reproduce material in some tables and illustrations.

Contents

Thermodynamics of irreversible processes

1

1.1 Fundamental concepts

1.1.1 Space and time scales of observation

The thermodynamics of irreversible processes relies on two fundamental hypotheses: the continuum hypothesis and the local equilibrium hypothesis. Their use amounts to an implicit setting up of the space and time scales of observation of the processes occurring within the physical system under study.

The system is described as a flowing continuum where state variables such as energy, pressure, electric potential and composition are functions of position $\vec{r}$ and time t. Every point $\vec{r}$ in space is occupied by a volume element dV (or fluid particle) that is small enough so that infinitesimal calculus can be applied to it but still macroscopic in the sense that it is constituted by a large number of molecules. In this (Eulerian) description of the fluid motion, we do not follow any volume element along its trajectory, but we rather look at the different volume elements (i.e. composed of different molecules) that occupy a given position $\vec{r}$ at different times. The fluid motion is described by the velocity field $\vec{v}(\vec{r}, t)$, where it should be stressed that $\vec{r}$ and t are independent variables; that is, the expression $d\vec{r}/dt$ is meaningless and $\vec{v} \neq d\vec{r}/dt$. Velocities are referred to the (inertial) laboratory reference frame unless otherwise stated.

The velocity $\vec{v}$ is the mass-average or barycentric velocity of the volume element that occupies position $\vec{r}$ at time t. This velocity has to be determined from the balance equation of the linear momentum (see Section 1.2.5) and it can be related to the velocities of the fluid components as follows. If we denote by $\vec{v}_i$ and dm_i the velocity and mass of component i in a volume element of total mass dm, the barycentric velocity is

$$\vec{v} \equiv \sum_i w_i \vec{v}_i, \tag{1.1}$$

where $w_i \equiv dm_i/dm$ is the local mass fraction of component i, and $\sum_i w_i = 1$.

Since the system under study is not in (thermodynamic) equilibrium, it might be argued that equilibrium thermodynamic relations should not be applied. However, the observed changes in composition, electric potential and/or electric current passing through the system take place over a time scale much larger than that of molecular motion. Thus, although the volume elements are not in (thermodynamic) equilibrium with each other, it can be assumed[1] that the molecular processes occurring within them are so fast that they guarantee the establishment of internal equilibrium (within the time scale of observation). The relations between the local values of the thermodynamic functions are then the same as in a state of complete equilibrium.

1.1.2 Local thermodynamic equations

Thermodynamics of irreversible processes is a field theory in which the system is described as a continuum and the thermodynamic functions and state variables are field quantities. The pressure p, the mass fractions w_i, and the (molar) chemical potentials μ_i vary with position, and therefore the usual thermodynamic equation for the Gibbs potential of a homogeneous system

$$G = \sum_i \mu_i n_i = U - TS + pV \tag{1.2}$$

cannot be used. Instead, the governing equations must be expressed in a local form that applies to every point within the continuum.

Imagine that we divide the system volume in elements $\mathrm{d}V$ so small that they can be considered as homogeneous subsystems. The Gibbs potential of a volume element is

$$\mathrm{d}G = \sum_i \mu_i \mathrm{d}n_i = \mathrm{d}U - T\mathrm{d}S + p\mathrm{d}V, \tag{1.3}$$

and dividing it by the volume $\mathrm{d}V$ we find the thermodynamic (Euler) equation that applies at the location of the volume element

$$g = \sum_i \mu_i c_i = u - Ts + p. \tag{1.4}$$

This is a fundamental equation that relates the local values of the state variables, T and p, and the volume densities of the thermodynamic functions: the internal energy density $u \equiv \mathrm{d}U/\mathrm{d}V$, the Gibbs potential density $g \equiv \mathrm{d}G/\mathrm{d}V$, and the entropy density $s \equiv \mathrm{d}S/\mathrm{d}V$. In eqns (1.2)–(1.4), n_i is the number of moles of component i and $c_i \equiv \mathrm{d}n_i/\mathrm{d}V$ is its molar concentration.

The local equilibrium hypothesis states that the change of the Gibbs potential follows the Gibbs equation[2]

$$\delta g = -s\delta T + \delta p + \sum_i \mu_i \delta c_i, \tag{1.5}$$

[1] This assumption imposes an upper bound on the size of the volume elements.

[2] The symbol δ denotes the change in a variable (when the system undergoes an infinitesimal process) and the symbol d is used for quantities referred to a volume element.

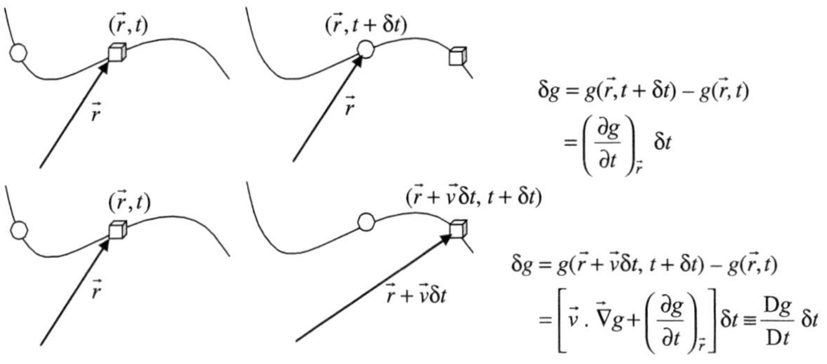

Fig. 1.1.
In the upper drawing, we compare the Gibbs potential density g in two different fluid particles that occupy the same position $\vec{r}$ at times t and $t+\delta t$. In the lower drawing, we compare the values of g in the same fluid particle at two different locations and times, $(\vec{r}+\vec{v}\delta t, t+\delta t)$ and $(\vec{r}, t)$, where $\vec{v}$ is the velocity of the fluid particle. In the first case, the time variation of g is given by its partial time derivative. In the second one, it is given by the substantial time derivative of g.

and the combination of eqns (1.4) and (1.5) yields the Gibbs–Duhem equation

$$\sum_i c_i \delta\mu_i = -s\delta T + \delta p \tag{1.6}$$

and the Gibbs equation

$$T\delta s = \delta u - \sum_i \mu_i \delta c_i. \tag{1.7}$$

Although many equations presented in this chapter are also valid for compressible fluids, the mechanical expansion work is often negligible in condensed phases, and therefore we restrict our attention to incompressible fluids.

The field quantities may vary from one volume element to another as well as with time. There are three types of variations that deserve comment (Fig. 1.1):

i) When we compare the values that a function takes in neighbouring volume elements (at fixed time), the symbol δ can be replaced by the gradient operator $\vec{\nabla}$. Thus, for instance, the Gibbs–Duhem equation can be written for isothermal systems as[3]

$$\sum_i c_i \vec{\nabla}\mu_{i,T} = \vec{\nabla} p. \tag{1.8}$$

ii) In other cases we are interested in comparing the values that a given function takes in the different volume elements that passed over a given location in space at different times. Then, the change of, e.g. s is given by $\delta s = (\partial s/\partial t)_{\vec{r}}\delta t$, and eqn (1.7) can be written as

$$T\frac{\partial s}{\partial t} = \frac{\partial u}{\partial t} - \sum_i \mu_i \frac{\partial c_i}{\partial t}. \tag{1.9}$$

[3] Since we restrict discussion to isothermal systems, the subscript T is often omitted for the sake of clarity.

iii) From a physical point of view, it is also very interesting to look at a given volume element and follow it along its trajectory. The change of, e.g. $g(\vec{r}, t)$ when this volume element undergoes a displacement $\delta\vec{R} \equiv \vec{v}\delta t$ in a time interval δt is then evaluated as $\delta g = (\mathrm{D}g/\mathrm{D}t)\delta t$, where

$$\frac{\mathrm{D}g}{\mathrm{D}t} \equiv \vec{v} \cdot \vec{\nabla} g + \left(\frac{\partial g}{\partial t}\right)_{\vec{r}} \tag{1.10}$$

is the material or substantial time derivative of $g(\vec{r}, t)$.

1.1.3 Electrolyte solutions

The equations in Section 1.1.2 involve sums over all the fluid components and we have made no reference to their charge state because the thermodynamic functions g and u introduced there do not contain any electrostatic energy contribution. The electrostatic energy is discussed in Section 1.2.8, but it seems convenient to introduce here some fundamental ideas about the thermodynamic description of electrolyte solutions.

According to eqn (1.5), the elementary change in the Gibbs potential at constant temperature and pressure due to a change in the local composition is

$$\delta g_{T,p} = \sum_i \mu_i \delta c_i. \tag{1.11}$$

If the change in the local composition affects several charged species in such a way that

$$\delta\rho_\mathrm{e} \equiv F \sum_i z_i \delta c_i = 0, \tag{1.12}$$

the electrostatic energy of the system is not affected; in eqn (1.12) z_i is the charge number of species i, F is the Faraday constant, and ρ_e is the electric charge density. Consider for example the solution of a strong binary electrolyte $\mathrm{A}_{\nu_1}\mathrm{C}_{\nu_2}$ that dissociates into ν_1 ions A^{z_1} and ν_2 ions C^{z_2}, where their charge numbers z_1 and z_2 satisfy the stoichiometric relation $z_1\nu_1 + z_2\nu_2 = 0$. A change δc_{12} in the local stoichiometric electrolyte concentration is equivalent to the changes $\delta c_1 = \nu_1 \delta c_{12}$ and $\delta c_2 = \nu_2 \delta c_{12}$ in the local ionic concentrations, and hence the associated change in the Gibbs potential (at constant temperature T, pressure p, and solvent concentration c_0) can be written as

$$\delta g_{T,p,c_0} = \sum_i \mu_i \delta c_i = \mu_1 \delta c_1 + \mu_2 \delta c_2 = \mu_{12} \delta c_{12}, \tag{1.13}$$

where

$$\mu_{12} \equiv \nu_1 \mu_1 + \nu_2 \mu_2 \tag{1.14}$$

is the chemical potential of the electrolyte. Similarly, in multi-ionic systems the local changes in composition involve changes in the Gibbs potential that can be written in terms of chemical potentials of neutral components or electroneutral combinations of charged components.

This last statement holds true even when the local change in composition does not satisfy eqn (1.12) because $\delta g_{T,p,c_0}$ does not account for electrostatic energy changes. For instance, in the case of a strong binary electrolyte $A_{\nu_1}C_{\nu_2}$, the Gibbs potential change is still described by eqn (1.13), $\delta g_{T,p,c_0} = \mu_{12}\delta c_{12}$, when $\delta\rho_e \neq 0$. Hence, the changes in the electrostatic energy of the system need to be described by additional terms. The energy required to bring an electrical charge $\delta\rho_e \mathrm{d}V$ from infinity (where we choose the origin of potential $\phi = 0$) to a volume element $\mathrm{d}V$ where the local electrical potential is ϕ can be evaluated as $\phi\delta\rho_e \mathrm{d}V$. Therefore, the sum of the chemical and electrical contributions to the change in the energy density is

$$\begin{aligned}\delta g_{T,p,c_0} + \phi\delta\rho_e &= \sum_i \mu_i \delta c_i + \phi F \sum_i z_i \delta c_i \\ &= \sum_i (\mu_i + z_i F\phi)\delta c_i = \sum_i \tilde{\mu}_i \delta c_i, \end{aligned} \tag{1.15}$$

where

$$\tilde{\mu}_i \equiv \mu_i + z_i F\phi \tag{1.16}$$

is the (molar) *electrochemical potential* of species i.

a) Locally electroneutral solutions

The energy required to charge a macroscopic system is very high, and therefore it seems reasonable to assume that the volume elements are electrically neutral. The local electroneutrality assumption states that the local electrical charge density ρ_e vanishes everywhere within the system

$$\rho_e \equiv F\sum_i z_i c_i \approx 0. \tag{1.17}$$

This assumption has important implications on the thermodynamic description of transport processes in solution.

We note first that in a locally electroneutral solution the ionic concentrations cannot be varied independently and, therefore, ions are constituents of the solution but not components in the sense of the Gibbs phase rule. The definition of the chemical potential of an ionic species i

$$\mu_i \equiv \left(\frac{\partial g}{\partial c_i}\right)_{T,p,\{c_{j\neq i}\}} \tag{1.18}$$

is not operational because the concentration c_i cannot be varied while keeping all the other concentrations constant and, at the same time, satisfying the local electroneutrality condition. On the contrary, the chemical potential of the electrolyte can be defined from eqn (1.13) as $\mu_{12} \equiv (\partial g/\partial c_{12})_{T,p,c_0}$. This means that the changes in μ_{12} (but not in μ_i) are experimentally measurable in this case.

Another important consequence of eqns (1.15) and (1.17) is that the local electric potential ϕ may be relevant when describing the transport of the charged components separately but not for the electroneutral volume element as a whole. Equations (1.4)–(1.9) involve sums over all species, and the local electroneutrality assumption implies that these equations remain valid if the chemical potential is replaced by the electrochemical potential inside the sums. For instance, the volume density of the Gibbs potential is

$$g = \sum_i \tilde{\mu}_i c_i, \quad \text{if } \rho_e = 0. \tag{1.19}$$

Similarly, the Gibbs–Duhem equation can also be written as

$$\sum_i c_i \vec{\nabla} \tilde{\mu}_i = \vec{\nabla} p, \quad \text{if } \rho_e = 0 \tag{1.20}$$

in the case of electroneutral solutions. However, the use of eqns (1.19) and (1.20) is not recommended because they are valid in locally electroneutral solutions only [1]; eqns (1.4) and (1.8) are preferred instead due to their general validity.

b) Locally charged solutions

The local electroneutrality condition is a reasonable assumption for the description of most transport processes but it is not a strict requirement. In fact, many electrochemical systems are not strictly electroneutral. Whenever the electric field varies with position (e.g. when a porous membrane separates two solutions with different concentrations of the same binary electrolyte or in an electrical double layer), there are deviations from local electroneutrality [2].

The small deviations from electroneutrality, although irrelevant when specifying the chemical composition of the solution, are crucial for the electrical contribution to the electrochemical potential of charged species. Hence, we should be cautious when extrapolating the conclusions derived for locally electroneutral solutions to real electrochemical systems. For instance, we have already mentioned that eqn (1.20) is not valid and the Gibbs–Duhem equation should be written either in terms of electrochemical potentials as

$$\sum_i c_i \vec{\nabla} \tilde{\mu}_i = \vec{\nabla} p - \rho_e \vec{\nabla} \phi, \tag{1.21}$$

or as shown in eqn (1.8). Similarly, we could inquire whether $\sum_i \tilde{\mu}_i c_i = g + \rho_e \phi$ represents the sum of the local Gibbs potential and the local electrostatic energy in the case of locally charged solutions. This latter enquiry is equivalent to finding out whether is it possible in a charged solution to define locally the electrochemical potential $\widetilde{\mu}_i$ without first introducing the electric potential. The answer to these questions can be found in Section 1.2.8 but now we provide some hints.

In relation to eqn (1.15), we said that the electrostatic energy required to bring an electrical charge $\delta\rho_e dV$ from infinity could be evaluated as $\phi\delta\rho_e dV$. In fact, this is only valid if the local electric potential is not affected by the modification of the electric charge density ρ_e. Moreover, the local term $\phi\delta\rho_e$ does not describe completely the changes in electrostatic energy density because, due to the long range of the electrostatic field, the addition of electric charge to a particular location in the fluid produces changes in the electrostatic energy throughout the space, even outside the volume of the fluid.

In the case of fluids at rest in thermodynamic equilibrium, it is possible to formulate a global definition of the electrochemical potential without first introducing the electric potential. Such a global definition is equivalent to the local definition in eqn (1.16) [3, 4]. This can be proved by considering a system that contains not only the multicomponent fluid under study (of finite volume) but also the electrostatic field (extending to infinity) and all the electrical charges. In non-equilibrium systems, however, it does not seem possible to define the electrochemical potential without first introducing the electric potential [4–10].

The electric potential is determined from the Poisson equation of electrostatics, which relates this potential to the electric charge density. This equation is not used explicitly in Section 1.2 because the system under consideration contains only the fluid and its thermodynamic functions do not incorporate electrostatic contributions. However, Poisson's equation is needed for the solution of transport problems in locally charged solutions.

1.2 Balance equations

1.2.1 Introduction

The description of transport processes is based on two fundamental principles. First, some physical quantities like the total mass, the electric charge, the total energy, and the total linear momentum[4] must satisfy principles of conservation. Second, the evolution of the system towards equilibrium must satisfy the second law of thermodynamics. This implies that the entropy of the system is not conserved and its rate of variation follows a balance equation that contains a positive term describing the entropy production due to the (irreversible) transport processes. In Sections 1.2.2–1.2.6, we explain the physical meaning of a balance equation and derive the most relevant ones, particularly the entropy-balance equation, which is the starting point for the phenomenological transport equations. We outline first some important ideas.

Every volume element of the fluid has an amount dB of an arbitrary extensive quantity B. We aim at describing the (time) change in dB due to the interaction between the volume element and its surroundings as well as the processes taking place inside it. The first contribution can be evaluated in terms of its flux density,

[4] The term total here means that the energy and the linear momentum of both the electrostatic field and the fluid are involved in the formulation of the conservation laws.

that is, in terms of the rate at which the quantity B crosses the boundary of the volume element. The second contribution is associated with the production rate of B due to the processes that occur inside the volume element. The analysis of these changes in dB leads to the balance equation

$$\frac{\partial b}{\partial t} + \vec{\nabla} \cdot \vec{j}_b = \pi_b, \tag{1.22}$$

where b is the local volume density of B, $\vec{j}_b$ is the flux density of B, and π_b is the local density of the production rate of B; for the sake of simplicity, however, we often refer to π_b as the production rate. For conservative quantities, such as the total mass and the electric charge, π_b is zero. On the contrary, non-conservative quantities such as internal energy and entropy can be produced or consumed in the transport processes.

The local equilibrium hypothesis (for an incompressible fluid)

$$T\frac{\partial s}{\partial t} = \frac{\partial u}{\partial t} - \sum_i \widetilde{\mu}_i \frac{\partial c_i}{\partial t} \tag{1.23}$$

is used as the starting point to derive the entropy balance equation. Thus, our first aim is to derive the balance equations for the amount of component i and for the internal energy. According to the second law of thermodynamics, the entropy production rate π_s must be positive and therefore the dissipation function $\theta \equiv T\pi_s$ must be positive-definite. In fact, this is the fundamental characteristics of irreversible processes. We aim below to show that the entropy-balance equation takes the form

$$\frac{\partial s}{\partial t} + \vec{\nabla} \cdot \vec{j}_s = \frac{1}{T}(\theta_{\text{ch}} + \theta_{\text{ed}} + \theta_\eta) \tag{1.24}$$

where the three terms in the right-hand side are the contributions of chemical reactions, electrodiffusion, and viscous flow to the dissipation function, respectively.

The derivation of eqn (1.24) and the expressions for the different contributions to the dissipation function is rather tedious and it seems convenient to explain here their importance in order to stimulate the lecture of the following sections. In equilibrium thermodynamics the expression of one thermodynamic potential in terms of its natural variables is known as the fundamental relation of the system. This relation contains all the thermodynamic information of the system and, therefore, we aim at deriving it. Statistical methods prove to be very useful in this task. In non-equilibrium thermodynamics we could say that the fundamental relation is the relation of the local entropy production with the thermodynamic driving forces and the flux densities. Balance equations allow us to derive such a relation and, once we know it, the transport equations can be formulated 'rigorously'.

The study of the balance equations provide us with much more than the expression for the dissipation function θ. First, the transport equations are not sufficient to analyse the transport processes and require to be complemented with the continuity equations, which are mass-balance equations. Second, since we have to choose a reference frame for the flux densities in the balance equations, the difference between diffusive and convective transport mechanisms becomes very clear. Convection requires a non-zero barycentric fluid velocity $\vec{v}$, while diffusion is associated with the exchange of quantity B in a reference frame bound to the moving fluid. In mathematical terms, the total flux density $\vec{j}_b$ is

$$\vec{j}_b = \vec{j}_b^m + b\vec{v}, \tag{1.25}$$

where the flux density relative to the fluid, $\vec{j}_b^m$, accounts for the diffusive contribution and the term $b\vec{v}$ describes the convective one. Third, the relations between the flux densities in different reference frames are also worked out. Fourth, and more important, a balance equation for the linear momentum is so different from a balance equation for entropy that their study provides us with a sound understanding of the difference between the transport equations for diffusion or electrodiffusion processes and the mechanical equation for macroscopic flow.

1.2.2 General form of the balance equations

In thermodynamics of irreversible processes, the governing equations are expressed in a local form that applies to every point within the continuum. The derivation of these equations, however, requires consideration of a finite system and we show below two ways of carrying out such a derivation. In the first case, we consider a finite system enclosed by a (real or imaginary) surface S that is fixed with respect to the laboratory reference frame. This is an open system in the sense that it can exchange matter with its surroundings. In the second one, we analyse a system enclosed by a boundary surface $S(t)$ that moves with the fluid so that the system is always composed by the same matter (Fig. 1.2).

The balance equation for an arbitrary extensive quantity B (such as mass, linear momentum, energy or entropy) can be derived by considering an open system of volume V enclosed by a fixed surface S. This volume V is divided into elements whose volume dV and location are time independent. The amount

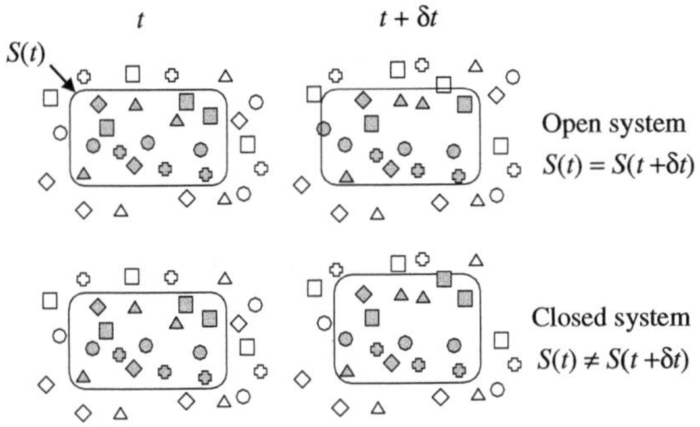

Fig. 1.2.
In the upper drawing, the surface is fixed in space and does not enclose the same fluid particles at different times. In the lower drawing, the surface moves through the space so that it always encloses the same fluid particles, i.e. the velocity of a surface element is equal to that of the particle at the position of this element.

of quantity B in a volume element is dB and the total amount in the system is

$$B = \iiint_V \mathrm{d}B = \iiint_V b\mathrm{d}V, \tag{1.26}$$

where $b(\vec{r}, t) \equiv \mathrm{d}B/\mathrm{d}V$ is the local volume density of B.

The time variation of B is

$$\frac{\delta B}{\delta t} = \frac{\delta}{\delta t}\iiint_V b\mathrm{d}V = \iiint_V \frac{\partial b}{\partial t}\mathrm{d}V, \tag{1.27}$$

and this must be equal to sum of the rates at which the quantity B crosses the surface S or is produced due to the processes taking place within V. The mathematical formulation of this requirement is the balance equation

$$\frac{\delta B}{\delta t} = -\oiint_S \vec{j}_b \cdot \mathrm{d}\vec{S} + \iiint_V \pi_b \mathrm{d}V. \tag{1.28}$$

In order to evaluate the amount of B that enters V through the surface S, we divide the latter into surface elements and label them by vectors $\mathrm{d}\vec{S}$ whose magnitude is equal to the area of the surface element and whose direction is normal to the surface, $\mathrm{d}\vec{S} = \mathrm{d}S\ \hat{n}$, where $\hat{n}$ is the outward normal unit vector (Fig. 1.3). The amount of quantity B that enters V through $\mathrm{d}\vec{S}$ in a time δt is $-\vec{j}_b \cdot \mathrm{d}\vec{S}\,\delta t$, where $\vec{j}_b$ is the vector flux density of B at the surface element.[5] Thus, the net influx rate can be obtained by integration over the surface S as

$$-\oiint_S \vec{j}_b \cdot \mathrm{d}\vec{S} = -\iiint_V \vec{\nabla}\cdot\vec{j}_b\,\mathrm{d}V \tag{1.29}$$

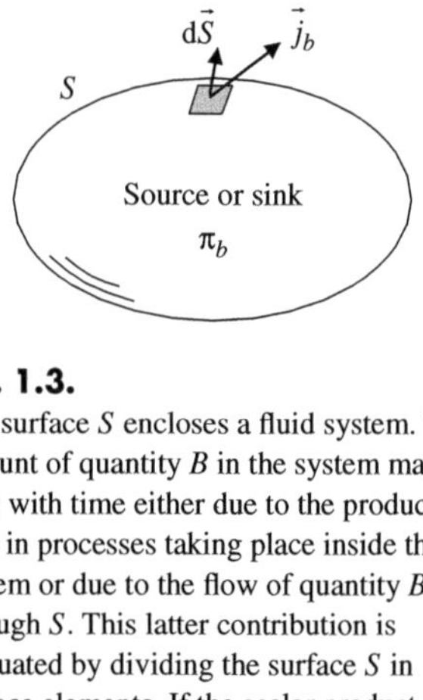

Fig. 1.3.
The surface S encloses a fluid system. The amount of quantity B in the system may vary with time either due to the production of B in processes taking place inside the system or due to the flow of quantity B through S. This latter contribution is evaluated by dividing the surface S in surface elements. If the scalar product of the flux density of $B, \vec{j}_b$, and the unit vector normal to S (and directed outwards) is positive at a given location on the surface, there is an outflow of B.

where the Gauss–Ostrogradski divergence theorem has been used. This is the first term in the right-hand side of eqn (1.28). Similarly, the contribution of the sources and sinks of B is given by the second term in the right-hand side of eqn (1.28), where $\pi_b(\vec{r}, t)$ is the local density of the production rate of B. This quantity can be zero, negative, or positive. It is zero if B is a conservative quantity. It is negative at those points where B is consumed by the local processes and positive where B is generated.

Combining eqns (1.27)–(1.29), the balance equation can be stated in terms of a single volume integral over V. Since this equation must be valid for any arbitrary volume V, it is concluded that the integrand must vanish everywhere within V, and therefore

$$\frac{\partial b}{\partial t} + \vec{\nabla}\cdot\vec{j}_b = \pi_b, \tag{1.30}$$

which is the local form of the balance equation of quantity B.

[5] The derivation of the general balance equation considers vectorial fluxes only although the description of chemical reactions and viscous flow also involves scalar and tensorial 'fluxes', respectively.

Alternatively, this local balance equation can be obtained by considering a volume $V(t)$ enclosed by a boundary surface $S(t)$ that moves with the fluid; i.e. the volume $V(t)$ that does not exchange matter with its surroundings. The total amount of B in this volume is

$$B = \iiint\limits_{V(t)} \mathrm{d}B = \iiint\limits_{V(t)} b\mathrm{d}V. \tag{1.31}$$

The displacement of every surface element $\mathrm{d}\vec{S}$ in a time interval δt must then be equal to the displacement $\delta\vec{R} = \vec{v}\delta t$ of the fluid particle that occupies the position of the surface element. A straightforward consequence of the motion of the surface elements is that the time variation of the system volume is given by

$$\frac{\delta V}{\delta t} = \frac{\delta}{\delta t}\iiint\limits_{V(t)} \mathrm{d}V = \oiint\limits_{S(t)} \vec{v}\cdot\mathrm{d}\vec{S} = \iiint\limits_{V(t)} \vec{\nabla}\cdot\vec{v}\,\mathrm{d}V. \tag{1.32}$$

Similarly, the time variation of the total amount B in the system is

$$\frac{\delta B}{\delta t} = \frac{\delta}{\delta t}\iiint\limits_{V(t)} b\mathrm{d}V = \iiint\limits_{V(t)} \frac{\partial b}{\partial t}\mathrm{d}V + \oiint\limits_{S(t)} b\vec{v}\cdot\mathrm{d}\vec{S}, \tag{1.33}$$

where the Reynolds transport theorem [11, 12] has been employed. Using the Gauss–Ostrogradski divergence theorem, this can be further transformed to

$$\frac{\delta B}{\delta t} = \iiint\limits_{V(t)} \left[\frac{\partial b}{\partial t} + \vec{\nabla}\cdot(b\vec{v})\right]\mathrm{d}V. \tag{1.34}$$

The flux density of B across a surface element that moves with the fluid is denoted as $\vec{j}_b^m$. Then, the amount of quantity B that enters $V(t)$ through an element $\mathrm{d}\vec{S}$ of $S(t)$ in a time δt is $-\vec{j}_b^m\cdot\mathrm{d}\vec{S}\,\delta t$. In close similarity to eqn (1.28), the balance equation is obtained as

$$\frac{\partial b}{\partial t} + \vec{\nabla}\cdot(\vec{j}_b^m + b\vec{v}) = \pi_b, \tag{1.35}$$

or, in terms of the substantial derivative, as

$$\frac{\mathrm{D}b}{\mathrm{D}t} + b\vec{\nabla}\cdot\vec{v} + \vec{\nabla}\cdot\vec{j}_b^m = \pi_b. \tag{1.36}$$

In incompressible fluids (see Section 1.2.3) this simplifies to

$$\frac{\mathrm{D}b}{\mathrm{D}t} + \vec{\nabla}\cdot\vec{j}_b^m = \pi_b. \tag{1.37}$$

The equivalence between eqns (1.30) and (1.35) requires that

$$\vec{j}_b = \vec{j}_b^m + b\vec{v}. \tag{1.38}$$

Somehow this equation is trivial, because the flux density of B across a surface element that moves with the fluid, $\vec{j}_b^m$, must differ from the flux density $\vec{j}_b$ across a fixed surface element at the same location in an amount proportional to the fluid velocity $\vec{v}$. Yet, eqn (1.38) is of primary importance because it tells us that in a flowing fluid there are two transport mechanisms that contribute to the flux of B. The so-called 'diffusive' term $\vec{j}_b^m$ is the flux density in a reference frame that moves (with respect to the laboratory) with the local barycentric velocity, which is known as the barycentric reference frame. The term $b\vec{v}$ describes the convective contribution. Thus, convection is associated with a non-zero barycentric velocity, while 'diffusion' is associated with the exchange of quantity B in a reference frame bound to the moving fluid.

Interestingly, eqn (1.38) reduces to $\vec{j}_m = \rho\vec{v}$ when applied to mass because there is no fluid motion across $S(t)$ and hence the mass flux density relative to the fluid is zero, $\vec{j}_m^m = 0$. This is equivalent to stating that, by definition, the only transport mechanism for (total) mass is convection. Note also that the term 'diffusion' is used above also to describe processes other than (component) mass diffusion. For instance, we show in Section 1.2.4 that eqn (1.38) takes the form $\vec{I} = \vec{I}^m + \rho_e\vec{v}$ when applied to electric charge. The 'diffusive' contribution $\vec{I}^m$ may then describe ohmic electric conduction.

Sometimes, reference frames other than the barycentric and the laboratory ones are convenient (Table 1.1). In a reference frame that moves (with respect to the laboratory) with the volume-average velocity

$$\vec{v}_\upsilon \equiv \sum_i c_i \upsilon_i \vec{v}_i, \tag{1.39}$$

where υ_i is the partial molar volume of component i, the flux density of B is

$$\vec{j}_b^\upsilon = \vec{j}_b - b\vec{v}_\upsilon. \tag{1.40}$$

This is known as the volume-average or Fick's reference frame and, by definition, the volume flux density in this reference frame is zero, $\vec{j}_\upsilon^\upsilon = 0$. Note also that the volume flux density in the laboratory reference frame is equal to the volume-average velocity, $\vec{j}_\upsilon = \vec{v}_\upsilon$, because $b = 1$ for the case of volume.

Table 1.1. Flux density of B in different reference frames.

Reference frame	Frame velocity with respect to the laboratory	Flux density of B
Laboratory	$\vec{0}$	$\vec{j}_b$
Barycentric (mass-average)	$\vec{v} \equiv \sum_i w_i \vec{v}_i$	$\vec{j}_b^m = \vec{j}_b - b\vec{v}$
Fick's (volume-average)	$\vec{v}_\upsilon \equiv \sum_i c_i \upsilon_i \vec{v}_i$	$\vec{j}_b^\upsilon = \vec{j}_b - b\vec{v}_\upsilon$
Hittorf's (solvent-fixed)	$\vec{v}_0$	$\vec{j}_b^{\mathrm{H}} = \vec{j}_b - b\vec{v}_0$

The similarity between eqns (1.38) and (1.40) is apparent and, in principle, $\vec{j}_b^{\,v}$ and $b\vec{v}_v$ could also be denoted as the diffusive and convective contributions to the flux density of B. However, there is a fundamental difference between the reference velocities $\vec{v}$ and $\vec{v}_v$: only the former is determined by the conservation law of linear momentum, and this makes the barycentric velocity preferred from a theoretical point of view.

Likewise, the flux density of B in the solvent-fixed or Hittorf's reference frame is

$$\vec{j}_b^{\,\mathrm{H}} = \vec{j}_b - b\vec{v}_0, \tag{1.41}$$

where $\vec{v}_0$ is the solvent velocity (with respect to the laboratory). Thus, the molar flux density of solvent in the solvent-fixed reference frame is zero, $j_0^{\mathrm{H}} = 0$. Hittorf's reference frame has a wide practical interest because some measurements (e.g. of transport numbers) are carried out with respect to water.

1.2.3 Total and component mass-balance equations

When the quantity B is the total mass of the system, b is the mass density ρ, and $\vec{j}_m = \rho\vec{v}$ is the mass flux density in the laboratory reference frame. Since the mass-production rate is zero because the total mass is conserved, eqn (1.30) reduces to

$$\frac{\partial \rho}{\partial t} + \vec{\nabla} \cdot (\rho \vec{v}) = 0. \tag{1.42}$$

This equation constitutes the principle of conservation of mass and is valid even in the presence of chemical reactions.[6]

When the fluid is incompressible, ρ is a constant, and the mass conservation is described by the simple equation

$$\vec{\nabla} \cdot \vec{v} = 0. \tag{1.43}$$

It is then clear from eqn (1.32) that an incompressible system that does not exchange matter with its surroundings conserves its volume.

The mass-balance equation can also be applied to the system components. The mass density of component i is $\rho_i = \rho w_i = M_i c_i$, where M_i is its molar mass, and its balance equation is

$$\frac{\partial (\rho w_i)}{\partial t} + \vec{\nabla} \cdot \vec{j}_{w_i} = \pi_{w_i}, \tag{1.44}$$

where the production rate π_{w_i} is non-zero if component i is involved in homogeneous chemical reactions. Heterogeneous reactions do not contribute to this production rate and they are only relevant for the statement of the boundary conditions of the transport equations.

[6] Nuclear reactions do not conserve mass and this principle of conservation must then be extended to include the energy released in the nuclear reaction.

In a system with N components, there are N mass-balance equations like eqn (1.44). Equation (1.42) can be considered as the result of summing them all because $\sum_i w_i = 1$, $\sum_i \pi_{w_i} = 0$, and $\sum_i \vec{j}_{w_i} = \rho\vec{v}$. This latter equation is another form of the definition of the barycentric velocity, eqn (1.1), because the mass flux density of component i is $\vec{j}_{w_i} = \rho w_i \vec{v}_i$.

Equation (1.44) can also be written in terms of the molar concentration $c_i = \rho w_i / M_i$ as

$$\frac{\partial c_i}{\partial t} + \vec{\nabla} \cdot \vec{j}_i = \pi_i, \tag{1.45}$$

where $\pi_i = \pi_{w_i}/M_i$ and $\vec{j}_i = \vec{j}_{w_i}/M_i = c_i\vec{v}_i$ is the molar flux density of component i. Equation (1.45) is known as the *continuity equation for component* i. In the absence of chemical reactions, the amount of component i is conserved and the balance equation takes the form of a conservation law

$$\frac{\partial c_i}{\partial t} + \vec{\nabla} \cdot \vec{j}_i = 0. \tag{1.46}$$

Furthermore, under steady-state conditions this equation reduces to

$$\vec{\nabla} \cdot \vec{j}_i = 0, \tag{1.47}$$

which states that the amount of component i entering the system through some boundaries is equal to the amount that exits through other system boundaries. Thus, for instance, in one-dimensional systems the steady-state molar flux density j_i is independent of position (in the absence of chemical reactions).

When component i is involved in homogeneous chemical reactions the change in its concentration is evaluated as a sum of contributions from the different reactions. If $\nu_{i,r}$ denotes the stoichiometric coefficient of component i in reaction r and ξ_r is the local reaction co-ordinate of this reaction, then the change in the molar concentration of component i due to reaction r is $\delta c_{i,r} = \nu_{i,r}\delta\xi_r$. The coefficient $\nu_{i,r}$ is positive for products and negative for reactants. Thus, the total production rate of component i is

$$\pi_i = \sum_r \nu_{i,r} \frac{\partial \xi_r}{\partial t}, \tag{1.48}$$

where $\partial\xi_r/\partial t$ is the rate of reaction r. Since every reaction conserves the total mass, the relations $\sum_i \nu_{i,r} M_i = 0$ and $\sum_i \pi_{w_i} = \sum_i \pi_i M_i = 0$ are satisfied.

1.2.4 Electric charge-balance equation

The homogeneous chemical reactions also conserve the electric charge, and hence it is satisfied that $\sum_i \nu_{i,r} z_i = 0$ for every reaction r and $\sum_i \pi_i z_i = 0$. The combination of this equation and eqn (1.45), leads to the mathematical formulation of the conservation of the electric charge

$$\frac{\partial \rho_e}{\partial t} + \vec{\nabla} \cdot \vec{I} = 0, \tag{1.49}$$

where

$$\rho_e \equiv F \sum_i z_i c_i \tag{1.50}$$

is the electric charge density and

$$\vec{I} \equiv F \sum_i z_i \vec{j}_i \tag{1.51}$$

is the conduction electric current density.

Alternatively, eqn (1.49) can be derived from Maxwell's equations. Ampere's law states that the rotational of the magnetic field $\vec{H}$ is equal to the sum of the conduction current density $\vec{I}$ and the displacement current density $\vec{I}_d \equiv \partial\vec{D}/\partial t$, where $\vec{D}$ is the electric displacement,

$$\vec{\nabla} \times \vec{H} = \vec{I}_d + \vec{I}. \tag{1.52}$$

Then, the total current density $\vec{I}_T \equiv \vec{I}_d + \vec{I}$ has zero divergence

$$\vec{\nabla} \cdot \vec{I}_T = \vec{\nabla} \cdot \vec{I}_d + \vec{\nabla} \cdot \vec{I} = 0, \tag{1.53}$$

and the time derivative of Poisson's equation $\vec{\nabla} \cdot \vec{D} = \rho_e$ leads to eqn (1.49). In other words, eqns (1.49) and (1.53) are two forms of the same conservation equation for the electric charge. Equation (1.53) implies that the net flux of total current across the system boundaries is zero. In particular, in a one-dimensional system I_T is independent of position and is equal to the current density exchanged with the surroundings, that is, the current that enters the system through one boundary and exits through the other.

In electroneutral solutions the electric charge density ρ_e and the displacement current $\vec{I}_d$ vanish so that conduction is the only transport mechanism for the electric charge. Moreover, the electric current density is then independent of the reference frame. On the contrary, in charged solutions the conduction current density depends on the reference frame because a convective transport mechanism is possible. Thus, for instance, the conduction current density can be written as $\vec{I} = \vec{I}^m + \rho_e\vec{v}$ where $\vec{I}_m$ and $\rho_e\vec{v}$ are the 'diffusive' and convective contributions, respectively. Note that these two contributions are due to ionic motions, while the displacement current density is not. The current density $\vec{I}_m$ is the conduction current density in the barycentric reference frame. Similarly, the conduction current densities in the Hittorf, $\vec{I}^H$, and in the volume-average, $\vec{I}^v$, reference frames satisfy the relation

$$\vec{I} = \vec{I}^H + \rho_e\vec{v}_0 = \vec{I}^v + \rho_e\vec{v}_v. \tag{1.54}$$

In summary, in electroneutral solutions the electric current density $\vec{I} \equiv F\sum_i z_i\vec{j}_i$ is due to the motion of ionic species, is independent of the reference frame, and has zero divergence. In charged solutions, however, there can be different contributions to the current density and Table 1.2 summarizes them.

Table 1.2. Contributions to the electric current density in charged (i.e. non-electroneutral) solutions.

Current density		Comment
'Diffusive'	$\vec{I}^m \equiv F \sum_i z_i \vec{j}_i^m$	Due to the motion of ionic species in a barycentric reference frame.
Convective	$\rho_e \vec{v}$	Due to the fluid motion. In an electroneutral fluid, this contribution vanishes and the conduction current is independent of the reference frame.
'Conduction'	$\vec{I} \equiv F \sum_i z_i \vec{j}_i = \vec{I}^m + \rho_e \vec{v}$	Sum of the 'diffusive' and convective contributions. Due to the motion of ionic species in the laboratory reference frame.
Displacement	$\vec{I}_d \equiv \frac{\partial \vec{D}}{\partial t}$	Significant in very fast transients (usually, on the scale of ns).
Total	$\vec{I}_T \equiv \vec{I} + \vec{I}_d$	This is the current that crosses the system boundaries and satisfies $\vec{\nabla} \cdot \vec{I}_T = 0$.

1.2.5 Linear momentum-balance equation

In eqns (1.30) and (1.37), the balance equation was formulated for a scalar quantity b but the linear momentum of the moving fluid is a vectorial quantity. In particular, the volume density of linear momentum is the product of the mass density and the barycentric velocity, $\rho\vec{v}$. The vectorial character makes the mathematical statement of its balance equation difficult because a vector balance equation involves a flux density $\overleftrightarrow{j}_v^m$ that is a second-order tensor. To avoid this difficulty, we can apply eqn (1.37) to the three components of the linear momentum separately and end up with three balance equations that involve three vectors $\vec{j}_{v_j}^m$ $(j = x, y, z)$ describing the flux density of these components. In this section we derive the vector balance equation for the simple case of a non-viscous fluid as well as the scalar balance equation of a viscous fluid that moves in the x direction. The linear momentum-balance equation for a viscous fluid moving arbitrarily is worked out in Section 1.2.7.

a) Linear momentum-balance equation for a non-viscous fluid

Newton's second law states that the rate of change of the linear momentum of a system is equal to the net force acting on it. The production rate of the linear momentum is then equal to the density of the net external force. The net force on a fluid volume element is the (vector) sum of surface forces that act on its surfaces and volume forces that act on every point inside it. We consider here that the only volume force acting on the fluid is the electrical one, and its density is $\rho_e\vec{E}$, where ρ_e is the electric charge density and $\vec{E}$ is the electric field. The surface forces can be pressure forces normal to the surfaces and shear forces parallel to them. In non-viscous fluids only the pressure forces are present and our next task is to describe their contribution to the transport of linear momentum.

Consider a parallelepipedic volume element with dimensions Δx, Δy, and Δz in a Cartesian co-ordinate system. This volume element has six faces and there are six surface forces normal to them. The forces on the x direction are exerted on the faces normal to the x axis. We identify these faces by their position co-ordinates x and $x+\Delta x$. The net force on them is $p(x)\Delta y\Delta z - p(x+\Delta x)\Delta y\Delta z$ and the force density can be approximated by $-\partial p/\partial x$. Similarly, the other components of the force density are $-\partial p/\partial y$ and $-\partial p/\partial z$, and the net force density can be written in vector form as $-\vec{\nabla}p$. The pressure gradient $\vec{\nabla}p$ describes the transport of linear momentum across the boundaries of the volume element. Therefore, we conclude that the linear momentum balance equation is

$$\rho\frac{\mathrm{D}\vec{v}}{\mathrm{D}t} = -\vec{\nabla}p + \rho_{\mathrm{e}}\vec{E}, \tag{1.55}$$

which is the *Navier–Stokes equation for a non-viscous fluid.* If the fluid is electroneutral, $\rho_{\mathrm{e}} = 0$ and this equation becomes the conservation law of linear momentum.

b) Linear momentum-balance equation for a viscous fluid that moves along direction *x*

When the fluid moves in the x direction only, the continuity equation, $\vec{\nabla}\cdot\vec{v} = 0$, enforces the velocity v_x to be independent of position co-ordinate x, although it can vary in other directions. For instance, in rectangular channel flow v_x may depend on the distance to the channel walls, while in a cylindrical channel flow it may depend on the radial position co-ordinate. The variation of velocity v_x with position in the direction normal to the channel walls is evidence of the fact that linear momentum is being transferred in this direction due to the fluid viscosity. Then, we conclude that the barycentric velocity $\vec{v} = (v_x, 0, 0)$ and the flux density of the linear momentum $\vec{j}^{\,m}_{v_x}$ are vector quantities that do no have the same direction.

In the absence of external forces, linear momentum is conserved ($\pi_{v_x} = 0$) and the balance equation, eqn (1.35) for $b = \rho v_x$, reduces to

$$\rho\frac{\partial v_x}{\partial t} + \vec{\nabla}\cdot\vec{j}^{\,m}_{v_x} = 0, \tag{1.56}$$

where we have used that $(\vec{v}\cdot\vec{\nabla})\,v_x = 0$. Since the mechanical equilibrium is established in a much shorter time than the distribution (or diffusional) equilibrium, stationary flow is often assumed. The term $\rho\partial v_x/\partial t$ then vanishes, and the balance equation reduces to $\vec{\nabla}\cdot\vec{j}^{\,m}_{v_x} = 0$, which is applied next to some typical flows.

c) Stationary Couette flow in planar geometry

Consider an incompressible fluid that occupies the space between two horizontal, parallel plates such that the upper one is moving in the positive x direction and the lower one is fixed. Due to the fluid viscosity, the fluid moves in the positive x direction and the velocity v_x varies linearly in the direction normal to the plates from 0 at the position of the lower plate to its maximum value (the plate velocity) at the position of the upper plate. The lower plate acts as a sink and the upper one as a source of linear momentum. If y denotes the position co-ordinate in the direction normal to the plates, the balance equation $\vec{\nabla}\cdot\vec{j}^{m}_{v_x} = 0$ reduces to

$$dj^{m}_{v_x,y}/dy = 0, \tag{1.57}$$

where $j^{m}_{v_x,y}$ is the component y of the linear momentum flux density $\vec{j}^{m}_{v_x}$. The physical meaning of eqn (1.57) is that momentum is transferred through the fluid without losses. Note that $\vec{j}^{m}_{v_x}$ is the flux density in the barycentric reference frame, i.e. bound to the fluid motion, and therefore $j^{m}_{v_x,x} = 0$.[7]

The net surface force along the x direction is due to the viscous friction between elements, which acts on its surfaces normal to the y direction (Fig. 1.4). The force on every surface can be evaluated as the product of the contact area and the viscous stress component xy on this surface. This stress component σ'_{xy} is given by Newton's law of viscosity

$$\sigma'_{xy} = \eta\frac{dv_x}{dy}, \tag{1.58}$$

where η is the dynamic viscosity of the fluid. Since every volume element must experience a zero net force in order to move with a stationary velocity v_x, it is required that $\sigma'_{xy}(y) = \sigma'_{xy}(y + \Delta y)$ or, equivalently, that $d\sigma'_{xy}/dy = 0$. This turns out to be the balance equation for the component x of the linear momentum, eqn (1.57); note that $\vec{j}^{m}_{v_x}$ have the dimensions of a surface stress or pressure. The velocity profile is then linear, as shown in Fig. 1.5, and the vector $\vec{j}^{m}_{v_x}$ is independent of position and directed from the moving to the fixed plate.

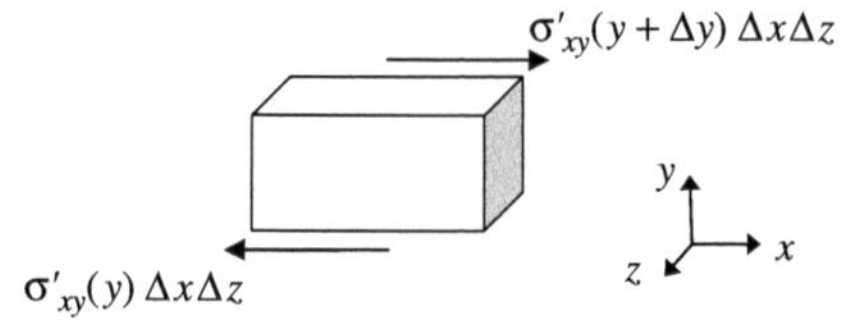

Fig. 1.4.
Shear forces acting on the surfaces normal to the y direction of a Cartesian volume element in the stationary Couette flow between parallel plates.

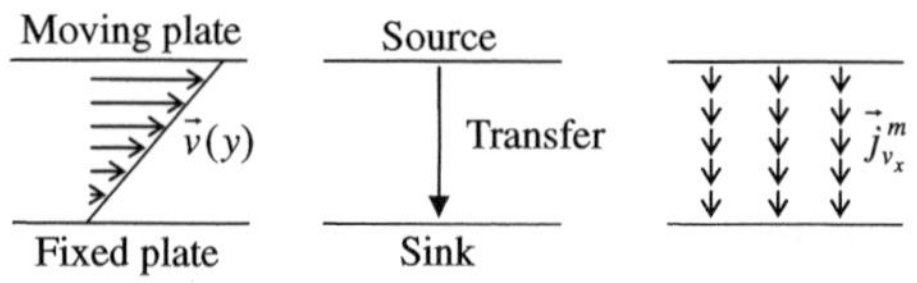

Fig. 1.5.
Velocity profile, schematic drawing of the flow of linear momentum (from the source to the sink), and vector field for the linear momentum flux density in the stationary Couette flow between parallel plates.

[7] In the laboratory reference frame the component x of the linear momentum flux density $j_{v_x,x}$ is non-zero but it is independent of position x.

d) Stationary Poiseuille flow in planar geometry

In the Poiseuille flow in planar geometry, an incompressible fluid occupies the space between two horizontal, parallel plates such that both of them are fixed and a pressure gradient is applied in the x direction, $\mathrm{d}p/\mathrm{d}x < 0$. The fluid then moves in the positive x direction and, due to the fluid viscosity, the velocity v_x varies in the y direction normal to the plates from 0 (in contact with them) to its maximum value at the channel centre. Both plates act here as sinks of linear momentum, while the external pump that applies the pressure gradient is the source of momentum. Linear momentum is transferred without losses through the fluid.

Once again we can use the idea that for any volume element to move with a stationary velocity v_x, the net force on it must be zero. Along direction x, there are four forces. They act on the four surfaces (of the fluid element) that are normal to the directions x and y (Fig. 1.6). Newton's law of viscosity, eqn (1.58), can be used to evaluate the stress on the surfaces normal to the y direction and the net force density due to their imbalance is $\mathrm{d}\sigma'_{xy}/\mathrm{d}y$. This force density is compensated by another one due to the pressure gradient acting on the surfaces of the volume element normal to the x direction, $-\mathrm{d}p/\mathrm{d}x$, and therefore the force balance requires that

$$-\frac{\mathrm{d}p}{\mathrm{d}x} + \frac{\mathrm{d}\sigma'_{xy}}{\mathrm{d}y} = -\frac{\mathrm{d}p}{\mathrm{d}x} + \eta\frac{\mathrm{d}^2 v_x}{\mathrm{d}y^2} = 0. \tag{1.59}$$

This statement of mechanical equilibrium is known as the *Stokes equation* and constitutes the local balance equation for the component x of the linear momentum, eqn (1.56), because in this flow the components of the flux density of linear momentum are $j^m_{v_x,x} = p$ and $j^m_{v_x,y} = -\sigma'_{xy} = -\eta \mathrm{d}v_x/\mathrm{d}y$. Equation (1.59) implies that the velocity profile is parabolic, as shown in Fig. 1.7, and the components x and y of the vector $\vec{j}^m_{v_x}$ vary linearly with position co-ordinates x and y, respectively.

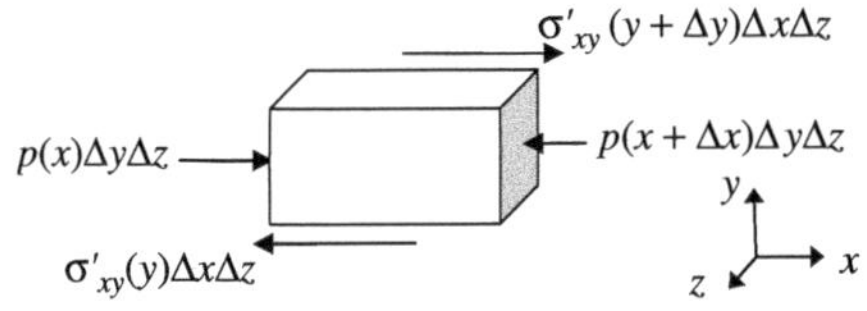

Fig. 1.6.
Pressure and shear forces acting on the surfaces normal to the y direction of a Cartesian volume element in the stationary Poiseuille flow between parallel plates.

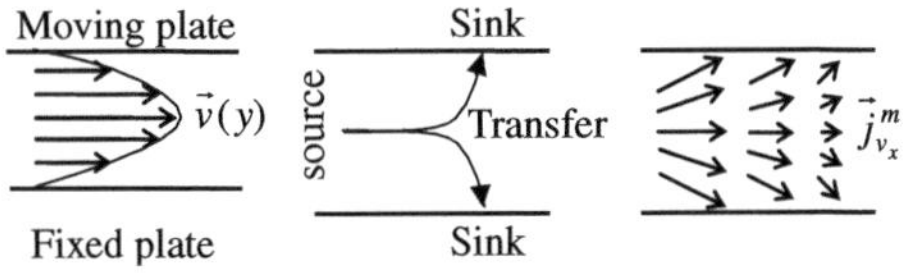

Fig. 1.7.
Velocity profile, schematic drawing of the flow of linear momentum (from the source to the sink), and vector field for the linear momentum flux density in the stationary Poiseuille flow between parallel plates.

1.2.6 Energy- and entropy-balance equations in a non-viscous fluid

Scalar multiplication of the linear momentum balance equation for a non-viscous fluid, eqn (1.55), and the barycentric velocity, leads to the balance equation for the translational kinetic energy density $e_\mathrm{k} \equiv \rho v^2/2$,

$$\frac{\partial e_\mathrm{k}}{\partial t} + \vec{\nabla} \cdot [(p + e_\mathrm{k})\vec{v}] = \frac{\mathrm{D}e_\mathrm{k}}{\mathrm{D}t} + \vec{\nabla} \cdot (p\vec{v}) = \rho_\mathrm{e}\vec{E} \cdot \vec{v}, \tag{1.60}$$

where the mass continuity equation for incompressible fluids, $\vec{\nabla} \cdot \vec{v} = 0$, has been used. The source term $\rho_\mathrm{e}\vec{E} \cdot \vec{v}$ is the rate of conversion of electric energy to kinetic energy of the fluid. This conversion can only take place in charged solutions ($\rho_\mathrm{e} \neq 0$).

The comparison of eqn (1.60) with the general form of the balance equation in eqn (1.30) evidences that the flux density of kinetic energy in a non-viscous fluid is

$$\vec{j}_{e_\mathrm{k}} = \vec{j}^{\,m}_{e_\mathrm{k}} + e_\mathrm{k}\vec{v} = (p + e_\mathrm{k})\vec{v}. \tag{1.61}$$

The 'diffusive' term $\vec{j}^{\,m}_{e_\mathrm{k}} = p\vec{v}$ is then associated with the mechanical power of pressure forces.

The total energy density of the fluid is the sum of the internal and the translational kinetic energy densities, $e = e_\mathrm{k} + u$. Other contributions such as the rotational kinetic energy are neglected. The electric field is considered to be external to the system and the internal energy u does not contain electric contributions. The total energy is not a conservative quantity because of the interaction of the fluid with the electric field. The production rate of total energy π_e is equal to the net power of the electrical forces[8] (i.e. the sum of the products of force density $Fz_ic_i\vec{E}$ acting on every charged species i and its velocity $\vec{v}_i$)

$$\pi_e = F\sum_i z_ic_i\vec{v}_i \cdot \vec{E} = \vec{I} \cdot \vec{E}. \tag{1.62}$$

The balance equation for the total energy of the fluid is then

$$\frac{\partial e}{\partial t} + \vec{\nabla} \cdot \vec{j}_e = \vec{I} \cdot \vec{E}. \tag{1.63}$$

As was stated in Section 1.2.1, the main outcome from Chapter 1 is the balance equation for the entropy. Irreversible transport processes are characterized by a positive entropy production $\pi_s \geq 0$ and we aim to evaluate it or, equivalently, the dissipation function $\theta \equiv T\pi_s$ because it is used as the starting point

[8] In the barycentric reference frame this power can be evaluated as $\sum_i Fz_ic_i\vec{E} \cdot \vec{v}_i^{\,m} = \vec{I}^m \cdot \vec{E}$, but it must then be taken into account that charged fluids can also increase their kinetic energy due to the power $\rho_\mathrm{e}\vec{E} \cdot \vec{v}$. That is, not only the ions but also the solvent obtains energy from the field due to the ion–solvent interactions. As expected, these two contributions add up to $\pi_e = \vec{I} \cdot \vec{E}$.

for Chapter 2. The entropy production is determined from the entropy-balance equation

$$\frac{\partial s}{\partial t} + \vec{\nabla} \cdot \vec{j}_s = \pi_s = \frac{\theta}{T}, \tag{1.64}$$

where the flux density of entropy is

$$\begin{aligned}\vec{j}_s = \vec{j}_s^m + s\vec{v} &= \frac{1}{T}\left(\vec{j}_u^m - \sum_i \mu_i \vec{j}_i^m\right) + \frac{1}{T}\left(u + p - \sum_i \mu_i c_i\right)\vec{v} \\ &= \frac{1}{T}\left(\vec{j}_u + p\vec{v} - \sum_i \mu_i \vec{j}_i\right).\end{aligned} \tag{1.65}$$

Subtracting eqn (1.60) from eqn (1.63), the balance of internal energy is obtained as

$$\frac{\partial u}{\partial t} + \vec{\nabla} \cdot \vec{j}_u = \vec{I}^m \cdot \vec{E}, \tag{1.66}$$

where $\vec{j}_u = \vec{j}_u^m + u\vec{v}$ and $\vec{j}_e = \vec{j}_{e_k} + \vec{j}_u$. The source term $\vec{I}^m \cdot \vec{E}$ represents the rate of conversion of electric energy to internal energy.

In the absence of chemical reactions, eqn (1.46) leads to

$$\begin{aligned}-\sum_i \mu_i \frac{\partial c_i}{\partial t} &= \sum_i \mu_i \vec{\nabla} \cdot \vec{j}_i \\ &= \vec{\nabla} \cdot \left(\sum_i \mu_i \vec{j}_i - p\vec{v}\right) - \sum_i \vec{j}_i^m \cdot \vec{\nabla}\mu_i,\end{aligned} \tag{1.67}$$

where the Gibbs–Duhem equation for isothermal fluids, $\sum_i c_i \vec{\nabla}\mu_i = \vec{\nabla}p$, and the continuity equation, $\vec{\nabla} \cdot \vec{v} = 0$, have been used. Combining eqns (1.9) and (1.64)–(1.67), it is concluded that the contribution of electrodiffusion to the dissipation function is

$$\theta_{\mathrm{ed}} = -\sum_i \vec{j}_i^m \cdot \vec{\nabla}\mu_i + \vec{I}^m \cdot \vec{E} = -\sum_i \vec{j}_i^m \cdot \vec{\nabla}\tilde{\mu}_i \geq 0, \tag{1.68}$$

which takes the form of a sum of products of the flux densities $\vec{j}_i^m$ and the negative gradients $-\vec{\nabla}\tilde{\mu}_i$. The latter are considered to be the driving forces for electrodiffusion.

The derivation of eqn (1.68) has been based on the consideration of ionic species. Alternatively, the contribution of electrodiffusion to the dissipation function can be written as

$$\theta_{\mathrm{ed}} = -\sum_K \vec{J}_K^m \cdot \vec{\nabla}\mu_K + \vec{I}^m \cdot \vec{E}_{\mathrm{ohm}} \geq 0, \tag{1.69}$$

where the sum extends over neutral components (i.e. dissociated electrolytes, neutral solutes, and solvent), $\vec{E}_{\text{ohm}} \equiv \vec{I}^m/\kappa$ is the ohmic electric field, and κ is the local electrical conductivity. The term $\theta_{\text{ohm}} = \vec{I}^m \cdot \vec{E}_{\text{ohm}} = (I^m)^2/\kappa \geq 0$ can be identified as the Joule power and accounts for the contribution of the irreversible electric conduction process to the dissipation function. The other term in eqn (1.69), $\theta_{\text{dif}} = -\sum_K \vec{J}_K^m \cdot \vec{\nabla}\mu_K$, is also positive-definite and describes the contribution of the diffusion of neutral components, also known as chemical diffusion, to the dissipation function. Note that we have used uppercase symbols for both the flux densities and the index that runs over the neutral components to make it clear the difference with the flux densities and the index that runs over ionic species.

Introducing the diffusional or internal electric field as $\vec{E}_{\text{dif}} \equiv \vec{E} - \vec{E}_{\text{ohm}}$ and comparing eqns (1.68) and (1.69), the contribution of chemical diffusion to the dissipation function can also be written as

$$\theta_{\text{dif}} = -\sum_i \vec{j}_i^m \cdot \vec{\nabla}\mu_i + \vec{I}^m \cdot \vec{E}_{\text{dif}}. \tag{1.70}$$

The relation between ionic and neutral chemical potential gradients is

$$\vec{\nabla}\mu_K = \sum_i \nu_{i,K} \vec{\nabla}\mu_i, \tag{1.71}$$

where $\nu_{i,K}$ is the stoichiometric coefficient of ionic species i in component K (which is zero if the dissociation of component K does not give rise to ionic species i in solution). Therefore, the equivalence between the ionic and the component formulations of θ_{dif} requires that

$$\vec{I}^m \cdot \vec{E}_{\text{dif}} = \sum_i \left(\vec{j}_i^m - \sum_K \nu_{i,K} \vec{J}_K^m \right) \cdot \vec{\nabla}\mu_i. \tag{1.72}$$

In the absence of electric current $\vec{I}^m = \vec{0}$, $\vec{E}_{\text{dif}} = \vec{E} \neq \vec{0}$, and the ionic flux densities are $\vec{j}_i^m = \sum_K \nu_{i,K} \vec{J}_K^m$. In the presence of electric current, the ionic flux density $\vec{j}_i^m$ is the sum of $\sum_K \nu_{i,K} \vec{J}_K^m$ and a term proportional to $\vec{I}^m$. In Sections 2.1.5 and 2.3.3 we pursue further the relation between the ionic and the component formalisms for the description of transport processes.

In the presence of chemical reactions, using eqn (1.48) and following a similar procedure, the entropy balance equation takes the form

$$\frac{\partial s}{\partial t} + \vec{\nabla} \cdot \vec{j}_s = \frac{1}{T}(\theta_{\text{ch}} + \theta_{\text{ed}}), \tag{1.73}$$

where the contribution of the chemical reactions to the dissipation function is

$$\theta_{\text{ch}} = \sum_r A_r \frac{\partial \xi_r}{\partial t} \geq 0, \tag{1.74}$$

and $A_r \equiv -\sum_i \nu_{i,r}\mu_i$ is the chemical affinity of reaction r. The chemical affinity A_r is zero when the reaction has reached equilibrium. If $A_r > 0$, the reaction proceeds towards equilibrium in such a way that $\partial\xi_r/\partial t > 0$, while $\partial\xi_r/\partial t < 0$ when $A_r < 0$, so that $\theta_{ch} \geq 0$ in all cases. The chemical affinity is considered to be the driving force for reaction r, and the reaction rate $\partial\xi_r/\partial t$ can be denoted as the reaction flux. Thus, θ_{ch} also takes the form of a sum of products of forces and fluxes. These forces and fluxes, however, are scalar.

Example: Transport in a binary electrolyte solution

Consider the one-dimensional transport in a 1:1 electrolyte solution under such conditions that $j_1^m = I/z_1F$ and $j_2^m = 0$. The dissipation function can be written as

$$\theta_{dif} + \theta_{ohm} = -J_{12}^m \frac{d\mu_{12}}{dx} + I^m E_{ohm} \quad \text{or} \quad \theta_{ed} = -j_1^m \frac{d\tilde{\mu}_1}{dx}.$$

In the first expression, the contributions of mass and electric charge transport are split into two terms. In the second expression, only one electrodiffusion term appears because electric charge is bound to ions, and hence mass and charge transport are not independent.

Using the relations

$$\frac{d\mu_1}{dx} = \frac{d\mu_2}{dx}, \quad J_{12}^m = \frac{D_2}{D_1 + D_2} j_1^m, \quad \text{and} \quad E = -\frac{1}{z_2F}\frac{d\mu_2}{dx},$$

which are fully justified in Chapter 2, we aim at discussing the relative importance of θ_{dif} and θ_{ohm} in terms of the ratio D_1/D_2 of ionic diffusion coefficients.

From the above relations, the dissipation function is

$$\theta_{ed} = -j_1^m \frac{d\tilde{\mu}_1}{dx} = -j_1^m\left(\frac{d\mu_1}{dx} - z_1FE\right) = -2j_1^m\frac{d\mu_1}{dx}.$$

Since the gradient of the electrolyte chemical potential is

$$\frac{d\mu_{12}}{dx} = \frac{d(\mu_1 + \mu_2)}{dx} = 2\frac{d\mu_1}{dx},$$

the contribution of chemical diffusion to the dissipation function is

$$\theta_{dif} = -J_{12}^m\frac{d\mu_{12}}{dx} = -\frac{D_2}{D_1 + D_2}2j_1^m\frac{d\mu_1}{dx} = \frac{D_2}{D_1 + D_2}\theta_{ed}.$$

From eqn (1.72), the diffusion electric field is (implicitly) given by

$$\vec{I}^m \cdot \vec{E}_{dif} = (j_1^m - J_{12}^m)\frac{d\mu_1}{dx} + (0 - J_{12}^m)\frac{d\mu_2}{dx} = \frac{D_1 - D_2}{D_1 + D_2}j_1^m\frac{d\mu_1}{dx},$$

and, therefore, the contribution of electric conduction to the dissipation function is

$$\theta_{\text{ohm}} = I^m E_{\text{ohm}} = I^m E - I^m E_{\text{dif}}$$
$$= -j_1^m \frac{\mathrm{d}\mu_1}{\mathrm{d}x} - \frac{D_1 - D_2}{D_1 + D_2} j_1^m \frac{\mathrm{d}\mu_1}{\mathrm{d}x} = \frac{D_1}{D_1 + D_2}\theta_{\text{ed}}$$

as should be expected, because $\theta_{\text{dif}} + \theta_{\text{ohm}} = \theta_{\text{ed}}$.

Although both θ_{dif} and θ_{ohm} are positive-definite, it is interesting to observe that mass and electric charge transport are not independent process because in the system under consideration only the ionic species 1 moves, and it is responsible for the transport of both mass and electric charge. Since species 2 does not move, the dissipation function θ_{ed} cannot depend on the ionic diffusion coefficient D_2. The fact that D_2 appears in the expressions for θ_{dif} and θ_{ohm} evidences that the decomposition of this process as the combination of chemical diffusion and electric conduction is somehow unnatural. Indeed, when $D_1/D_2 \gg 1$, the process should be considered as electric conduction because $\theta_{\text{dif}} \ll \theta_{\text{ohm}} \approx \theta_{\text{ed}}$, while in the opposite case $D_1/D_2 \ll 1$, the process should be considered as chemical diffusion of the electrolyte because $\theta_{\text{ohm}} \ll \theta_{\text{dif}} \approx \theta_{\text{ed}}$. In conclusion, depending on the value of D_1/D_2 the same process can be considered as electric conduction, chemical diffusion or a combination of them, and hence the name electrodiffusion is better suited in this situation.

1.2.7 Energy- and entropy-balance equations in a Newtonian viscous fluid

The viscosity of a flowing fluid causes a continuous degradation of kinetic energy into internal energy and, hence, entropy production. When this irreversible process is taken into account, the energy- and entropy-balance equations are

$$\frac{\partial e}{\partial t} + \vec{\nabla} \cdot \vec{j}_e = \vec{I} \cdot \vec{E} \tag{1.75}$$

$$\frac{\partial e_{\text{k}}}{\partial t} + \vec{\nabla} \cdot \vec{j}_{e_{\text{k}}} = \rho_{\text{e}} \vec{v} \cdot \vec{E} - \theta_\eta \tag{1.76}$$

$$\frac{\partial u}{\partial t} + \vec{\nabla} \cdot \vec{j}_u = \vec{I}^m \cdot \vec{E} + \theta_\eta \tag{1.77}$$

$$\frac{\partial s}{\partial t} + \vec{\nabla} \cdot \vec{j}_s = \frac{1}{T}(\theta_{\text{ch}} + \theta_{\text{ed}} + \theta_\eta), \tag{1.78}$$

where $\theta_\eta \geq 0$ is the contribution of viscous flow to the dissipation function.

We now aim at showing that the viscous contribution to the dissipation function θ_η also takes the form of the product of a 'force' and a 'flux density', which are second-order tensors. To avoid the mathematical complexity of algebraic and differential operations involving tensors, we describe first the case in which the fluid moves in the x direction and present the general case later in this section.

Consider, in particular, the stationary Poiseuille flow in planar geometry. An incompressible and electroneutral fluid occupies the space between two horizontal, parallel plates. Both plates are fixed and a pressure gradient $\mathrm{d}p/\mathrm{d}x < 0$ is applied in the x direction. The variation of the fluid velocity v_x with position y is given by the linear momentum-balance equation, eqn (1.59), and the components of the flux density of the linear momentum $\vec{j}^{\,m}_{v_x}$ are $j^m_{v_x,x} = p$ and $j^m_{v_x,y} = -\sigma'_{xy}$. Scalar multiplication of eqn (1.59) by the velocity v_x leads to the balance equation for the kinetic energy

$$\frac{\mathrm{d}(pv_x)}{\mathrm{d}x} - \frac{\mathrm{d}(\sigma'_{xy}v_x)}{\mathrm{d}y} + \sigma'_{xy}\frac{\mathrm{d}v_x}{\mathrm{d}y} = \vec{\nabla}\cdot(\vec{j}^{\,m}_{v_x}\,v_x) + \sigma'_{xy}\frac{\mathrm{d}v_x}{\mathrm{d}y} = 0. \tag{1.79}$$

The first term in the left-hand side of eqn (1.79) is the power of the pressure forces, i.e. it is the input rate of kinetic energy to the volume element. Since the velocity is independent of time, this power input must get out of the fluid element due to shear forces or dissipated inside it due to the fluid viscosity. The second term in the left-hand side of eqn (1.79), $-\mathrm{d}(\sigma'_{xy}v_x)/\mathrm{d}y$, represents the rate of energy transfer due to shear stresses and the third term, $\sigma'_{xy}\mathrm{d}v_x/\mathrm{d}y = \eta(\mathrm{d}v_x/\mathrm{d}y)^2 = \theta_\eta \geq 0$, represents the energy dissipation rate due to the fluid viscosity. In other words, under stationary conditions, eqn (1.76) reduces to $\vec{\nabla}\cdot\vec{j}_{e_k} = -\theta_\eta$, where $e_k = \rho v_x^2/2$. The comparison of this equation and eqn (1.79) shows that, in this case, the flux density of kinetic energy is $\vec{j}^{\,m}_{e_k} = \vec{j}^{\,m}_{v_x}v_x = (pv_x, -\sigma'_{xy}v_x, 0)$.

In other cases in which the fluid moves in the x direction only and there are no external forces, scalar multiplication of the linear momentum-balance equation, eqn (1.56), by the velocity v_x leads to the balance equation for the kinetic energy

$$\frac{\partial e_k}{\partial t} + \vec{\nabla}\cdot\vec{j}^{\,m}_{e_k} = \frac{\partial e_k}{\partial t} + \vec{\nabla}\cdot(\vec{j}^{\,m}_{v_x}\,v_x) = \vec{j}^{\,m}_{v_x}\cdot\vec{\nabla}v_x = -\theta_\eta. \tag{1.80}$$

Note that the divergence of the convective flux density of kinetic energy $\vec{\nabla}\cdot(e_k\vec{v})$ vanishes when the fluid moves in the x direction only.

Once we have discussed viscous dissipation under relatively simple flow conditions, we tackle now the difficult task of finding the general expression of θ_η. This requires the derivation of the balance equation for the kinetic energy, eqn (1.76), and the necessary preliminary step is the derivation of the general form of the balance equation of the linear momentum of a viscous fluid that was postponed in Section 1.2.5.

The linear momentum is a vector quantity and its flux density is a second-order tensor, so that the corresponding balance equation is mathematically rather complex. In the barycentric reference frame, the flux density of the linear momentum of the fluid $\overleftrightarrow{j}^{m}_{\vec{v}}$ is related to the surface forces and is equal to the negative of the stress tensor $\overleftrightarrow{\sigma}$. In the case of incompressible Newtonian fluids, the Cartesian component jk (j,$k = x, y, z$) of the stress tensor is

$$\sigma_{jk} = -p\delta_{jk} + \sigma'_{jk} = -p\delta_{jk} + \eta\left(\frac{\partial v_j}{\partial x_k} + \frac{\partial v_k}{\partial x_j}\right), \tag{1.81}$$

where δ_{jk} is the Kronecker delta ($\delta_{jk} = 1$ if $j = k$, $\delta_{jk} = 0$ if $j \neq k$). This equation can be written in tensorial form as

$$\overset{\leftrightarrow}{\sigma} = -p\overset{\leftrightarrow}{1} + \overset{\leftrightarrow}{\sigma}{}' = -p\overset{\leftrightarrow}{1} + 2\eta\overset{\leftrightarrow}{\gamma}{}' \tag{1.82}$$

where p is the pressure, $\overset{\leftrightarrow}{\sigma}{}'$ is the shear stress tensor, and $\overset{\leftrightarrow}{\gamma}{}'$ is the deformation rate tensor.

In the case of charged fluids, the interaction between the fluid and the electromagnetic field, through the electrostatic force $\rho_e\vec{E}$, implies that the linear momentum of the fluid is not conserved. The linear momentum density associated to the moving fluid is $\rho\vec{v}$ and its production rate is $\vec{\pi}_{\vec{v}} = \rho_e\vec{E}$. The balance equation for the component j of the linear momentum is then

$$\rho\frac{\mathrm{D}v_j}{\mathrm{D}t} = -\vec{\nabla}\cdot\vec{j}^{\,m}_{v_j} + \pi_{v_j} = -\frac{\partial p}{\partial x_j} + \eta\nabla^2 v_j + \rho_e E_j, \tag{1.83}$$

and the balance equation for the vector linear momentum can be written as

$$\rho\frac{\mathrm{D}\vec{v}}{\mathrm{D}t} = \vec{\nabla}\cdot\overset{\leftrightarrow}{\sigma} + \rho_e\vec{E} = -\vec{\nabla}p + \eta\nabla^2\vec{v} + \rho_e\vec{E}, \tag{1.84}$$

which is the *Navier–Stokes equation* in the presence of an electric force.

An alternative form of this equation is

$$\rho\frac{\partial\vec{v}}{\partial t} + \rho(\vec{v}\cdot\vec{\nabla})\vec{v} - \vec{\nabla}\cdot\overset{\leftrightarrow}{\sigma} = \rho\frac{\partial\vec{v}}{\partial t} + \vec{\nabla}\cdot(-\overset{\leftrightarrow}{\sigma} + \rho\vec{v}\vec{v}) = \rho_e\vec{E} \tag{1.85}$$

where the condition of incompressibility of the fluid has been used and the tensor $\vec{v}\vec{v}$ is the exterior (or dyadic) product of the fluid velocity and itself. The component jk of this tensor is $v_j v_k$. It is clear from this equation that the flux density of linear momentum in the laboratory reference frame is $\overset{\leftrightarrow}{j}_{\vec{v}} = -\overset{\leftrightarrow}{\sigma} + \rho\vec{v}\vec{v}$.

The mathematical modelling of convective transport processes often requires the solution of eqn (1.84). In practice, however, the Navier–Stokes equation is reduced to a much simpler form. Since the mechanical equilibrium is much faster than the distribution (or diffusional) equilibrium, stationary flow is assumed and $\partial\vec{v}/\partial t = \vec{0}$. In addition, most electrochemical techniques using convective flow involve such low Reynolds numbers that the convective acceleration $(\vec{v}\cdot\vec{\nabla})\vec{v}$ can be neglected. Thus, the Navier–Stokes equation is reduced to the statement of mechanical equilibrium (Stokes equation)

$$\vec{\nabla}p - \rho_e\vec{E} = \eta\nabla^2\vec{v}, \tag{1.86}$$

where the terms in the left-hand side generally represent the forces that induce the flow motion, and the term in the right-hand side represents the frictional force due to the solution viscosity. Nevertheless, the solution flow satisfying the Laplace equation $\vec{0} = \nabla^2\vec{v}$ can also by induced by shear stresses, such as in the Couette flow.

Scalar multiplication of the linear momentum-balance equation, eqn (1.85), by the barycentric velocity leads to the balance equation for the kinetic energy shown in eqn (1.76). In this equation, the flux density of kinetic energy density is

$$\vec{j}_{e_k} = -\overleftrightarrow{\sigma} \cdot \vec{v} + e_k\vec{v} = p\vec{v} - \overleftrightarrow{\sigma}' \cdot \vec{v} + e_k\vec{v}, \tag{1.87}$$

and the viscous contribution to the dissipation function is

$$\theta_\eta \equiv \sum_j \sum_k \sigma'_{jk} \frac{\partial v_j}{\partial x_k} = \sum_j \sum_k \sigma'_{jk}\gamma'_{kj} = \overleftrightarrow{\sigma}' : \overleftrightarrow{\gamma}' = 2\eta\overleftrightarrow{\gamma}' : \overleftrightarrow{\gamma}' \geq 0. \tag{1.88}$$

In deriving eqn (1.76), we have also used the relation $\vec{v}\cdot(\vec{\nabla}\cdot\overleftrightarrow{\sigma}') = \vec{\nabla}\cdot(\overleftrightarrow{\sigma}'\cdot\vec{v}) - \theta_\eta$.

In closing, we emphasize that the three contributions to the dissipation function θ_{ch}, θ_{ed}, and θ_η have the similar form of products of forces and fluxes. However, θ_{ch} involves products of scalars, θ_{ed} involves products of vectors, and $\theta_\eta \equiv \overleftrightarrow{\sigma}' : \overleftrightarrow{\gamma}'$ is the product of second order tensors. The fact that not only the total dissipation function θ but also the individual contributions θ_{ch}, θ_{ed}, and θ_η are positive-definitive is a consequence of the absence of coupling phenomena between irreversible processes of different tensorial degree (in a medium that is isotropic at equilibrium). The mathematical proof of this absence of coupling is known as the Curie theorem.

1.2.8 Electromagnetic energy and linear momentum

So far we have dealt with the properties of the fluid system only. In this section we consider the electric field and discuss its balance equations and its interaction with the fluid system [4,10,13]. We assume that there is no applied magnetic field and that the magnetization of the fluid is negligible because the magnetic field created by the electric currents in electrochemical systems is small. The equations presented in this section correspond to the *mathematical* limit in which the magnetic permeability of the fluid μ_m tends to zero. Therefore, the magnetic induction $\vec{B} = \mu_m\vec{H}$ is neglected and the electric field is assumed to be irrotational, $\vec{\nabla} \times \vec{E} = -\partial\vec{B}/\partial t \approx 0$. This implies that an electric potential ϕ can be defined from the equation $\vec{E} = -\vec{\nabla}\phi$. Note, however, that the magnetic field $\vec{H}$ is not neglected.

a) Electric energy-balance equation

The electric energy density is $e_{\mathrm{e}} = \vec{D} \cdot \vec{E}/2 = \varepsilon E^2/2$, where we have used the constitutive equation $\vec{D} = \varepsilon\vec{E}$. For the sake of simplicity, the dielectric permittivity ε is considered to be constant (i.e. it is not affected by the changes in the electric field and the local composition). To derive the balance equation of the electric energy, we note first that its time derivative is equal to the scalar product of the electric field and the displacement current density

$$\frac{\partial e_\mathrm{e}}{\partial t} = \frac{\partial \vec{D}}{\partial t} \cdot \vec{E} = \vec{I}_\mathrm{d} \cdot \vec{E} = \vec{I}_\mathrm{T} \cdot \vec{E} - \vec{I} \cdot \vec{E}. \tag{1.89}$$

The flux density of electromagnetic energy is given by the Poynting vector $\vec{E} \times \vec{H}$, and its divergence is

$$\vec{\nabla} \cdot (\vec{E} \times \vec{H}) = \vec{H} \cdot (\vec{\nabla} \times \vec{E}) - \vec{E} \cdot (\vec{\nabla} \times \vec{H}) \approx -\vec{I}_\mathrm{T} \cdot \vec{E}, \tag{1.90}$$

where we have neglected the magnetic energy density and used Ampere's law, $\vec{\nabla} \times \vec{H} = \vec{I}_\mathrm{T}$. Combining eqns (1.89) and (1.90), the balance equation for the electric energy is

$$\frac{\partial e_\mathrm{e}}{\partial t} + \vec{\nabla} \cdot (\vec{E} \times \vec{H}) = -\vec{I} \cdot \vec{E}. \tag{1.91}$$

It is interesting to note that the production rate is the negative of that in the energy balance equation of the fluid, eqn (1.75). That is, the sum of the energies of the fluid and the electric field is a conservative quantity, and their interaction is described by the product $\vec{I} \cdot \vec{E}$, which is the rate at which the electric field provides energy to the fluid. The mathematical statement of this conservation law is

$$\begin{aligned} &\frac{\partial (e + e_\mathrm{e})}{\partial t} + \vec{\nabla} \cdot (\vec{j}_e + \vec{E} \times \vec{H}) \\ &\approx \frac{\partial (e + e_\mathrm{e})}{\partial t} + \vec{\nabla} \cdot (\vec{j}_e + \phi \vec{I}_\mathrm{T}) = 0, \end{aligned} \tag{1.92}$$

where we have used that $\vec{\nabla} \cdot \vec{I}_\mathrm{T} = 0$ to introduce the approximation $\vec{\nabla} \cdot (\vec{E} \times \vec{H}) \approx \vec{\nabla} \cdot (\phi \vec{I}_\mathrm{T})$.

b) Electromagnetic linear momentum-balance equation

Equations (1.76) and (1.77) show that, from the power $\vec{I} \cdot \vec{E} = \vec{I}^m \cdot \vec{E} + \rho_\mathrm{e} \vec{v} \cdot \vec{E}$, a fraction $\vec{I}^m \cdot \vec{E}$ is converted to internal energy and a fraction $\rho_\mathrm{e} \vec{v} \cdot \vec{E}$ to kinetic energy of the fluid. This can be confirmed by analysing the balance of linear momentum of the field. The density of electromagnetic linear momentum is $\vec{D} \times \vec{B} = (\vec{E} \times \vec{H})/c^2$, where $\vec{B}$ is the magnetic induction and c is the speed of light in the medium. Making use of the Maxwell equations, $\vec{\nabla} \cdot \vec{D} = \rho_\mathrm{e}$, $\vec{\nabla} \cdot \vec{B} = 0$, $\vec{\nabla} \times \vec{E} = -\partial \vec{B}/\partial t$, and $\vec{\nabla} \times \vec{H} = \vec{I}_\mathrm{T}$, and neglecting the electric polarization energy and the magnetic energy, the time derivative of the density of electromagnetic linear momentum is approximately given by

$$\frac{\partial (\vec{D} \times \vec{B})}{\partial t} \approx \vec{\nabla} \cdot \overleftrightarrow{T} - \rho_\mathrm{e} \vec{E}, \tag{1.93}$$

where $\overleftrightarrow{T} \equiv \vec{D}\vec{E} - (1/2)\varepsilon_0 E^2 \overleftrightarrow{1}$ is the Maxwell stress tensor, $\vec{D}\vec{E}$ is a second-order tensor whose component jk is $D_j E_k$, and $\overleftrightarrow{1}$ is the unit second-order tensor represented by the matrix diag(1,1,1). This equation shows that (the volume density of) the production rate of linear momentum of the field is $-\rho_e \vec{E}$. Since eqn (1.85) showed that this is the rate of transference of linear momentum from the field to the fluid, it is concluded that the sum of the linear momenta of the fluid and the field is also a conservative quantity. Furthermore, the mechanical power done by the field on the fluid is $\rho_e \vec{v} \cdot \vec{E}$ and this is the increase rate of the kinetic energy density of the fluid.

c) Thermodynamic potentials revisited

As we noticed in Section 1.1.3, the thermodynamic functions of the fluid, e.g. u and g, do not contain any electric contribution and that is why the equations in Section 1.1.2 involve the chemical potentials rather than the electrochemical potentials. We briefly analyse here the difficulties that arise when attempting to incorporate the electric energy in the thermodynamic functions.

Replacing the chemical potentials by the electrochemical potentials, the local equilibrium equation takes the form

$$\frac{\partial u}{\partial t} + \phi \frac{\partial \rho_e}{\partial t} = T \frac{\partial s}{\partial t} + \sum_i \tilde{\mu}_i \frac{\partial c_i}{\partial t}, \tag{1.94}$$

which suggests to analyse the contribution $\phi \partial \rho_e / \partial t$ to the time variation of the electric energy density. The internal energy balance equation can be written as

$$\frac{\partial u}{\partial t} + \phi \frac{\partial \rho_e}{\partial t} + \vec{\nabla} \cdot (\vec{j}_u + \phi \vec{I}) = -\rho_e \vec{v} \cdot \vec{E} + \theta_\eta, \tag{1.95}$$

and its combination with the kinetic energy balance leads to

$$\frac{\partial e}{\partial t} + \phi \frac{\partial \rho_e}{\partial t} + \vec{\nabla} \cdot (\vec{j}_e + \phi \vec{I}) = 0. \tag{1.96}$$

Although this resembles eqn (1.92), it should be noticed that eqn (1.96) does not describe a conservation law and that the difference between these equations is

$$\phi \frac{\partial \rho_e}{\partial t} - \frac{\partial e_e}{\partial t} = \vec{\nabla} \cdot (\phi \vec{I}_d). \tag{1.97}$$

Some of the difficulties associated with the thermodynamic description of non-electroneutral solutions come from the fact that $\phi \delta \rho_e \neq \delta e_e$.[9] We know from classical thermodynamics that heat and (mechanical) work are not state functions but energies in transfer. In general, their differentials $T \delta s$ and $p \delta (1/\rho)$ depend on the process undergone by the system and, therefore, they are not exact. Similarly, the differential $\phi \delta \rho_e$ is not exact because it depends

[9] Due to the long range of electrical interactions, changes in the local electric charge density at a given position can provoke changes in the electric potential and charge density at other positions in the system or even outside it.

on the constraints imposed to the process, and this is the reason why it is not convenient to write the local thermodynamic equations in terms of a kind of 'electrochemical Gibbs potential' $\widetilde{g} \equiv g + \rho_e \phi = \sum_i \widetilde{\mu}_i c_i$. Moreover, this also rules out the possibility of defining the electrochemical potential of an ionic species as $\widetilde{\mu}_i \equiv (\partial \widetilde{g}/\partial c_i)_{T,p,\{c_{j\neq i}\}}$ because the change in the local concentration of species i implies a change in the local electric potential that cannot be evaluated in general. In an attempt to overcome this problem, some authors write $\widetilde{\mu}_i \equiv (\partial \widetilde{g}/\partial c_i)_{T,p,\{c_{j\neq i}\},\phi}$ but it should be emphasized that the electric potential has not appeared in the previous sections as a thermodynamic variable characterizing the state of the system and, more importantly, that the constraint of constant ϕ while changing the local electric charge density is hardly realizable in practice.

Finally, it is interesting to observe that the classical expression for the energy density of a continuous distribution of electric charges is neither $e_e = \varepsilon E^2/2$ nor $\rho_e \phi$, but $\rho_e \phi/2$. The relation between them is

$$e_e = \frac{1}{2}\rho_e \phi - \frac{1}{2}\vec{\nabla} \cdot (\phi \vec{D}), \tag{1.98}$$

which is obtained from the mathematical identity $\vec{\nabla} \cdot (\phi \vec{D}) = \phi \vec{\nabla} \cdot \vec{D} + \vec{D} \cdot \vec{\nabla} \phi$.

d) Example 1: Discharge of a capacitor over a resistance

Consider a dielectric film between the plates of a parallel capacitor of capacitance C that is initially charged under a potential difference $\Delta\phi(0)$. At time $t = 0$, it is allowed to discharge over a resistance R. The transient current is given by $I(t) = I(0)\,\mathrm{e}^{-t/\tau}$, where $\tau = RC$ is the electrical relaxation time, $I(0) = Q(0)/\tau$ is the initial current, and $Q(0) = C\Delta\phi(0)$ is the initial charge on the plates. The total current density $I_T = I + I_d$ is constant along the circuit, as required by the principle of conservation of charge, $\vec{\nabla} \cdot \vec{I}_T = 0$, but inside the resistance there is only conduction current and inside the capacitor there is only displacement current. In the discharge process, the energy $Q(0)\Delta\phi(0)/2 = Q(0)^2/2C$ initially stored in the capacitor is dissipated as Joule heat inside the resistance. The electric energy flux density in the capacitor is $\phi\vec{I}_d$ and the energy that flows out of it can be evaluated by integrating the divergence of this flux over the volume of the dielectric film

$$\iiint_C \vec{\nabla} \cdot (\phi \vec{I}_d)\,\mathrm{d}V = \oiint_C \phi \vec{I}_d \cdot \mathrm{d}\vec{S} = \Delta\phi I A,$$

where A is the area of the plates. Thus, inside the capacitor there are no (free) electric charges and $\phi \partial \rho_e/\partial t = 0$, but the electric energy density changes and, in agreement with eqn (1.97), $\partial e_e/\partial t = -\vec{\nabla} \cdot (\phi \vec{I}_d) < 0$.

Similarly, the electrostatic energy flux density in the resistance is $\phi\vec{I}$ and the energy that flows through it can be evaluated by integrating the divergence of this flux over the volume of the resistance

$$-\iiint_R \vec{\nabla}\cdot(\phi\vec{I})\,\mathrm{d}V = -\oiint_R \phi\vec{I}\cdot\mathrm{d}\vec{S} = \Delta\phi IA,$$

where the minus sign comes from the fact that we are evaluating the flow of energy towards the inside of the resistance. In the resistance, electric energy is converted to internal energy due to the Joule dissipation at a rate $-\vec{I}\cdot\vec{E}$ and the total electric energy converted is

$$-\iiint_R \vec{I}\cdot\vec{E}\,\mathrm{d}V = \iiint_R \vec{\nabla}\cdot(\phi\vec{I})\,\mathrm{d}V = \oiint_R \phi\vec{I}\cdot\mathrm{d}\vec{S} = -\Delta\phi IA.$$

Therefore, there is no change in the electric energy of the resistance because all the electric energy that enters is converted to internal energy. In fact, this balance also holds locally in the resistance because $\vec{\nabla}\cdot(\phi\vec{I}) = -\vec{I}\cdot\vec{E}$, and therefore $\partial e_\mathrm{e}/\partial t = 0$; note that in the resistance both the space charge density and the displacement current density are zero. In conclusion, the above equations describe well the conversion of electric energy into internal energy that takes place in this electrical relaxation process.

e) Example 2: Electrical double-layer formation at an isolated metal electrode

Consider that a piece of metal, with no electrical contacts, is introduced instantaneously at time $t = 0$ in an electrolyte solution containing the redox couple Fe^{2+}/Fe^{3+} [14]. The molar concentrations of these ions in the solution are $c_{Fe^{2+}}$ and $c_{Fe^{3+}}$. Initially, the metal is electrically neutral and the electrical potential difference between the metal and the solution is zero, $\Delta_\mathrm{s}^\mathrm{m}\phi(0) = 0$. The metal is not in equilibrium with the solution because there is a tendency for the redox couple to exchange electrons with the piece of metal until the Nernst equilibrium potential

$$\Delta_\mathrm{s}^\mathrm{m}\phi_\mathrm{eq} = \Delta\phi^\circ + \frac{RT}{F}\ln\frac{c_{Fe^{3+}}}{c_{Fe^{2+}}}$$

is reached. That is, the potential difference $\Delta_\mathrm{s}^\mathrm{m}\phi = \phi^\mathrm{m} - \phi^\mathrm{s}$ evolves monotonically with time from 0 to $\Delta_\mathrm{s}^\mathrm{m}\phi_\mathrm{eq}$.

The value of the Nernst equilibrium potential can be positive or negative depending on the solution composition. Similarly, the affinity for the iron oxidation, $Fe^{2+} \rightleftarrows Fe^{3+} + e^-$, which can be written as $A = -\sum_i \nu_i\widetilde{\mu}_i = F(\Delta_\mathrm{s}^\mathrm{m}\phi - \Delta_\mathrm{s}^\mathrm{m}\phi_\mathrm{eq})$, can also be positive or negative. Oxidation occurs when $A > 0$, and reduction when $A < 0$.

The iron (cathodic) reduction takes place when $\Delta_\mathrm{s}^\mathrm{m}\phi_\mathrm{eq} > \Delta_\mathrm{s}^\mathrm{m}\phi(t) > 0$. The conduction electric current then flows from the solution towards the electrode, and hence the electrode becomes positively charged; the solution adjacent to the electrode bears a compensating negative charge density. Still, the metal behaves as a cathode. The electric field is then directed from the electrode towards the solution, and hence the conduction current density and the electric field are directed in opposite directions in the solution adjacent to the electrode, $\vec{I} \cdot \vec{E} < 0$. Similar arguments lead to the conclusion that $\vec{I} \cdot \vec{E} < 0$ also when $\Delta_{\mathrm{s}}^{\mathrm{m}}\phi_{\mathrm{eq}} < \Delta_{\mathrm{s}}^{\mathrm{m}}\phi(t) < 0$ and iron (anodic) oxidation takes place.

The fact that the power $\vec{I} \cdot \vec{E} = \vec{I}^m \cdot \vec{E}$ is negative might be surprising but it is not in contradiction either with the second law of thermodynamics or with Ohm's law, $\vec{I} = \kappa \vec{E}_{\mathrm{ohm}}$. In the process described, $\vec{I}^m \cdot \vec{E} \neq \vec{I}^m \cdot \vec{E}_{\mathrm{ohm}} = \theta_{\mathrm{ohm}} \geq 0$ because the concentrations are non-uniform and the diffusion electric field, $\vec{E}_{\mathrm{dif}} = \vec{E} - \vec{E}_{\mathrm{ohm}}$, is important. Moreover, $\theta_{\mathrm{ed}} = \sum_i \vec{j}_i^m \cdot \vec{\nabla}\mu_i + \vec{I}^m \cdot \vec{E} = -\sum_i \vec{j}_i^m \cdot \vec{\nabla}\tilde{\mu}_i \geq 0$ because the ionic motions take place mainly as a result of diffusion (although the effect of the electric field cannot be neglected) and the contribution $-\sum_i \vec{j}_i^m \cdot \vec{\nabla}\mu_i$ is always positive and compensates for any eventual negative values of $\vec{I}^m \cdot \vec{E}$.

Since the metal is not connected to any external circuit, the total current density is zero and the conduction and displacement current densities in the solutions are opposite to each other, $\vec{I} = -\vec{I}_{\mathrm{d}}$. Thus, in agreement with eqn (1.89), the formation of the electrical double layer implies an increase of the electric energy of the system

$$\frac{\partial e_{\mathrm{e}}}{\partial t} = \vec{I}_{\mathrm{d}} \cdot \vec{E} = -\vec{I} \cdot \vec{E} > 0,$$

at the expense of decreasing the internal energy. Hence, we conclude that the equations derived in Section 1.2.8 also describe satisfactorily the conversion of internal energy into electric energy that takes place in this process driven by a chemical affinity.

f) Example 3: Electrical transient of the formation of the liquid-junction potential

Consider two dilute NaCl aqueous solutions at (slightly) different concentrations separated by an impermeable wall. At time $t = 0$, the wall is removed and the solutions are allowed to mix. The mixing, however, is constrained to a region of thickness d centred at the wall. That is, the solution concentrations at $x = 0$ and $x = d$ are kept constant, and the system evolves from an initial situation in which the concentration is uniform (although with different values) in the regions $0 < x < d/2$ and $d/2 < x < d$, to a final stationary situation in which the concentration distribution is linear throughout the constraint diffusion region $0 < x < d$. We are interested here in the initial times of this process in which the most interesting electrical phenomena take place. This is the so-called electrical relaxation process [15].

In the initial state the solutions are electroneutral, and internal energy is the only contribution to the total energy. After removing the wall, the solutions start to mix. During this evolution, the system is electrically isolated and the total current $\vec{I}_{\mathrm{T}}$ is zero. The open-circuit condition $\vec{I}_{\mathrm{T}} = \vec{0}$ implies that the conduction current carried by the ions moving initially at different velocities is 'compensated' by the displacement current associated with the time-dependent electric field, $\vec{I}_d = -\vec{I}$. In their tendency towards making uniform the concentration, Na^+ and Cl^- ions try to move independently. At the initial stages, when no electric field exists, Cl^- ions move faster than Na^+ ions because of their larger diffusion coefficient, and there is a non-zero conduction current density $\vec{I} = \vec{I}^m$ from the low to the high concentration region. A space-charge region (i.e. a non-electroneutral region) with intense electric fields appears at the junction. The electric field thus created tends to slow down the Cl^- ions and to speed up the Na^+ ions, until they eventually move at the same velocity, much as if transport of neutral NaCl molecules were then taking place. However, the electric field has not disappeared completely. An electric field is needed to keep Cl^- and Na^+ ions moving at the same velocity. When this electrical relaxation process is concluded (which requires a time of the order of nanoseconds [15]), the conduction electric current $\vec{I}^m$ is negligible, and no more charge separation occurs. The space-charge region simply spreads out due to the diffusional relaxation. This time evolution of the space-charge density is illustrated in Fig. 1.8, which has been obtained from the solution of the Nernst–Planck and Poisson equations (see Ref. [15] for details) in a system where the diffusional relaxation time, $\tau_d = d^2/[\pi^2(D_{Na^+} + D_{Cl^-})]$, is 100 times larger than the electrical relaxation time τ_e.

The formation of a space charge-region with intense electric fields requires energy (Fig. 1.9). Since the total energy is conserved, the electric energy associated to this space-charge region is taken from the internal energy. When, in the initial stages, Cl^- ions move faster than Na^+ ions, there is a non-zero conduction current density $\vec{I} = \vec{I}^m$ from the low to the high concentration solution. The conduction current has the same direction throughout the region, takes its maximum value at the position of the initial junction and decreases rapidly with the distance to the junction. The electric field created at the junction has the opposite direction and the electric power $\vec{I}^m \cdot \vec{E}$ is negative. Thus, according to

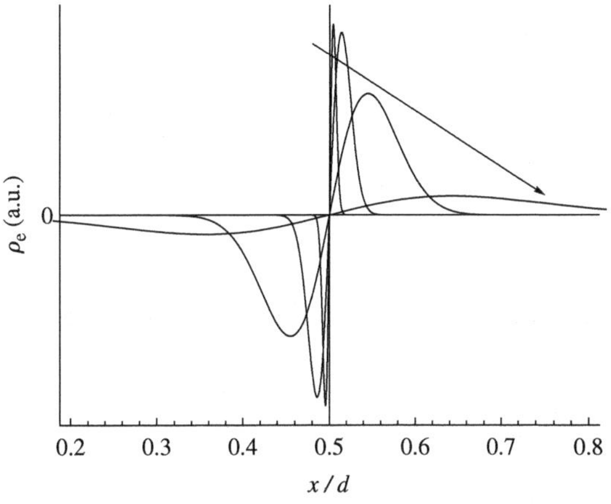

Fig. 1.8.
Space-charge density (in arbitrary units) vs. position at times $t/\tau_e = 0.01, 0.1, 1$, and 10 (the arrow indicates the direction of increasing time) during the formation of the liquid junction potential between two NaCl solutions of slightly different concentrations that meet at $x = d/2$, where d is the thickness of the constrained diffusion region [15].

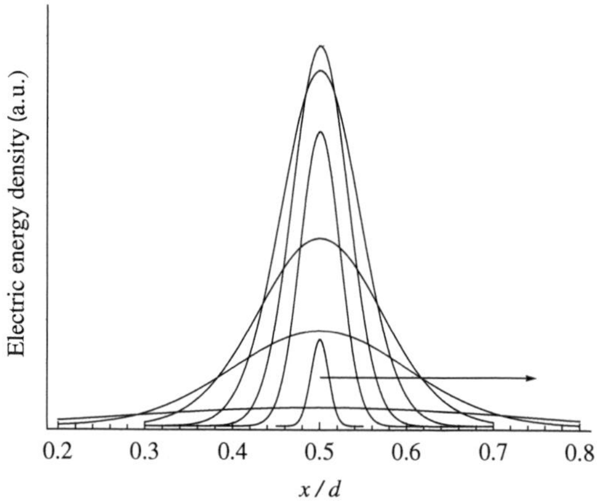

Fig. 1.9.
Electric energy density (in arbitrary units) vs. position at times $t/\tau_e = 0.1, 0.5, 1, 2, 5, 10$, and 50 (the arrow indicates the direction of increasing time) for the system described in Fig. 1.8.

the balance equations eqns (1.66) and (1.91), the internal energy decreases and the electric energy increases at the junction. After the electrical relaxation time, when both ions move at the same velocity and the liquid-junction potential has reached its steady-state value, the conduction current density $\vec{I}^m$ is negligible, and so is the power $\vec{I}^m \cdot \vec{E}$, which means that no more energy conversion takes place.

As a final comment, it should be mentioned that the internal energy of dilute electrolyte solutions depends on the temperature but not so much on other state variables. This means that changes in internal energy involve changes in temperature, whereas thermal conduction has been neglected in our description. Similarly, we have neglected the transport of solvent or, equivalently, the solution motion due to the (very small) Lorentz force that acts on the space-charge region. All this is justified because they are second-order effects that are not essential for the understanding of the formation of the liquid-junction potential.

Exercises

1.1 The local Gibbs equation describes the change in the Gibbs potential density g with temperature, pressure, and composition. In the aqueous solution of a strong binary electrolyte at constant temperature, pressure, and solvent concentration, this change is given by $\delta g_{T,p,c_0} = \mu_{12}\delta c_{12}$, where c_{12} is the stoichiometric electrolyte concentration. Consider now a weak binary electrolyte solution, in equilibrium with respect to the dissociation reaction $A_{\nu_1}C_{\nu_2} \rightleftarrows \nu_1 A^{z_1} + \nu_2 C^{z_2}$, and show that the local Gibbs equation takes the form

$$\delta g_{T,p,c_0} = \mu_{12}\delta c_{12,\mathrm{T}},$$

where $c_{12,\mathrm{T}} = c_{12} + c_{12,\mathrm{u}}$ is the total electrolyte molar concentration (dissociated plus undissociated). Therefore, the only chemical potential experimentally measurable under these conditions is

$$\mu_{12} = \left(\frac{\partial g}{\partial c_{12,\mathrm{T}}}\right)_{T,p,c_0}.$$

1.2 Use the equation of continuity of mass in a compressible fluid

$$\frac{\partial \rho}{\partial t} + \vec{\nabla} \cdot (\rho \vec{v}) = 0$$

to show the equivalence between the following forms of the general balance equation

$$\frac{\partial b}{\partial t} + \vec{\nabla} \cdot \vec{j}_b = \frac{\mathrm{D}b}{\mathrm{D}t} - \frac{b}{\rho}\frac{\mathrm{D}\rho}{\mathrm{D}t} + \vec{\nabla} \cdot \vec{j}_b^m = \pi_b.$$

1.3 Show that in a locally electroneutral solution, the conduction electric current density is independent of the reference frame in which it is evaluated.

1.4 Prove that the conservation of electrical charge in homogeneous chemical reactions, $\sum_i \nu_{i,r} z_i = 0$, implies that the chemical affinity can also be written as $A_r = -\sum_i \nu_{i,r} \tilde{\mu}_i$.

1.5 Write down the energy- and entropy-balance equations in an electroneutral, isothermal, non-viscous, and chemically inert solution in the absence of external forces.

1.6 The Couette flow between two parallel plane walls located at $y = 0$ and $y = h$ is due to the application of shear stresses in opposite directions on the plates. The motion is described with respect to the wall at $y = 0$ (which is then fixed) and the wall at $y = h$ moves at velocity U along direction x. The (barycentric) velocity distribution is then

$$v_x(y) = Uy/h.$$

The velocity gradient is U/h and, according to Newton's law of viscosity, the shear stress is $\eta U/h$. This stress is uniform inside the fluid. The shear stress externally applied on the wall at $y = h$ to cause the motion is $\eta U/h$, while that applied on the wall at $y = 0$ to keep it fixed is $-\eta U/h$. Thus, the net force is zero, as should be expected in a stationary flow. The external power on the fluid, however, is not zero because the fixed plate makes no power and the moving one gives energy to the fluid at a rate $A(\eta U/h)U$, where A is the plate area. This energy is dissipated uniformly within the fluid volume Ah.

(a) Evaluate the components of the deformation rate tensor

$$\gamma'_{jk} = \frac{1}{2}\left(\frac{\partial v_j}{\partial x_k} + \frac{\partial v_k}{\partial x_j}\right),$$

and the shear stress tensor, $\sigma'_{jk} = 2\eta\gamma'_{jk}$.

(b) Show that the viscous dissipation function $\theta_\eta \equiv \overleftrightarrow{\sigma}' : \overleftrightarrow{\gamma}' = \sum_j \sum_k \sigma'_{jk}\gamma'_{kj}$ can be written as

$$\theta_\eta = JX,$$

where $J = \langle v_x \rangle = U/2$ is the volume flux density and $X = [(\eta U/h) - (-\eta U/h)]/h = 2\eta U/h^2$ is the shear stress gradient.

(c) Show that the expression $\theta_\eta = -\vec{\nabla} \cdot \vec{j}_{e_k}$, which is valid for stationary flows, leads to the same result for θ_η.

1.7 The Poiseuille flow in a cylindrical, horizontal tube of (inner) radius R is due to the application of pressure gradient along the tube axis, $\vec{\nabla} p = (\mathrm{d}p/\mathrm{d}x, 0, 0) = (-\Delta p/L, 0, 0)$, where $\Delta p > 0$ is the pressure drop over a length L of the tube.

Under steady-state conditions, the velocity distribution (with respect to the tube) can be expressed as

$$v_x(r) = v_x(0)\left[1 - \left(\frac{r}{R}\right)^2\right] = \frac{\Delta p}{4\eta L}(R^2 - r^2),$$

where r is the radial position co-ordinate. The average velocity across the tube Section is $\langle v_x \rangle = v_x(0)/2 = R^2\Delta p/(8\eta L)$, which is also equal to the average volume flow density $J \equiv \langle v_x \rangle$, and the corresponding 'driving force' is the negative pressure gradient $X \equiv \Delta p/L$.

(a) Evaluate the components of the deformation rate tensor

$$\gamma'_{jk} = \frac{1}{2}\left(\frac{\partial v_j}{\partial x_k} + \frac{\partial v_k}{\partial x_j}\right)$$

and the shear stress tensor, $\sigma'_{jk} = 2\eta\gamma'_{jk}$.

(b) Show that the viscous dissipation function is $\theta_\eta \equiv \sum_j \sum_k \sigma'_{jk}\gamma'_{kj} = (r^2/4\eta)(\Delta p/L)^2$ and its average value can be written as

$$\langle\theta_\eta\rangle = JX,$$

where J and X are the flow and the driving force defined above.

(c) Show that the expression $\theta_\eta = -\vec{\nabla}\cdot\vec{j}_{e_k}$, which is valid for stationary flows, leads to the same result for θ_η.

References

[1] T.S. Sørensen and V. Compañ, 'On the Gibbs–Duhem equation for thermodynamic systems of mixed Euler order with special reference to gravitational and nonelectroneutral systems', *Electrochim. Acta*, 42 (1997) 639–649.

[2] M. Planck, 'Ueber die Erregung von Electricität und Wärme in Electrolyten', *Ann. Phys. Chem. N.F.*, 39 (1890) 161–186.

[3] R. Defay and P. Mazur, 'Sur la definition locale dels potentials chimiques dans les systemes electrochimiques', *Bull. Soc. Chim. Belg.*, 63 (1954) 562–579.

[4] A. Sandfeld, *Thermodynamics of Charged and Polarized Layers*, Wiley-Interscience, London, 1968, p. 7 and 27.

[5] I. Prigogine, *Introduction to Thermodynamics of Irreversible Processes*, Interscience Publishers, New York, 1961.

[6] S.R. de Groot and P. Mazur, *Non-equilibrium Thermodynamics*, North-Holland, Amsterdam, 1962.

[7] R. Haase, *Thermodynamics of Irreversible Processes*, Addison-Wesley, Reading, MA, 1969.

[8] I. Gyarmarti, *Non-equilibrium Thermodynamics. Field Theory and Variational Principles*, Springer-Verlag, Berlin, 1970.

[9] S. Wisniewski, B. Staniszewski, and E. Szymanik, *Thermodynamics of Nonequilibrium Processes*, D. Reidel Pub. Co., Dordrecht, 1976.

[10] Y.L. Yao, *Irreversible Thermodynamics*, Science Press, Beijing, 1981.

[11] D.S. Chandrasekhariah and L. Debnath, *Continuum Mechanics*, Academic Press, San Diego, 1994.

[12] J.H. Spurk, *Fluid Mechanics*, Springer, Berlin, 1997.

[13] G.D.C. Kuiken, *Thermodynamics of Irreversible Processes. Applications to Diffusion and Rheology*, John Wiley & Sons, Chichester, 1994.

[14] W.D. Murphy, J.A. Manzanares, S. Mafé, and H. Reiss, 'A numerical study of the equilibrium and nonequilibrium diffuse double layer in electrochemical cells', *J. Phys. Chem.*, 96 (1992) 9983–9991.

[15] S. Mafé, J.A. Manzanares, and J. Pellicer, 'The charge separation process in nonhomogeneous electrolyte solutions', *J. Electroanal. Chem.*, 241 (1988) 57–77.

2 Transport equations

2.1 Linear phenomenological equations

2.1.1 Introduction

Irreversible processes are described in terms of generalized fluxes and forces. These quantities vanish at equilibrium and, therefore, they somehow quantify the departure from equilibrium. The fluxes and forces are denoted by J and X, respectively, and can be scalars, vectors or tensors of second order, depending on the process under consideration. The dissipation function is given by the sum of their products

$$\theta = \sum_s J_s X_s + \sum_v \vec{J}_v \cdot \vec{X}_v + \Sigma_t \overleftrightarrow{J}_t : \overleftrightarrow{X}_t, \tag{2.1}$$

where the indices s, v, and t stand for scalar, vector, and tensor. Indeed, we know from eqns (1.68), (1.74), and (1.88) that

$$\theta = \theta_{\mathrm{ch}} + \theta_{\mathrm{ed}} + \theta_\eta = \sum_r \frac{\mathrm{d}\xi_r}{\mathrm{d}t} A_r + \sum_i \vec{j}_i^{\,m} \cdot (-\vec{\nabla}\tilde{\mu}_i) + \overleftrightarrow{\sigma}' : \overleftrightarrow{\gamma}', \tag{2.2}$$

so that the generalized fluxes and forces can be selected as follows $J_r = \mathrm{d}\xi_r/\mathrm{d}t$, $X_r = A_r$, $\vec{J}_i = \vec{j}_i^{\,m}$, $\vec{X}_i = -\vec{\nabla}\tilde{\mu}_i$, $\overleftrightarrow{J} = \overleftrightarrow{\gamma}'$, and $\overleftrightarrow{X} = \overleftrightarrow{\sigma}'$.

Assuming that the fluxes are functions of the forces only, we can write

$$\mathrm{d}\xi_r/\mathrm{d}t = f_r(A_1, A_2, \ldots) \tag{2.3}$$

$$\vec{j}_i^{\,m} = \vec{g}_i(\vec{\nabla}\tilde{\mu}_1, \vec{\nabla}\tilde{\mu}_2, \ldots) \tag{2.4}$$

$$\overleftrightarrow{\gamma}' = \overleftrightarrow{h}(\overleftrightarrow{\sigma}'). \tag{2.5}$$

These equations represent the general form of the phenomenological equations, which are the starting point for the description of transport processes. It has to be realized that eqns (2.3)–(2.5) do not exclude, e.g. the possible mutual coupling of chemical reactions and diffusion via mass balance.[1]

[1] Direct mutual coupling is forbidden by the Curie–Prigogine principle because they are of different tensorial degree [1]. However, this principle only applies as long as the system is isotropic at equilibrium. The coupling of transport processes and chemical reactions occurs, e.g. in biological systems.

The next task is to determine the functions f_r, $\vec{g}_i$ and $\overleftrightarrow{h}$. In Section 1.2.7 we already used an equation similar to eqn (2.5): the linear relation $\overleftrightarrow{\sigma}' = 2\eta\overleftrightarrow{\gamma}'$ that defines a Newtonian fluid. When describing the balance of electrostatic energy in Section 1.2.8 we also used Ohm's law $\vec{I}^m = \kappa\vec{E}_{\text{ohm}}$, which is related to eqn (2.4) because $\vec{I}^m = F\sum_i z_i\vec{j}_i^m$. Similarly, Fick's first law $\vec{j}_i^v = -D_i\vec{\nabla}c_i$ for diffusion and Fourier's law $\vec{j}_u^m = \vec{j}_q^m = -\kappa_T\vec{\nabla}T$ for thermal conduction are also linear transport equations.

Linear relations can be understood as the first-order term in the Taylor series expansions of the functions in eqns (2.3)–(2.5) and this justifies their validity when the system is not too far from equilibrium. In fact, it has been confirmed by statistical mechanics studies that as far as electrodiffusional transport processes are concerned, the linear equations are as general as the local formulation of the thermodynamics of irreversible processes. On the contrary, linear equations apply to chemical reactions only in the very close vicinity of chemical equilibrium. Nevertheless, the phenomenological equations for homogeneous chemical reactions are not of interest in this book because chemical relaxation times are usually much smaller than diffusional relaxation times. Furthermore, it can often be assumed that the chemical equilibrium has already been reached when describing electrodiffusional processes.

2.1.2 Ionic transport equations

In the linear approximation, the phenomenological equations for electrodiffusion state that all the thermodynamic driving forces contribute to every flux density so that the latter can be evaluated as

$$\vec{j}_i^m = -\sum_k l_{i,k}\vec{\nabla}\tilde{\mu}_k, \tag{2.6}$$

where the $l_{i,k}$s are the phenomenological coefficients. These coefficients can depend on the local state variables (temperature, pressure, and concentrations) but not on fluxes and forces (because otherwise we would observe a non-linear behaviour). It has been experimentally verified that the coefficients satisfy the Onsager reciprocal theorem

$$l_{i,k} = l_{k,i}, \tag{2.7}$$

which is also supported by microscopic considerations [2].

In principle, the sum in eqn (2.6) and that in the contribution of electrodiffusion to the dissipation function

$$\theta_{\text{ed}} = -\sum_i \vec{j}_i^m \cdot \vec{\nabla}\tilde{\mu}_i \tag{2.8}$$

extend over all the system components. We denote the number of components, including the solvent, by $N + 1$. However, all flux densities $\vec{j}_i^m$ are not independent because they are linked through the definition of the barycentric velocity $\vec{v} \equiv \sum_i w_i \vec{v}_i$. This definition can be rewritten as

$$\sum_i M_i \vec{j}_i^m = \vec{0}, \tag{2.9}$$

and it is then clear that the number of independent flux densities cannot exceed N. Similarly, the gradients of the electrochemical potentials must satisfy the Gibbs–Duhem equation (for isothermal and incompressible systems)

$$\sum_i c_i \vec{\nabla} \tilde{\mu}_i = \vec{\nabla} p - \rho_e \vec{E}. \tag{2.10}$$

This all means that if we are using the barycentric velocity and the pressure gradient to describe the transport of linear momentum, the description of electrodiffusion can only involve N fluxes and N forces, where N is the number of components excluding the solvent.

By elimination of the solvent flux density from eqn (2.9) as

$$\vec{j}_0^m = c_0(\vec{v}_0 - \vec{v}) = -\frac{1}{M_0} \sum_{i \neq 0} M_i \vec{j}_i^m, \tag{2.11}$$

where the sum excludes the solvent ($i = 0$), the electrodiffusional contribution to the dissipation function can be written as

$$\theta_{\text{ed}} = -\sum_i \vec{j}_i^m \cdot \vec{\nabla} \tilde{\mu}_i = -\sum_{i \neq 0} \vec{j}_i^m \cdot \left(\vec{\nabla} \tilde{\mu}_i - \frac{M_i}{M_0} \vec{\nabla} \mu_0 \right). \tag{2.12}$$

Similarly, eliminating the solvent chemical potential gradient from eqn (2.10), θ_{ed} can be written as

$$\begin{aligned} \theta_{\text{ed}} &= -\vec{j}_0^m \cdot \vec{\nabla} \mu_0 - \sum_{i \neq 0} \vec{j}_i^m \cdot \vec{\nabla} \tilde{\mu}_i \\ &= -(\vec{v}_0 - \vec{v}) \cdot (\vec{\nabla} p - \rho_e \vec{E}) - \sum_{i \neq 0} \vec{j}_i^{\text{H}} \cdot \vec{\nabla} \tilde{\mu}_i, \end{aligned} \tag{2.13}$$

where

$$\vec{j}_i^{\text{H}} \equiv c_i (\vec{v}_i - \vec{v}_0) \tag{2.14}$$

is the molar flux density of species i in a reference frame bound to the solvent. This is known as the Hittorf reference frame and eqn (2.14) allows us to evaluate θ_{ed} when the flux densities are referred to this frame.

The flux densities can also be referred to the Fick reference frame, which is bound to the volume flux density

$$\vec{v}_\upsilon \equiv \sum_i \upsilon_i \vec{j}_i = \vec{v}_\upsilon^m + \vec{v} = \vec{v}_\upsilon^{\mathrm{H}} + \vec{v}_0, \tag{2.15}$$

where υ_i is the partial molar volume of species i and $\sum_i c_i \upsilon_i = 1$. When the flux densities are referred to the Fick reference frame

$$\vec{j}_i^\upsilon = \vec{j}_i - c_i \vec{v}_\upsilon = \vec{j}_i^m - c_i \vec{v}_\upsilon^m \tag{2.16}$$

the electrodiffusional contribution to the dissipation function takes the form

$$\begin{aligned} \theta_{\mathrm{ed}} &= -\vec{v}_\upsilon^m \cdot \sum_i c_i \vec{\nabla} \tilde{\mu}_i - \sum_i \vec{j}_i^\upsilon \cdot \vec{\nabla} \tilde{\mu}_i \\ &= -(\vec{v}_\upsilon - \vec{v}) \cdot (\vec{\nabla} p - \rho_{\mathrm{e}} \vec{E}) - \sum_{i=1} \vec{j}_i^\upsilon \cdot \left(\vec{\nabla} \tilde{\mu}_i - \frac{\upsilon_i}{\upsilon_0} \vec{\nabla} \mu_0 \right), \end{aligned} \tag{2.17}$$

where we have used the condition $\sum_i \upsilon_i \vec{j}_i^\upsilon = \vec{0}$ to eliminate the solvent flux density.

The above equations are most often applied to electroneutral solutions in the absence of a pressure gradient. Table 2.1 shows the corresponding form of the electrodiffusional contribution to the dissipation function for electroneutral solutions in different reference frames; note that the sums in the dissipation function exclude the solvent, $i = 0$.

It should be stressed that the flux densities are coupled through the definition of the reference velocity and the driving forces are coupled through the Gibbs–Duhem equation. As far as the following sections are concerned, the different reference frames introduced are of secondary importance since most of them are restricted to dilute solutions and then

$$\vec{j}_i^m \approx \vec{j}_i^\upsilon \approx \vec{j}_i^{\mathrm{H}}, \quad \vec{v} \approx \vec{v}_0 \approx \vec{v}_\upsilon, \tag{2.18}$$

and

$$\theta_{\mathrm{ed}} = -\sum_{i \neq 0} \vec{j}_i^{\mathrm{H}} \cdot \vec{\nabla} \tilde{\mu}_i. \tag{2.19}$$

Table 2.1. Electrodiffusional contribution to the dissipation function in different reference frames

Reference frame	Reference velocity	Dissipation function
Barycentric	$\vec{v} \equiv \frac{1}{\rho} \sum_i M_i \vec{j}_i = \sum_i w_i \vec{v}_i$	$\theta_{\mathrm{ed}} = -\sum_{i \neq 0} \vec{j}_i^m \cdot \left(\vec{\nabla} \tilde{\mu}_i - \frac{M_i}{M_0} \vec{\nabla} \mu_0 \right)$
Fick's	$\vec{v}_\upsilon \equiv \sum_i \upsilon_i \vec{j}_i = \sum_i c_i \upsilon_i \vec{v}_i$	$\theta_{\mathrm{ed}} = -\sum_{i \neq 0} \vec{j}_i^\upsilon \cdot \left(\vec{\nabla} \tilde{\mu}_i - \frac{\upsilon_i}{\upsilon_0} \vec{\nabla} \mu_0 \right)$
Hittorf's	$\vec{v}_0$	$\theta_{\mathrm{ed}} = -\sum_{i \neq 0} \vec{j}_i^{\mathrm{H}} \cdot \vec{\nabla} \tilde{\mu}_i$

From the above dissipation function, the phenomenological transport equations in a multicomponent system can be written as

$$\vec{j}_i^{\mathrm{H}} = -\sum_k l_{i,k} \vec{\nabla}\tilde{\mu}_k, \tag{2.20}$$

or, in matrix form,

$$\begin{pmatrix} \vec{j}_1^{\mathrm{H}} \\ \vec{j}_2^{\mathrm{H}} \\ \vdots \\ \vec{j}_N^{\mathrm{H}} \end{pmatrix} = -\begin{pmatrix} l_{1,1} & l_{1,2} & \cdots & l_{1,N} \\ l_{1,2} & l_{2,2} & \cdots & l_{2,N} \\ \vdots & \vdots & \ddots & \vdots \\ l_{1,N} & l_{2,N} & \cdots & l_{N,N} \end{pmatrix} \begin{pmatrix} \vec{\nabla}\tilde{\mu}_1 \\ \vec{\nabla}\tilde{\mu}_2 \\ \vdots \\ \vec{\nabla}\tilde{\mu}_N \end{pmatrix}, \tag{2.21}$$

where the Onsager reciprocal relations, eqn (2.7), have already been used and, therefore, there are only $N(N+1)/2$ independent ionic phenomenological transport coefficients, $l_{i,k}$. Note that the index k in eqn (2.20) excludes the solvent, even though no explicit reference to this fact is shown hereinafter.

2.1.3 Binary electrolyte solution

In this section we consider an electrodiffusion process in a binary electrolyte solution. The electrolyte $A_{\nu_1}C_{\nu_2}$ is assumed to be completely dissociated into ν_1 ions A^{z_1} and ν_2 ions B^{z_2} whose charge numbers z_1 and z_2 satisfy the stoichiometric relation $z_1\nu_1 + z_2\nu_2 = 0$. The dissipation function is

$$-\theta_{\mathrm{ed}} = \vec{j}_1^{\mathrm{H}} \cdot \vec{\nabla}\tilde{\mu}_1 + \vec{j}_2^{\mathrm{H}} \cdot \vec{\nabla}\tilde{\mu}_2, \tag{2.22}$$

and, therefore, the ionic phenomenological transport equations can be written as

$$\begin{pmatrix} \vec{j}_1^{\mathrm{H}} \\ \vec{j}_2^{\mathrm{H}} \end{pmatrix} = -\begin{pmatrix} l_{1,1} & l_{1,2} \\ l_{1,2} & l_{2,2} \end{pmatrix} \begin{pmatrix} \vec{\nabla}\tilde{\mu}_1 \\ \vec{\nabla}\tilde{\mu}_2 \end{pmatrix}. \tag{2.23}$$

Other choices of fluxes and forces are also possible provided that the sum of their products leads to the same dissipation function, that is, the dissipation function must remain invariant [2]. The choice can be made, for instance, on the basis of the simplicity of the transport equations. We show next the transformation rules for forces, fluxes, and phenomenological coefficients by comparing the transport equations for neutral components and those for ionic species in a binary solution.

The dissipation function in eqn (2.22) can also be written as

$$-\theta_{\mathrm{ed}} = \vec{J}_{12}^{\mathrm{H}} \cdot \vec{\nabla}\mu_{12} + \vec{I} \cdot \vec{\nabla}\phi_{\mathrm{ohm}}, \tag{2.24}$$

and an alternative set of phenomenological transport equations is

$$\begin{pmatrix} \vec{J}^{\mathrm{H}}_{12} \\ \vec{I} \end{pmatrix} = - \begin{pmatrix} L_{12,12} & L_{12,\phi} \\ L_{12,\phi} & L_{\phi,\phi} \end{pmatrix} \begin{pmatrix} \vec{\nabla}\mu_{12} \\ \vec{\nabla}\phi_{\mathrm{ohm}} \end{pmatrix}, \tag{2.25}$$

where $\vec{J}^{\mathrm{H}}_{12}$, $\vec{I}$, $\vec{\nabla}\mu_{12}$, and $\vec{\nabla}\phi_{\mathrm{ohm}}$ are the electrolyte flux density, the conduction current density and the gradients of the electrolyte chemical potential and of the ohmic potential, respectively. These fluxes and forces are related to those used in eqn (2.22) by the following equations

$$\begin{pmatrix} \vec{J}^{\mathrm{H}}_{12} \\ \vec{I} \end{pmatrix} = \begin{pmatrix} t_2/\nu_1 & t_1/\nu_2 \\ z_1 F & z_2 F \end{pmatrix} \begin{pmatrix} \vec{j}^{\mathrm{H}}_{1} \\ \vec{j}^{\mathrm{H}}_{2} \end{pmatrix} \tag{2.26}$$

$$\begin{pmatrix} \vec{\nabla}\tilde{\mu}_1 \\ \vec{\nabla}\tilde{\mu}_2 \end{pmatrix} = \begin{pmatrix} t_2/\nu_1 & z_1 F \\ t_1/\nu_2 & z_2 F \end{pmatrix} \begin{pmatrix} \vec{\nabla}\mu_{12} \\ \vec{\nabla}\phi_{\mathrm{ohm}} \end{pmatrix}, \tag{2.27}$$

where

$$t_1 \equiv \frac{z_1(z_1 l_{1,1} + z_2 l_{1,2})}{z_1^2 l_{1,1} + 2 z_1 z_2 l_{1,2} + z_2^2 l_{2,2}}, \tag{2.28}$$

and $t_2 = 1 - t_1$ are the migrational transport numbers of the ionic species. It is then required that the two sets of transport coefficients satisfy the relation [2]

$$\begin{pmatrix} L_{12,12} & L_{12,\phi} \\ L_{12,\phi} & L_{\phi,\phi} \end{pmatrix} = \begin{pmatrix} t_2/\nu_1 & t_1/\nu_2 \\ z_1 F & z_2 F \end{pmatrix} \begin{pmatrix} l_{1,1} & l_{1,2} \\ l_{1,2} & l_{2,2} \end{pmatrix} \begin{pmatrix} t_2/\nu_1 & z_1 F \\ t_1/\nu_2 & z_2 F \end{pmatrix}. \tag{2.29}$$

Notice that the transformation matrices between the fluxes and forces in eqns (2.26) and (2.27) are the transpose of each other.

Interestingly, the matrix multiplication in eqn (2.29) leads to the result

$$L_{12,\phi} = 0, \tag{2.30}$$

and hence the transport equations in eqn (2.25) simplify to

$$\vec{J}^{\mathrm{H}}_{12} = -L_{12,12}\vec{\nabla}\mu_{12} \tag{2.31}$$

$$\vec{I} = -\kappa\vec{\nabla}\phi_{\mathrm{ohm}}, \tag{2.32}$$

where $\kappa \equiv L_{\phi,\phi} \equiv F^2(z_1^2 l_{1,1} + 2z_1 z_2 l_{1,2} + z_2^2 l_{2,2})$ is the electrical conductivity of the solution. These equations can be considered as generalizations of Fick's first law and Ohm's law, respectively, and they show that the mass and charge transport are decoupled [3, 4]. This important conclusion is not exclusive of the binary case under consideration and it is a natural consequence of the fact that the fluxes of the electrically neutral components cannot depend on the ohmic potential gradient.

2.1.4 Electric conduction

In Section 1.2.6 it was stated that the electric conduction is an irreversible process whose contribution to the dissipation function is

$$\theta_{\text{ohm}} = -\vec{I} \cdot \vec{\nabla}\phi_{\text{ohm}} = I^2/\kappa \geq 0. \tag{2.33}$$

This equation was not proved there and the ohmic potential gradient was defined through the generalized Ohm's law

$$\vec{I} = -\kappa \vec{\nabla}\phi_{\text{ohm}}. \tag{2.34}$$

In Section 2.1.3 we have shown that eqns (2.33) and (2.34) are valid in binary solutions. These equations can also be applied to multi-ionic solutions provided that the electrical conductivity and the ohmic potential gradient are defined as

$$\kappa \equiv F^2 \sum_i \sum_k z_i z_k l_{i,k}, \tag{2.35}$$

$$\vec{\nabla}\phi_{\text{ohm}} = \frac{1}{F} \sum_i \frac{t_i}{z_i} \vec{\nabla}\tilde{\mu}_i, \tag{2.36}$$

where t_i is the transport number of species i. Introducing the contribution of species i to the electrical conductivity of the solution as

$$\kappa_i \equiv F^2 z_i \sum_k z_k l_{i,k}, \tag{2.37}$$

the definition of the transport number becomes

$$t_i \equiv \kappa_i/\kappa, \tag{2.38}$$

and the ohmic conduction contribution to the flux density of species i is

$$\vec{j}_{i,\text{ohm}} = \frac{t_i \vec{I}}{z_i F}. \tag{2.39}$$

Note that $\kappa = \sum_i \kappa_i$, $\sum_i t_i = 1$, and $\vec{I} = F \sum_i z_i \vec{j}_{i,\text{ohm}}$. Equations (2.33)–(2.39) describe electric conduction even in the presence of concentration gradients.

2.1.5 Component transport equations

The contribution of the diffusion of dissociated electrolytes and neutral solutes to the dissipation function, $\theta_{\text{dif}} \equiv \theta_{\text{ed}} - \theta_{\text{ohm}}$, is determined from eqns (2.19) and (2.36) as

$$\theta_{\mathrm{dif}} = -\sum_i \vec{j}_i^{\mathrm{H}} \cdot \vec{\nabla}\tilde{\mu}_i + \vec{I} \cdot \vec{\nabla}\phi_{\mathrm{ohm}} = -\sum_i \left(\vec{j}_i^{\mathrm{H}} - \frac{t_i\vec{I}}{z_iF}\right) \cdot \vec{\nabla}\tilde{\mu}_i$$

$$= -\sum_i \left(\vec{j}_i^{\mathrm{H}} - \frac{t_i\vec{I}}{z_iF}\right) \cdot \vec{\nabla}\mu_i. \tag{2.40}$$

In these sums, the index i runs over the ionic species, but it should be apparent that the electrolyte diffusion can be described more naturally in terms of the neutral electrolyte components. If we use uppercase symbols for both the flux densities and the index that runs over the neutral components, the ionic flux density $\vec{j}_i^{\mathrm{H}}$ can be expressed in terms of the component flux densities $\vec{J}_K^{\mathrm{H}}$ as[2]

$$\vec{j}_i^{\mathrm{H}} = \sum_K \nu_{i,K}\vec{J}_K^{\mathrm{H}} + \frac{t_i\vec{I}}{z_iF}, \tag{2.41}$$

where $\nu_{i,K}$ is the stoichiometric coefficient of ion i in component K (which is zero if the dissociation of component K does not produce species i in solution).

By noting that the chemical potential gradient of electrolyte K is

$$\vec{\nabla}\mu_K = \sum_i \nu_{i,K}\vec{\nabla}\tilde{\mu}_i = \sum_i \nu_{i,K}\vec{\nabla}\mu_i, \tag{2.42}$$

where the relation $\sum_i z_i\nu_{i,K} = 0$ has been used in the last equality, the diffusion contribution to the dissipation function can also be written as

$$\theta_{\mathrm{dif}} = -\sum_K \vec{J}_K^{\mathrm{H}} \cdot \vec{\nabla}\mu_K. \tag{2.43}$$

The phenomenological transport equations for the electrolytes are then

$$\vec{J}_I^{\mathrm{H}} = -\sum_K L_{I,K}\vec{\nabla}\mu_K. \tag{2.44}$$

The equivalence between the descriptions of transport processes based on either the ionic species or the neutral components requires that they involve the same number of transport coefficients. In an electroneutral solution with N ionic species there are $N-1$ independent neutral electrolytes. In the ionic approach, the number of phenomenological transport coefficients is $N(N+1)/2$. In the component approach there are $N(N-1)/2$ independent coefficients $L_{I,K}$, $(N-1)$ independent transport numbers t_i (which are involved in the definition of the ohmic potential gradient), and the electrical conductivity κ. This also makes a total of $N(N+1)/2$ coefficients.

[2] The indices i and K also run over the non-dissociated neutral components, but there is no difference then between the ionic and the component formulations.

2.1.6 Ternary electrolyte solutions

Consider a ternary electrolyte solution formed by two binary electrolytes with a common ion. The common ion is denoted by index $i = 3$, and the other two by indices 1 and 2. The electrolytes $A_{\nu_1}C_{\nu_{3,1}}$ and $D_{\nu_2}C_{\nu_{3,2}}$ are denoted by indices 13 and 23, respectively; and are completely dissociated according to the equilibria

$$A_{\nu_1}C_{\nu_{3,1}} \leftrightharpoons \nu_1 A^{z_1} + \nu_{3,1}C^{z_3} \tag{2.45}$$

$$D_{\nu_2}C_{\nu_{3,2}} \leftrightharpoons \nu_2 D^{z_2} + \nu_{3,2}C^{z_3}. \tag{2.46}$$

The electrolyte molar concentrations are c_{13} and c_{23}, and the stoichiometric relations[3] $z_1\nu_1 + z_3\nu_{3,1} = 0$ and $z_2\nu_2 + z_3\nu_{3,2} = 0$ are satisfied.

The phenomenological transport equations can be written in matrix form as

$$\begin{pmatrix} \vec{j}^{\,H}_{13} \\ \vec{j}^{\,H}_{23} \\ \vec{I} \end{pmatrix} = - \begin{pmatrix} L_{13,13} & L_{13,23} & 0 \\ L_{13,23} & L_{23,23} & 0 \\ 0 & 0 & \kappa \end{pmatrix} \begin{pmatrix} \vec{\nabla}\mu_{13} \\ \vec{\nabla}\mu_{23} \\ \vec{\nabla}\phi_{\text{ohm}} \end{pmatrix}, \tag{2.47}$$

where the Onsager reciprocal relation has been applied. Since the driving forces and the fluxes satisfy the equations

$$\begin{pmatrix} \vec{\nabla}\mu_{13} \\ \vec{\nabla}\mu_{23} \\ \vec{\nabla}\phi_{\text{ohm}} \end{pmatrix} = \begin{pmatrix} \nu_1 & 0 & \nu_{3,1} \\ 0 & \nu_2 & \nu_{3,2} \\ t_1/z_1F & t_2/z_2F & t_3/z_3F \end{pmatrix} \begin{pmatrix} \vec{\nabla}\tilde{\mu}_1 \\ \vec{\nabla}\tilde{\mu}_2 \\ \vec{\nabla}\tilde{\mu}_3 \end{pmatrix} \tag{2.48}$$

$$\begin{pmatrix} \vec{j}^{\,H}_{1} \\ \vec{j}^{\,H}_{2} \\ \vec{j}^{\,H}_{3} \end{pmatrix} = \begin{pmatrix} \nu_1 & 0 & t_1/z_1F \\ 0 & \nu_2 & t_2/z_2F \\ \nu_{3,1} & \nu_{3,2} & t_3/z_3F \end{pmatrix} \begin{pmatrix} \vec{j}^{\,H}_{13} \\ \vec{j}^{\,H}_{23} \\ \vec{I} \end{pmatrix}, \tag{2.49}$$

we conclude that the relation between the phenomenological transport coefficients in the ionic and component formalisms is

$$\begin{pmatrix} l_{1,1} & l_{1,2} & l_{1,3} \\ l_{1,2} & l_{2,2} & l_{2,3} \\ l_{1,3} & l_{2,3} & l_{3,3} \end{pmatrix} = \begin{pmatrix} \nu_1 & 0 & t_1/z_1F \\ 0 & \nu_2 & t_2/z_2F \\ \nu_{3,1} & \nu_{3,2} & t_3/z_3F \end{pmatrix} \begin{pmatrix} L_{13,13} & L_{13,23} & 0 \\ L_{13,23} & L_{23,23} & 0 \\ 0 & 0 & \kappa \end{pmatrix} \begin{pmatrix} \nu_1 & 0 & \nu_{3,1} \\ 0 & \nu_2 & \nu_{3,2} \\ t_1/z_1F & t_2/z_2F & t_3/z_3F \end{pmatrix}.$$

[3] The stoichiometric coefficients ν_1, ν_2, $\nu_{3,1}$, and $\nu_{3,2}$ correspond to $\nu_{1,13}$, $\nu_{2,23}$, $\nu_{3,13}$, and $\nu_{3,23}$, respectively, in the notation of the previous section.

2.2 The Fickian approach

2.2.1 Introduction

The Fickian approach provides an alternative description of the diffusion of neutral components. In the phenomenological approach, the driving forces are the chemical potential gradients. In the Fickian approach, the concentration gradients are used as driving forces and the transport equations are

$$\vec{j}_I^{\rm H} = -\sum_J D_{I,J}\vec{\nabla}c_J, \tag{2.50}$$

where the sum runs over all the neutral components and $D_{I,K}$ are known as the *Fickian diffusion coefficients*.[4] Note that the flux density of any component is influenced by the concentration gradients of all components.

The diffusion coefficients can be related to the phenomenological transport coeficients as follows. Neglecting the influence of the pressure gradient on the electrolyte diffusion, the gradient of the chemical potential of component K can be written as

$$\vec{\nabla}\mu_K \approx \vec{\nabla}\mu_K^{\rm c} = \sum_J \frac{\partial \mu_K^{\rm c}}{\partial c_J}\vec{\nabla}c_J, \tag{2.51}$$

where the superscript c on $\mu_K^{\rm c}$ indicates that this is only the composition-dependent contribution to the chemical potential. Substituting eqn (2.51) in eqn (2.44) and comparing with eqn (2.50), the relation between diffusion and phenomenological coefficients is

$$D_{I,J} = \sum_K L_{I,K}\frac{\partial \mu_K^{\rm c}}{\partial c_J}. \tag{2.52}$$

Equation (2.52) shows that the diffusion coefficients of the neutral component depend on the composition. In fact, one of the practical problems of the Fickian approach is that this dependence is strong in multicomponent systems. The simplified approach based on the Nernst–Planck approximation (see Section 2.3.1 below) then proves to be very useful because it involves measurable ionic fluxes and ionic diffusion coefficients (that are proportional to the ionic mobilities), which show a much weaker dependence on the composition.

2.2.2 Fick's law

In an aqueous solution of a neutral (non-dissociable) solute and in the absence of a pressure gradient, eqn (2.50) reduces to[5]

$$\vec{j}_1^{\rm H} = -D_1\vec{\nabla}c_1, \tag{2.53}$$

[4] They are named Fickian diffusion coefficients to make clear the difference with the Stefan–Maxwell diffusion coefficients introduced in Sections 2.4.2 and 2.4.4. These adjectives refer to the formalism and not to the reference frame.

[5] In this case, the diffusion coefficient is written as D_1 instead of $D_{1,1}$ for the sake of simplicity.

which is known as *Fick's first law*. From eqn (2.52) and assuming ideal behaviour, the diffusion coefficient can be related to the phenomenological coefficient as

$$L_{1,1} = \frac{D_1}{\partial \mu_1^c / \partial c_1} \approx \frac{D_1 c_1}{RT}. \tag{2.54}$$

Although the diffusion coefficient D_1 is not a constant either, eqn (2.54) shows that the phenomenological coefficient $L_{1,1}$ depends on the local composition (and temperature). The integration of a transport equation as simple as $\vec{J}_1^{\mathrm{H}} = -L_{1,1}\vec{\nabla}\mu_1$ then becomes a difficult task unless we know the dependence of $L_{1,1}$ and μ_1 on the concentration c_1. The integration of Fick's first law, $\vec{J}_1^{\mathrm{H}} = -D_1\vec{\nabla}c_1$, on the contrary, is much easier under the assumption that the diffusion coefficient is a constant.

From an experimental point of view, Fick's first law should be stated in the Fick reference frame as

$$\vec{J}_1^{\upsilon} = -D_1\vec{\nabla}c_1, \tag{2.55}$$

and the diffusion coefficient should include a superscript υ to stress its dependence on the reference frame. However, in dilute solutions such a difference is negligible; note that the transport equations have been deduced from the approximate expression of θ_{ed} in eqn (2.19).

The diffusion coefficient D_1 is not a property of the solute only, but of the system solvent–solute and, in fact, it should be denoted as $D_{1,0}$ (see Section 2.4.2). To show this in a simple way, we start from the Euler equation $c_0\upsilon_0 + c_1\upsilon_1 = 1$, which simply states that the sum of the fractions of volume occupied by the solvent and the solute is one. Taking the gradient of this equation and assuming that the partial molar volumes υ_i ($i = 0, 1$) remain approximately constant when the composition varies, we obtain the relation

$$\upsilon_0\vec{\nabla}c_0 + \upsilon_1\vec{\nabla}c_1 = \vec{0}. \tag{2.56}$$

Since the volume-average velocity is zero in Fick's reference frame, we have that

$$\upsilon_0\vec{J}_0^{\upsilon} + \upsilon_1\vec{J}_1^{\upsilon} = \vec{0} \tag{2.57}$$

and eqn (2.55) can be rewritten as

$$\vec{J}_0^{\upsilon} = -D_1\vec{\nabla}c_0, \tag{2.58}$$

so that D_1 is also the diffusion coefficient of the solvent. This result should be expected because diffusion cannot take place in a monocomponent system. In a binary solution the diffusion of the solute and the solvent are not two different processes but a single process, which is characterized by a single diffusion coefficient.

2.2.3 Diffusion-conduction equations

When the Fickian transport equations for the neutral components are substituted into eqn (2.41), the ionic flux density $\vec{j}_i^{\rm H}$ can be expressed as

$$\vec{j}_i^{\rm H} = \underbrace{-\sum_J \sum_K \nu_{i,K} D_{K,J} \vec{\nabla} c_J}_{\text{chemical diffusion}} + \underbrace{\frac{t_i \vec{I}}{z_i F}}_{\text{ohmic conduction}} \quad . \tag{2.59}$$

This is the *diffusion–conduction equation* for the ionic species *i*. It must be stressed that it involves diffusion of neutral components (not of ions) and *ohmic conduction*. The term *chemical diffusion* is sometimes used to make explicit the difference from the ionic diffusion [5].

The simplest particular case of these equations corresponds to a strong binary electrolyte solution where the ionic flux densities are

$$\vec{j}_i^{\rm H} = -\nu_i D_{12} \vec{\nabla} c_{12} + \frac{t_i \vec{I}}{z_i F}, \quad i = 1, 2. \tag{2.60}$$

Similarly, in the ternary electrolyte case described in Section 2.1.6, the Fickian transport equations for the components are

$$-\vec{J}_{13}^{\rm H} = D_{13,13} \vec{\nabla} c_{13} + D_{13,23} \vec{\nabla} c_{23}, \tag{2.61}$$

$$-\vec{J}_{23}^{\rm H} = D_{23,13} \vec{\nabla} c_{13} + D_{23,23} \vec{\nabla} c_{23}, \tag{2.62}$$

and the diffusion–conduction equations are

$$\vec{j}_1^{\rm H} = -\nu_1 (D_{13,13} \vec{\nabla} c_{13} + D_{13,23} \vec{\nabla} c_{23}) + \frac{t_1 \vec{I}}{z_1 F}, \tag{2.63}$$

$$\vec{j}_2^{\rm H} = -\nu_2 (D_{23,13} \vec{\nabla} c_{13} + D_{23,23} \vec{\nabla} c_{23}) + \frac{t_2 \vec{I}}{z_2 F}, \tag{2.64}$$

$$\begin{aligned} \vec{j}_3^{\rm H} = & -\nu_{3,1} (D_{13,13} \vec{\nabla} c_{13} + D_{13,23} \vec{\nabla} c_{23}) \\ & -\nu_{3,2} (D_{23,13} \vec{\nabla} c_{13} + D_{23,23} \vec{\nabla} c_{23}) + \frac{t_3 \vec{I}}{z_3 F}, \end{aligned} \tag{2.65}$$

where the cross-diffusion coefficients $D_{13,23}$ and $D_{23,13}$ are not equal to each other. Although these equations look relatively simple, it must be observed that the electrolyte diffusion coefficients and the ionic transport numbers[6] are not constant, and we require expressions to evaluate them before these transport equations can be integrated. Such expressions are derived in Section 2.3.7, making use of the Nernst–Planck approximation.

[6] The transport numbers can be related to the ionic phenomenological coefficients as shown in eqn (2.38), but in the component approach, either Fickian or phenomenological, they are independent transport coefficients that cannot be related to the electrolyte diffusion coefficients.

2.3 The Nernst–Planck approximation

2.3.1 Introduction

The transport equations include cross-coefficients that couple the transport of the different components. In fact, one of the major achievements of the thermodynamics of irreversible processes was the explanation of coupled transport phenomena. The phenomenological and the Fickian approaches, however, cannot provide estimates for the cross-coefficients. Moreover, these approaches do not help much in understanding the physical basis of such coupled transport phenomena. Such understanding is only possible when we realize that the electrolytes are dissociated into ions in solution. Ions are charged species that interact with each other as well as with the solvent. These interactions are quite complex and difficult to model, but a simple approach proposed by the end of nineteenth century has proved to be able to provide a satisfactory explanation to most transport processes in ionic solutions. It has become essential for the development of electrochemical transport processes, membrane separation processes, and even semiconductor devices.

The fundamental idea behind this approach is the so-called *principle of independence* of the ionic fluxes. This principle states that, in a first approximation, cross-phenomenological coefficients can be neglected in the transport equations for ionic species, so that

$$l_{i,k} \approx 0 \text{ if } i \neq k, \tag{2.66}$$

and

$$\vec{j}_i^{\mathrm{H}} \approx -l_{i,i}\vec{\nabla}\tilde{\mu}_i = -\frac{t_i\kappa}{z_i^2F^2}\vec{\nabla}\tilde{\mu}_i. \tag{2.67}$$

The flux density of a species i is then determined by its electrochemical potential gradient only, and not by the electrochemical potential gradients of other species. The coefficient $l_{i,i}$ is related to the short-range interaction between the ionic species i and the solvent. The approximation $l_{i,k} \approx 0$ $(i \neq k)$ is somehow equivalent to assuming that there are no short-range interactions among ions.[7] In dilute solutions, the probability that two ions get close is relatively small and this picture works nicely. But in concentrated solutions, an ion is surrounded not only by solvent molecules but also by other ions. Short-range interactions become then important and eqn (2.66) is no longer valid. Either the full set of phenomenological equations, eqn (2.20), or the Stefan–Maxwell approach, are then better suited.

[7] However, long-range electrostatic interactions among the ionic species in solution are not negligible. These are satisfactorily described in dilute solutions within the Nernst–Planck approach, as explained in Section 2.3.3.

2.3.2 Nernst–Planck equation

If we introduce the ionic diffusion coefficient D_i from the relation $l_{i,i} = D_i c_i/RT$, and approximate the electrochemical potential gradient by $\vec{\nabla}\tilde{\mu}_i \approx RT\vec{\nabla}\ln c_i + z_i F\vec{\nabla}\phi$, eqn (2.67) can be transformed to

$$\vec{j}_i^{\rm H} = \underbrace{-D_i\vec{\nabla}c_i}_{\text{ionic diffusion}} + \underbrace{z_i D_i c_i f\vec{E}}_{\text{ionic migration}} = -D_i(\vec{\nabla}c_i + z_i c_i f\,\vec{\nabla}\phi) \tag{2.68}$$

where $f \equiv F/RT$. This is known as the classical *Nernst–Planck equation* and shows that the ionic flux density (in Hittorf's reference frame) has one contribution proportional to the concentration gradient and another one proportional to the electric field. The first one is related to *ionic diffusion* and the second one to *ionic migration* (or drift) and for this reason the Nernst–Planck equation is also known as the *diffusion-migration equation* (or drift-diffusion equation).

In the absence of concentration changes, an ionic species of charge number z_i under the influence of an electric field $\vec{E}$ moves with respect to the solvent with an average velocity

$$\vec{v}_i^{\rm H} = u_i z_i F\vec{E}, \tag{2.69}$$

where u_i is the (electrochemical or mechanical) mobility of species i.[8] The comparison of the flux density

$$\vec{j}_i^{\rm H} = c_i\vec{v}_i^{\rm H} = c_i u_i z_i F\vec{E} = -u_i z_i c_i F\vec{\nabla}\phi, \tag{2.70}$$

with eqn (2.68) shows that the transport coefficients D_i and u_i must satisfy the *Nernst–Einstein relation*, $D_i = u_i RT$.

For the sake of simplicity, we drop the superscript H on the flux densities hereinafter. In the presence of convective flow we write the Nernst–Planck equation as

$$\vec{j}_i = -D_i(\vec{\nabla}c_i + z_i c_i f\,\vec{\nabla}\phi) + c_i\vec{v}, \tag{2.71}$$

where, in principle, $\vec{v}$ should be the barycentric velocity. In practice, this equation is also used taking $\vec{v}$ as the solvent velocity $\vec{v}_0$. This approximation is considered to be consistent with the other ones used in the derivation of eqn (2.67), which are:

1) cross-phenomenological coefficients are neglected, $l_{i,k} \approx 0$ if $i \neq k$,
2) deviations from the Nernst–Einstein relation (due, e.g. to electrophoretic contributions) are neglected, i.e. the electrical and the diffusion mobilities, u_i and D_i/RT, are the same and the phenomenological coefficient is $l_{i,i} = D_i c_i/RT = u_i c_i$,
3) the gradient of the activity coefficient is neglected (i.e. the activity coefficient is independent of concentration), and

[8] The electric or electrophoretic mobility, $u_i z_i F$, is used in some other references.

4) the difference between the barycentric and the solvent velocity is neglected. They all can be accepted as reasonable approximations for dilute solutions. Accordingly, the coefficients D_i are usually given the (constant) values corresponding to infinite dilution, which do not depend on the other ions present in solution.

Finally, it is interesting to note that neglecting the cross-phenomenological coefficients in the Nernst–Planck approximation simplifies significantly the description of electric conduction. Thus, for instance, the ionic transport number becomes

$$t_i \equiv \frac{\lambda_i c_i}{\kappa} \equiv \frac{\kappa_i}{\kappa} = \frac{z_i^2 D_i c_i}{\sum_k z_k^2 D_k c_k}, \tag{2.72}$$

where[9] $\lambda_i \equiv \kappa_i / c_i = z_i^2 F^2 D_i / RT$ is the molar conductivity of species i and

$$\kappa = \sum_i \kappa_i = \frac{F^2}{RT} \sum_i z_i^2 D_i c_i \tag{2.73}$$

is the electrical conductivity of the solution.

2.3.3 Electrical coupling between the ionic fluxes

We have explained in Section 2.3.1 that the Nernst–Planck approach is based on the principle of independence of the ionic fluxes. The name of this principle is rather unfortunate because the migration term in eqn (2.68) couples the motions of all charged species and they are not independent. In fact, the flux density of species i can also be written in electroneutral solutions as

$$-\vec{j}_i = \sum_k D_{i,k} \vec{\nabla} c_k - \frac{t_i \vec{I}}{z_i F}, \tag{2.74}$$

which involves the concentration gradients of all the ionic species, $k = 1, 2, \ldots, N$, where N is the number of ionic species. The diffusion coefficients in eqn (2.74) are

$$D_{i,k} \equiv D_i \delta_{ik} + \frac{t_i}{z_i} z_k (D_i - D_k), \tag{2.75}$$

where δ_{ik} is the Kronecker delta ($\delta_{ik} = 1$ when $i = k$ and $\delta_{ik} = 0$ when $i \neq k$).

Equation (2.74) resembles eqn (2.59) and it is indeed another form of the diffusion–conduction equation. The difference between them is that the former involves ionic species, while the latter involves neutral components. In order for them to be really equivalent, they should also involve the same number

[9] This is another form of the *Nernst–Einstein relation*.

of transport coefficients. On the one hand, the sum in eqn (2.74) runs over all ionic species and hence there are N^2 ionic diffusion coefficients $D_{i,k}$. On the other hand, the sum in eqn (2.59) runs over neutral components and hence there are $(N-1)^2$ Fickian diffusion coefficients $D_{I,K}$ because in an electroneutral solution with N ionic species there are $N-1$ independent neutral components. This suggests that we can use the local electroneutrality assumption to eliminate in eqn (2.74) the concentration of, e.g. species $i = N$ in terms of the others. In this case the number of ionic diffusion coefficients $D_{i,k}$ is also reduced to $(N-1)^2$. The flux density of species $i = N$ is calculated from the equation for the electric current as $z_N\vec{j}_N = (\vec{I}/F) - \sum_{k \neq N} z_k\vec{j}_k$, where index k runs from 1 to $N-1$. The flux densities of species $i = 1, 2, \ldots, N-1$ are given by

$$-\vec{j}_i = \sum_{k \neq N} D'_{i,k}\vec{\nabla}c_k - \frac{t_i\vec{I}}{z_iF}, \tag{2.76}$$

where the cross-diffusion coefficients $D'_{i,k}$ are defined as

$$D'_{i,k} \equiv D_i\delta_{ik} + \frac{t_i}{z_i}z_k(D_N - D_k). \tag{2.77}$$

The $(N-1)^2$ diffusion coefficients $D'_{i,k}$ are not independent, and so neither are the Fickian coefficients $D_{I,K}$. In fact, in the Nernst–Planck approach the number of independent transport coefficients is simply N (the diffusion coefficients D_i), while in the Fickian (and in the phenomenological) approach there are $N(N+1)/2$ independent coefficients. The difference between these two numbers is $N(N-1)/2$, which is just equal to the number of ionic cross-phenomenological coefficients $l_{i,k}$ that are neglected in the Nernst–Planck approach.

The diffusion coefficients $D_{i,k}$ and $D'_{i,k}$ can be conveniently written in terms of the ionic molar conductivities λ_i as

$$D_{i,k} = \frac{RT}{z_i^2F^2}\left(\lambda_i\delta_{ik} + t_i\frac{z_k^2\lambda_i - z_i^2\lambda_k}{z_iz_k}\right), \tag{2.78}$$

$$D'_{i,k} = \frac{RT}{z_i^2F^2}\left(\lambda_i\delta_{ik} + z_it_i\frac{z_k^2\lambda_N - z_N^2\lambda_k}{z_N^2z_k}\right). \tag{2.79}$$

Note finally that, since the ionic cross-diffusion coefficients $D_{i,k} (i \neq k)$ are proportional to $(D_i - D_k)$, in solutions that do not contain significant concentrations of ions with relatively high or low diffusion coefficients (like hydrogen ions or large organic ions, respectively), the transport equations decouple and take the approximate form

$$-\vec{j}_i = D\vec{\nabla}c_i - \frac{t_i\vec{I}}{z_iF}, \tag{2.80}$$

where D is an average diffusion coefficient.

2.3.4 Diffusion potential

We have shown in Sections 2.2.3 and 2.3.2 that the flux density of an ionic species i can be written either as the sum of two contributions describing chemical diffusion and ohmic conduction or as the sum of two contributions describing ionic diffusion and ionic migration

$$\vec{j}_i = \vec{j}_{i,\,\text{chem dif}} + \vec{j}_{i,\,\text{ohm}} = \vec{j}_{i,\,\text{ion dif}} + \vec{j}_{i,\,\text{mig}}. \tag{2.81}$$

It is important to observe that the ionic migration contribution to the flux density

$$\vec{j}_{i,\,\text{mig}} \equiv -z_i D_i c_i f \vec{\nabla}\phi = -\frac{t_i \kappa}{z_i F}\vec{\nabla}\phi \tag{2.82}$$

is not equal to the electric conduction contribution

$$\vec{j}_{i,\,\text{ohm}} = \frac{t_i \vec{I}}{z_i F} = -\frac{t_i \kappa}{z_i F}\vec{\nabla}\phi_{\text{ohm}} \tag{2.83}$$

because the total electric potential gradient and the ohmic potential gradient are not the same. Their difference is the diffusion potential gradient[10]

$$\vec{\nabla}\phi_{\text{dif}} \equiv \vec{\nabla}\phi - \vec{\nabla}\phi_{\text{ohm}} = -\frac{1}{F}\sum_k \frac{t_k}{z_k}\vec{\nabla}\mu_k, \tag{2.84}$$

where we have used eqn (2.36).

An interesting consequence of the above equations is that under open-circuit conditions ($I = 0$) there can be ionic migration but not conduction, because the field involved in the electric conduction is only the 'external' or ohmic electric field, $-\vec{\nabla}\phi_{\text{ohm}} \equiv \vec{I}/\kappa$. On the contrary, the electric field involved in the migration term is the sum of the 'internal' field ($-\vec{\nabla}\phi_{\text{dif}}$) and the ohmic field ($-\vec{\nabla}\phi_{\text{ohm}}$) imposed externally to force the flow of electric current. Thus, the difference between the migration and the conduction terms in the ionic flux equation is

$$\vec{j}_{i,\,\text{mig}} - \vec{j}_{i,\,\text{ohm}} = -\frac{t_i \kappa}{z_i F}\vec{\nabla}\phi_{\text{dif}}, \tag{2.85}$$

and, according to eqn (2.81), this is also the difference between the chemical diffusion and the ionic diffusion terms

$$\vec{j}_{i,\,\text{chem dif}} - \vec{j}_{i,\,\text{ion dif}} = -\frac{t_i \kappa}{z_i F}\vec{\nabla}\phi_{\text{dif}}. \tag{2.86}$$

These equations evidence that the diffusion potential is responsible for the electrical coupling between the ionic fluxes.

[10] In the presence of a pressure gradient, there is another contribution to the electric potential gradient.

The origin of the diffusion potential gradient can be clarified by making use of the local electroneutrality assumption

$$\sum_i z_i c_i = 0. \tag{2.87}$$

Taking the chemical diffusion term from eqn (2.74)[11] with the cross-diffusion coefficients defined in eqn (2.75), and substituting it in eqn (2.86), it is obtained that the diffusion potential gradient in electroneutral solutions can also be evaluated as

$$\vec{\nabla}\phi_{\text{dif}} = \frac{F}{\kappa}\sum_k z_k(D_i - D_k)\vec{\nabla}c_k. \tag{2.88}$$

Hence, the diffusion potential originates from the differences in the ionic diffusion coefficients (and the need to maintain the local electroneutrality). In fact, since the ionic diffusion coefficients are usually of similar order of magnitude (with the exceptions of H^+ and OH^- ions), the diffusion potential gradient can be relatively small in the presence of electric current. Actually, in electrochemical cells with supporting electrolytes and applied current, the electric field is mostly ohmic. Nevertheless, we can assure that the diffusion potential gradient is important whenever there is electrostatic coupling between different ionic flux densities, and in the absence of electric current the electric field has a purely diffusional origin.

Finally, it is also interesting to comment that any ohmic potential gradient involves Joule dissipation because $\theta_{\text{ohm}} = -\vec{I}\cdot\vec{\nabla}\phi_{\text{ohm}} = (\vec{\nabla}\phi_{\text{ohm}})^2/\kappa \geq 0$. On the contrary, a non-zero diffusion potential gradient does not necessarily involve the existence of an irreversible process [6]. To clarify this point we must note first that eqn (2.84) is not restricted to electroneutral solutions. However, in non-electroneutral solutions, this potential gradient does not originate only from the coupling between the ion fluxes, and indeed the name 'diffusion potential gradient' becomes inappropriate. For instance, in the *equilibrium* electric double layer close to a charged electrode, there are no flux densities and no irreversible process takes place. However, in this layer there are concentration gradients and electric field. According to eqn (2.84) this field is $\vec{\nabla}\phi = \vec{\nabla}\phi_{\text{dif}} = -(1/F)\sum_k (t_k/z_k)\vec{\nabla}\mu_k$. Similarly, if we look at the Nernst–Planck equation, the flux density is zero but the ionic diffusion contribution and the ionic migration contribution (which is proportional to the diffusion potential gradient) are non-zero separately.

[11] Indeed, eqn (2.74) is obtained after inserting in eqn (2.68) the decomposition of the electric field $\vec{\nabla}\phi = \vec{\nabla}\phi_{\text{diff}} + \vec{\nabla}_{\text{ohm}}$, and making use of eqns (2.84) and (2.87).

2.3.5 Integration of the transport equations in multi-ionic solutions

In Section 2.3.3 we learned that the ionic fluxes are coupled because their driving forces are also coupled through the common electric field in the migration terms. This field is indirectly determined from the equation for the electric current density

$$\vec{I} = F \sum_i z_i \vec{j}_i, \tag{2.89}$$

and the need to satisfy the local electroneutrality condition.[12] Equation (2.74) shows this electrical coupling between the ionic fluxes. However, when we try to integrate it, we face the problem that the diffusion coefficients $D_{i,k}$ involve the transport numbers t_i, and these are functions of the unknown local ionic concentrations in multi-ionic solutions. Therefore, we have to find an alternative approach.

Making use of the local electroneutrality assumption in the form $\sum_i z_i \vec{\nabla} c_i = \vec{0}$, the Nernst–Planck equations of the different ionic species, eqn (2.68), can be combined to yield the following expression for the electric field

$$f\, \vec{\nabla}\phi = -\frac{\sum_k (z_k/D_k)\vec{j}_k}{\sum_j z_j^2 c_j}. \tag{2.90}$$

The individual Nernst–Planck equations can then be cast in the form

$$\frac{\vec{j}_i}{D_i} = -\vec{\nabla} c_i + \frac{z_i c_i}{\sum_j z_j^2 c_j} \sum_k z_k \frac{\vec{j}_k}{D_k}, \tag{2.91}$$

and solved in terms of sums such as

$$\sum_i \frac{\vec{j}_i}{D_i} = -\vec{\nabla} \sum_i c_i. \tag{2.92}$$

This procedure can be illustrated by considering a system with only two ionic classes, in which the charge number of all cations is z_+ and that of all anions is z_-. This system satisfies the relation $\sum_j z_j^2 c_j = -z_+ z_- c_{\mathrm{T}}$ where $c_{\mathrm{T}} \equiv \sum_i c_i$ is the total ionic concentration, and therefore eqns (2.90) and (2.92) lead to

$$f\, \vec{\nabla}\phi = \Gamma \vec{\nabla} \ln c_{\mathrm{T}}, \tag{2.93}$$

where

$$\Gamma \equiv -\frac{1}{z_+ z_-} \frac{\sum_k z_k j_k / D_k}{\sum_i j_i / D_i}. \tag{2.94}$$

[12] The Poisson equation of electrostatics must be used when this approximation fails.

Under steady-state conditions the flux densities, the ionic flux densities and Γ are constant, so that eqns (2.92) and (2.93) can be integrated providing c_T and ϕ are functions of position. Moreover, multiplying both sides of the Nernst–Planck equation for species i by $(c_T)^{z_i\Gamma}$ and making use of eqns (2.92) and (2.93), it can be transformed to[13]

$$\frac{j_i/D_i}{\sum_k j_k/D_k} \frac{\vec{\nabla}[(c_T)^{1+z_i\Gamma}]}{1+z_i\Gamma} = \vec{\nabla}[c_i(c_T)^{z_i\Gamma}], \tag{2.95}$$

which can be integrated easily because it is a linear equation with constant coefficients. This procedure is explained in further detail in Section 3.2.6.

2.3.6 Binary electrolyte solution

In a strong binary electrolyte the two ionic concentrations are coupled through the electroneutrality assumption, $z_1c_1 + z_2c_2 = 0$. Therefore, the solution composition can be specified by only one concentration variable, which can be either of the two ionic concentrations, the stoichiometric electrolyte concentration $c_{12} \equiv c_1/\nu_1 = c_2/\nu_2$, the mean electrolyte concentration $c_{\pm,12} \equiv (c_1^{\nu_1}c_2^{\nu_2})^{1/\nu_{12}}$, the total concentration $c_T \equiv c_1 + c_2$, the ionic strength $I \equiv (z_1^2c_1 + z_2^2c_2)/2$, the electrical conductivity $\kappa = (F^2/RT)(z_1^2D_1c_1 + z_2^2D_2c_2)$, etc. The transport equations obtained when employing any of these variables are all equivalent to each other.

In a binary solution, the local ionic transport numbers defined in eqn (2.72) are independent of the position and take the values

$$t_1 = \frac{\nu_2D_1}{\nu_2D_1 + \nu_1D_2} = \frac{z_1D_1}{z_1D_1 - z_2D_2} = 1 - t_2. \tag{2.96}$$

Furthermore, the sum in eqn (2.76) reduces to just one term, and the only diffusion coefficient is $D'_{1,1} = t_1D_2 + t_2D_1 \equiv D_{12}$. This is the Nernst–Hartley diffusion coefficient of the dissociated electrolyte and, with the help of eqn (2.96), it can be easily seen that

$$D_{12} \equiv t_1D_2 + t_2D_1 = \frac{\nu_{12}D_1D_2}{\nu_2D_1 + \nu_1D_2} = \frac{(z_1 - z_2)D_1D_2}{z_1D_1 - z_2D_2}, \tag{2.97}$$

where $\nu_{12} \equiv \nu_1 + \nu_2$. Since D_{12} and t_i are constant, the transport equation

$$\vec{j}_i = -\nu_iD_{12}\vec{\nabla}c_{12} + \frac{t_i\vec{I}}{z_iF} = -D_{12}\vec{\nabla}c_i + \frac{t_i\vec{I}}{z_iF} \tag{2.98}$$

can be integrated straightforwardly. It is important to stress that eqn (2.98) is the same as eqn (2.60), but the strength of the Nernst–Planck approach is that it also provides eqns (2.96) and (2.97) as estimates for the transport numbers

[13] In the case $z_i\Gamma = -1$, the term $\vec{\nabla}[(c_T)^{1+z_i\Gamma}]/(1+z_i\Gamma)$ should be replaced by $\vec{\nabla}\ln c_T$.

and the electrolyte diffusion coefficient, while similar equations were missing in the Fickian approach.

The electric field can be evaluated from eqn (2.90) as

$$f\,\vec{\nabla}\phi = -\frac{1}{z_1^2 c_1 + z_2^2 c_2}\left(\frac{z_1\vec{j}_1}{D_1} + \frac{z_2\vec{j}_2}{D_2}\right). \tag{2.99}$$

Substitution of eqn (2.99) into the Nernst–Planck equation, eqn (2.68), for any of the two ions gives the Fick first law

$$\vec{J}_{12} = -D_{12}\vec{\nabla}c_{12}, \tag{2.100}$$

where

$$\vec{J}_{12} \equiv \frac{t_2}{\nu_1}\vec{j}_1 + \frac{t_1}{\nu_2}\vec{j}_2 \tag{2.101}$$

is the electrolyte flux density. Under steady-state conditions, the flux densities are constant, and eqn (2.100) implies that the concentration profiles are linear. Similarly, the integration of eqn (2.99) shows that the electric potential varies logarithmically with position. Finally, as was shown in eqn (2.84), it must be observed that the electric field is the sum of the ohmic field and the diffusion electric field due to the difference in ionic mobilities,

$$\begin{aligned}\vec{\nabla}\phi = \vec{\nabla}\phi_{\text{dif}} + \vec{\nabla}\phi_{\text{ohm}} &= -\frac{1}{f}\left(\frac{t_1}{z_1} + \frac{t_2}{z_2}\right)\vec{\nabla}\ln c_{12} - \frac{\vec{I}}{\kappa} \\ &= -\frac{1}{f}\frac{D_1 - D_2}{z_1 D_1 - z_2 D_2}\vec{\nabla}\ln c_{12} - \frac{\vec{I}}{\kappa}.\end{aligned} \tag{2.102}$$

2.3.7 Ternary electrolyte solutions

In Section 2.3.6 we have shown that the Nernst–Planck approach provides estimates for the transport numbers and the electrolyte diffusion coefficient in a binary solution. In this section we derive approximate expressions for the phenomenological coefficients $L_{I,K}$ and the Fickian diffusion coefficients $D_{I,K}$ in a ternary system. Following the notation introduced in Sections 2.1.6 and 2.2.3, the electrolytes $\mathrm{A}_{\nu_1}\mathrm{C}_{\nu_{3,1}}$ and $\mathrm{D}_{\nu_2}\mathrm{C}_{\nu_{3,2}}$ are denoted by indices 13 and 23, respectively, and they are assumed to be completely dissociated according to the reactions (2.45) and (2.46). The electrolyte molar concentrations are c_{13} and c_{23}, the ionic molar concentrations are $c_1 = \nu_1 c_{13}$, $c_2 = \nu_2 c_{23}$, and $c_3 = \nu_{3,1}c_{13} + \nu_{3,2}c_{23}$, and the stoichiometric relations $z_1\nu_1 + z_3\nu_{3,1} = 0$ and $z_2\nu_2 + z_3\nu_{3,2} = 0$ are satisfied.

From eqns (2.41) and (2.47), the component flux densities can be written in terms of the ionic flux densities as

$$-\vec{J}_{13} = L_{13,13}\vec{\nabla}\mu_{13} + L_{13,23}\vec{\nabla}\mu_{23} = -\frac{1}{\nu_1}\left(\vec{j}_1 - \frac{t_1\vec{I}}{z_1 F}\right), \tag{2.103}$$

$$-\vec{J}_{23} = L_{13,23}\vec{\nabla}\mu_{13} + L_{23,23}\vec{\nabla}\mu_{23} = -\frac{1}{\nu_2}\left(\vec{j}_2 - \frac{t_2\vec{I}}{z_2 F}\right), \tag{2.104}$$

where the chemical potential of the electrolytes are

$$\mu_{13} = \nu_1\tilde{\mu}_1 + \nu_{3,1}\tilde{\mu}_3, \tag{2.105}$$

$$\mu_{23} = \nu_2\tilde{\mu}_2 + \nu_{3,2}\tilde{\mu}_3. \tag{2.106}$$

With the help of eqns (2.34), (2.36), and (2.67), eqn (2.103) is transformed to

$$\begin{aligned} -\vec{J}_{13} &= \frac{t_1\kappa}{z_1\nu_1 F^2}\left(\frac{1-t_1}{z_1}\vec{\nabla}\tilde{\mu}_1 - \frac{t_2}{z_2}\vec{\nabla}\tilde{\mu}_2 - \frac{t_3}{z_3}\vec{\nabla}\tilde{\mu}_3\right) \\ &= \frac{t_1(1-t_1)\kappa}{(Fz_1\nu_1)^2}\vec{\nabla}\mu_{13} - \frac{t_1 t_2\kappa}{F^2 z_1\nu_1 z_2\nu_2}\vec{\nabla}\mu_{23}, \end{aligned} \tag{2.107}$$

and therefore the phenomenological coefficients are identified as

$$L_{13,13} = \frac{t_1(1-t_1)\kappa}{(Fz_1\nu_1)^2} = \frac{(1-t_1)D_1c_{13}}{\nu_1 RT} > 0, \tag{2.108}$$

$$L_{13,23} = L_{23,13} = -\frac{t_1 t_2\kappa}{F^2 z_1\nu_1 z_2\nu_2} < 0. \tag{2.109}$$

Similarly, eqn (2.104) leads to

$$L_{23,23} = \frac{t_2(1-t_2)\kappa}{(Fz_2\nu_2)^2} = \frac{(1-t_2)D_2c_{23}}{\nu_2 RT} > 0. \tag{2.110}$$

From these equations we derive the following conclusions:

1) the Nernst–Planck approximation allows us to obtain simple (and rather accurate) expressions for the phenomenological coefficients $L_{I,K}$ of the components;
2) the phenomenological transport coefficients are functions of the local composition and cannot be considered as constant when integrating the phenomenological transport equations, eqn (2.44);
3) the cross-coefficient $L_{13,23} = L_{23,13}$ is of the same order of magnitude as $L_{13,13}$ and $L_{23,23}$ and, therefore, cross-effects cannot be neglected in component diffusion; and
4) the cross-coefficient $L_{13,23} = L_{23,13}$ is negative, which allows for the possibility of transport against concentration gradients (i.e. the so-called uphill transport that is discussed, e.g., in Section 4.1.8).

We aim now at obtaining the estimates for the Fickian diffusion coefficients. From the comparison of eqns (2.59) and (2.74) we have that

$$-\sum_J \sum_K \nu_{i,K} D_{K,J}\, \vec{\nabla} c_J = -\sum_k D_{i,k} \vec{\nabla} c_k = -\sum_J \sum_k D_{i,k} \nu_{k,J} \vec{\nabla} c_J, \tag{2.111}$$

and, therefore, the two sets of diffusion coefficients must satisfy the stoichiometric relation

$$\begin{pmatrix} \nu_1 & 0 \\ 0 & \nu_2 \\ \nu_{3,1} & \nu_{3,2} \end{pmatrix} \begin{pmatrix} D_{13,13} & D_{13,23} \\ D_{23,13} & D_{23,23} \end{pmatrix} = \begin{pmatrix} D_{1,1} & D_{1,2} & D_{1,3} \\ D_{2,1} & D_{2,2} & D_{2,3} \\ D_{3,1} & D_{3,2} & D_{3,3} \end{pmatrix} \begin{pmatrix} \nu_1 & 0 \\ 0 & \nu_2 \\ \nu_{3,1} & \nu_{3,2} \end{pmatrix}. \tag{2.112}$$

Using eqn (2.75) for the Nernst–Planck ionic coefficients $D_{i,k}$, we conclude that

$$D_{13,13} = (1 - t_1)D_1 + t_1 D_3, \tag{2.113}$$

$$D_{23,23} = (1 - t_2)D_2 + t_2 D_3, \tag{2.114}$$

$$D_{13,23} = \frac{\nu_{3,2}}{\nu_{3,1}} t_1 (D_3 - D_2), \tag{2.115}$$

$$D_{23,13} = \frac{\nu_{3,1}}{\nu_{3,2}} t_2 (D_3 - D_1). \tag{2.116}$$

Interestingly, if we use the coefficients $D'_{i,k}$ defined in eqn (2.77) with $N = 3$, it is found that $D'_{1,1} = D_{13,13}$, $D'_{2,2} = D_{23,23}$, $D'_{1,2} = (\nu_1/\nu_2)D_{13,23}$, and $D'_{2,1} = (\nu_2/\nu_1)D_{23,13}$.

It should be observed that $D_{13,23} \neq D_{23,13}$ and that all these coefficients are functions of the local concentration. It is also clear from eqns (2.115) and (2.116) that the cross-coefficients $D_{13,23}$ and $D_{23,13}$ can be negative and this allows for the description of uphill transport in the Fickian approach. Moreover, as mentioned in relation to the phenomenological coefficients, it must be noticed that the cross-diffusion coefficients are of the same order of magnitude as $D_{13,13}$ and $D_{23,23}$, so that cross-effects cannot be neglected when describing electrolyte diffusion.

Note, finally, that although the relation $l_{i,i} = D_i c_i / RT$ holds in a binary solution (within the Nernst–Planck approximation), similar equations do not hold in the ternary solution, e.g. $L_{13,13} \neq D_{13,13} c_{13}/RT$; moreover, $\vec{\nabla}\mu_{12} = (\nu_1 + \nu_2)\vec{\nabla} \ln c_{12}$ in a binary solution but $\vec{\nabla}\mu_{13} \neq (\nu_1 + \nu_{3,1})\vec{\nabla} \ln c_{13}$ in a ternary solution.

Example: Effective diffusion coefficient of a polyelectrolyte

Consider the solution formed by the complete dissolution of sodium chloride and an anionic polyelectrolyte in sodium form, and denote the ions P^{-z}, Cl^-, and Na^+ as species 1, 2, and 3, respectively. The charge numbers and stoichiometric coefficients are then $-z_1 = \nu_{3,1} = z$ and $-z_2 = z_3 = \nu_1 = \nu_2 = \nu_{3,2} = 1$. The flux density of the polyelectrolyte salt is

$$-\vec{J}_{13} = D_{13,13}\vec{\nabla}c_{13} + D_{13,23}\vec{\nabla}c_{23},$$

where the diffusion coefficients are given in eqns (2.113) and (2.115). In the presence of an excess of NaCl, the transport number of the polyelectrolyte ion vanishes, the cross-diffusion coefficient $D_{13,23}$ is negligible, and the polyelectrolyte flux density reduces to

$$-\vec{J}_{13} \approx D_1\vec{\nabla}c_{13} \quad , \text{ when } t_1 \to 0.$$

This equation shows that the effective diffusion coefficient of the polyelectrolyte in the limit of excess added salt is equal to that of the polyelectrolyte anion. In the opposite limit of absence of added salt, $c_{23} = 0$, the flux density of the polyelectrolyte salt is

$$-\vec{J}_{13} \approx D_{13}\vec{\nabla}c_{13},$$

where

$$D_{13} \equiv t_3D_1 + t_1D_3 = \frac{(z+1)D_1D_3}{zD_1 + D_3}$$

is the Nernst–Hartley diffusion coefficient of the polyelectrolyte salt. Since $D_3 > D_1$ (due to the smaller size of the sodium ion), the Nernst–Hartley diffusion coefficient of the polyelectrolyte salt is larger than that of the polyelectrolyte ion, $D_{13} > D_1$. This means that the polyelectrolyte diffusion proceeds faster than in the case of excess added salt; note, however, that the sodium ion is slowed down because it also moves with the electrolyte diffusion coefficient $D_{13} < D_3$. These limiting behaviours have been represented in Fig. 2.1 and are in agreement with experimental evidence [7]. An important conclusion of this example is that the Fickian diffusion coefficients can vary significantly with the local composition and they cannot be considered as constant in the integration of transport equations.

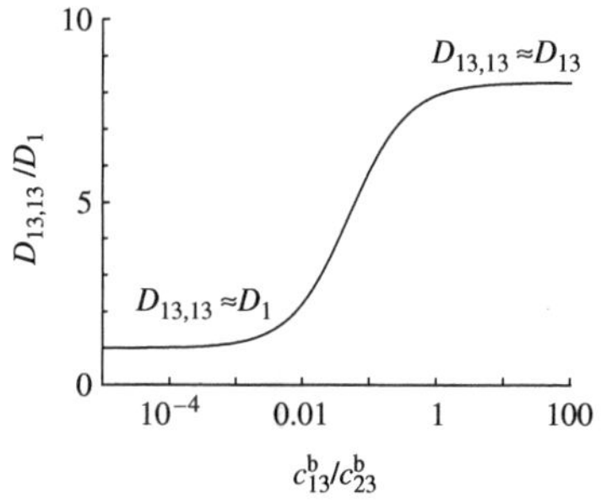

Fig. 2.1.
Variation of the diffusion coefficient $D_{13,13}$ with the concentration ratio c^b_{13}/c^b_{23}. In the presence of an excess added NaCl, $c^b_{13}/c^b_{23} \ll 1$, the effective diffusion coefficient of the polyelectrolyte salt is that of its ion. In the absence of added NaCl, $c^b_{13}/c^b_{23} \gg 1$, the polyelectrolyte salt diffuses with its Nernst–Hartley coefficient. This plot has been calculated using the values $z = 20$, $D_2/D_1 = 20$, and $D_2/D_3 = 2.0/1.3$.

2.3.8 Weak electrolytes

In weak electrolytes, ion transport is coupled to the homogeneous dissociation reaction

$$\mathrm{A}_{\nu_1}\mathrm{C}_{\nu_2} \rightleftarrows \nu_1\mathrm{A}^{z_1} + \nu_2\mathrm{C}^{z_2}. \tag{2.117}$$

The ionic concentrations are $c_1 = \nu_1c_{12}$ and $c_2 = \nu_2c_{12}$, where c_{12} is the concentration of dissociated electrolyte, and the stoichiometric coefficients satisfy the relation $z_1\nu_1 + z_2\nu_2 = 0$. The ionic flux densities are still given by eqn (2.98)

$$\vec{j}_i = -\nu_iD_{12}\vec{\nabla}c_{12} + \frac{t_i\vec{I}}{z_iF}, \tag{2.118}$$

where the transport numbers and the electrolyte diffusion coefficient are independent of the local composition, as shown in eqns (2.96) and (2.97). However, the ionic flux densities $\vec{j}_i$ $(i = 1, 2)$, and those of the dissociated

$$\vec{J}_{12} = -D_{12}\vec{\nabla}c_{12} \tag{2.119}$$

and the undissociated electrolyte

$$\vec{J}_{12,\text{u}} = -D_{12,\text{u}}\vec{\nabla}c_{12,\text{u}} \tag{2.120}$$

vary now with position because the dissociation degree α also does.

The continuity equation ensures that the total electrolyte flux density

$$\vec{J}_{12,\text{T}} = \vec{J}_{12,\text{u}} + \vec{J}_{12} = -D_{12,\text{T}}\vec{\nabla}c_{12,\text{T}} \tag{2.121}$$

is constant throughout the system. The effective diffusion coefficient of the weak electrolyte $D_{12,\text{T}}$ is a function of $c_{12,\text{T}}$ and has yet to be determined. Introducing the dissociation degree α through the relations $c_{12} = \alpha c_{12,\text{T}}$ and $c_{12,\text{u}} = (1-\alpha)c_{12,\text{T}}$, the diffusion coefficient $D_{12,\text{T}}$ can be expressed as

$$D_{12,\text{T}} \equiv \frac{\nu_{12}(1-\alpha)D_{12,\text{u}} + \alpha D_{12}}{\nu_{12}(1-\alpha)+\alpha} \tag{2.122}$$

where $\nu_{12} \equiv \nu_1 + \nu_2$. When the dissociation reaction is fast compared with the transport process, local chemical equilibrium can be assumed, and the relation between $c_{12,\text{T}}$ and α is

$$K = \frac{c_1^{\nu_1}c_2^{\nu_2}}{c_{12,\text{u}}} = (c_{12,\text{T}})^{\nu_{12}-1}\frac{\nu_1^{\nu_1}\nu_2^{\nu_2}\alpha^{\nu_{12}}}{1-\alpha}. \tag{2.123}$$

As expected, in the limiting cases of very weak, $K \ll (c_{12,\text{T}})^{\nu_{12}-1}$, and very strong dissociation, $K \gg (c_{12,\text{T}})^{\nu_{12}-1}$, the effective diffusion coefficient of the electrolyte reduces to the diffusion coefficient of the undissociated electrolyte $D_{12,\text{T}} \approx D_{12,\text{u}}$ and to the Nernst–Hartley diffusion coefficient $D_{12,\text{T}} \approx D_{12}$, respectively.

The integration of eqn (2.121) can be better carried out by evaluating the flux density of the total electrolyte as

$$-\vec{J}_{12,\text{T}} = D_{12,\text{u}}\vec{\nabla}\left[(1-\alpha)c_{12,\text{T}}\right] + D_{12}\vec{\nabla}\left[\alpha c_{12,\text{T}}\right]. \tag{2.124}$$

Since the diffusion coefficients $D_{12,\text{u}}$ and D_{12} are constant, the steady-state continuity equation $\vec{\nabla}\cdot\vec{J}_{12,\text{T}} = 0$ leads to a Laplace equation, $\nabla^2(D_{12,\text{u}}c_{12,\text{u}} + D_{12}c_{12}) = 0$, that can often be solved analytically.

It was mentioned above that the ionic flux density $\vec{j}_i$ is a function of position and this can complicate the integration of eqn (2.118). To solve this problem, it is convenient to introduce the concept of the total ionic constituent concentration. If we consider the amount of species 1 that is either dissociated or in the form of undissociated electrolyte, its total concentration is $c_{1,\mathrm{T}} = c_1 + \nu_1 c_{12,\mathrm{u}}$. Similarly, the total concentration of species 2 is $c_{2,\mathrm{T}} = c_2 + \nu_2 c_{12,\mathrm{u}}$. The flux density of the total ionic constituent i ($i = 1, 2$) is then

$$\vec{j}_{i,\mathrm{T}} \equiv \vec{j}_i + \nu_i \vec{j}_{12,\mathrm{u}} = \nu_i \vec{J}_{12,\mathrm{T}} + \frac{t_i \vec{I}}{z_i F}, \tag{2.125}$$

and the continuity equation ensures that $\vec{j}_{i,\mathrm{T}}$ is independent of position.[14]

2.3.9 Moderately concentrated solutions

When dealing with moderately concentrated solutions, the Nernst–Planck equations can be modified to account for the fact that the gradient of the electrochemical potential of species i actually is[15]

$$\vec{\nabla}\tilde{\mu}_i = RT\vec{\nabla}\ln a_i + z_i F\vec{\nabla}\phi = \left(1 + \frac{\mathrm{d}\ln\gamma_i}{\mathrm{d}\ln c_i}\right) RT\vec{\nabla}\ln c_i + z_i F\vec{\nabla}\phi. \tag{2.126}$$

This implies that the diffusion coefficients must be modified to

$$D_i^{\gamma} \equiv \left(1 + \frac{\mathrm{d}\ln\gamma_i}{\mathrm{d}\ln c_i}\right) D_i \equiv \beta_i D_i \tag{2.127}$$

where the superscript γ on D_i^{γ} denotes that it contains the activity coefficient correction. Note that the derivative $\mathrm{d}\ln\gamma_i/\mathrm{d}\ln c_i$ can be either positive or negative and that the correction does not apply to the migrational contribution to the flux density.

The only activity coefficient that can be measured is the mean activity coefficient of a binary electrolyte $\gamma_{12} = (\gamma_1^{\nu_1}\gamma_2^{\nu_2})^{1/\nu_{12}}$. However, eqn (2.127) involves the ionic activity coefficient, and hence some (arbitrary) convention is needed to calculate them from γ_{12}. This is usually done by employing the equation [8]

$$\sum_i \frac{c_i}{z_i}\ln\gamma_i = 0, \tag{2.128}$$

which retains the concept of the ionic strength in the sense of the Debye–Hückel theory. Thus, for example, in the case of 1:1 electrolytes eqn (2.128) implies $\gamma_{12} = \gamma_1 = \gamma_2$. Equations (2.127) and (2.128) then allow the activity correction to the diffusion coefficient to be determined from the measured thermodynamic data [9].

14 Strictly speaking, the continuity equation says that this flux density has zero divergence, but in one-dimensional systems these two statements are equivalent to each other.

15 In thermodynamics textbooks, the symbol used for the activity coefficient depends on the concentration scale used. In the molar concentration scale, the symbol most often used is y_i rather than the symbol γ_i employed here. The latter, however, is commonly used in books on transport phenomena where only one concentration scale is used.

In the presence of convection, it must be taken into account that the barycentric solution velocity and the solvent velocity may differ. Thus, since convective velocity is determined from the momentum balance as a barycentric velocity, the ionic flux densities must also be expressed in the barycentric reference frame. In concentrated solutions this implies that we must note, according to eqn (2.12), that the driving force for species i is $\vec{\nabla}\tilde{\mu}_i - (M_i/M_0)\vec{\nabla}\mu_0$. Thus, the Nernst–Planck equation takes the form [10, 11]

$$\begin{aligned}\vec{j}_i^m &= -l_{i,i}\left(\vec{\nabla}\tilde{\mu}_i - \frac{M_i}{M_0}\vec{\nabla}\mu_0\right)\\ &= -l_{i,i}\left[\left(1+\frac{d\ln\gamma_i}{d\ln c_i}\right)RT\vec{\nabla}\ln c_i + z_iF\vec{\nabla}\phi + \upsilon_i\vec{\nabla}p - \frac{M_i}{M_0}\vec{\nabla}\mu_0\right]\\ &= -D_i^{\gamma}\vec{\nabla}c_i - z_ic_iD_i\frac{F}{RT}\vec{\nabla}\phi - \frac{D_ic_i}{RT}\left(\upsilon_i\vec{\nabla}p - \frac{M_i}{M_0}\vec{\nabla}\mu_0\right),\end{aligned} \tag{2.129}$$

where the effect of the pressure gradient on $\vec{\nabla}\tilde{\mu}_i$ [not shown in eqn (2.126)] has also been taken into account. This equation is certainly more difficult to solve than the simple $\vec{j}_i^{\mathrm{H}} = -l_{i,i}\vec{\nabla}\tilde{\mu}_i$ that we can use in diluted solutions.

Finally, it is important to remember that the Nernst–Planck approach neglects the cross-phenomenological coefficients. Thus, although the activity correction makes the diffusion coefficient concentration dependent, the Nernst–Planck approach is only strictly applicable to dilute solutions. In other words, the accurate description of transport processes in concentrated solutions requires more transport coefficients than are available in the Nernst–Planck approach.

2.4 The Stefan-Maxwell approach

2.4.1 Introduction

The Nernst–Planck equation states that the molar flux density of species i, that is, the product of its molar concentration and its velocity, is proportional to the thermodynamic force, $\vec{j}_i = c_i\vec{v}_i = -l_{i,i}\vec{\nabla}\tilde{\mu}_i$. That is, in the time regime when this equation is valid, the thermodynamic force $-\vec{\nabla}\tilde{\mu}_i$ no longer produces an acceleration of these ions but keeps them moving at constant velocity $\vec{v}_i$. This must obviously be due to the fact that additional frictional forces are present. These are proportional to the relative velocities of the components (and hence to the flux densities in an appropriate reference frame or to differences of flux densities with appropriate factors). The transport equations can then be understood as a statement of equilibrium of (vector) forces

$$\text{driving force} + \text{frictional force} = 0.$$

This equation is the starting point for the frictional or Stefan–Maxwell approach for the description of transport processes, which was also proposed by Onsager [12, 13].

Although in some simple cases they can be shown to be equivalent to each other, the Stefan–Maxwell and the Fickian approaches differ in the transport coefficients that they use. In the Fickian approach, the velocity of the solute (with respect to the solvent) is considered to be proportional to the force as

$$\text{velocity} = \text{mobility} \times \text{driving force}.$$

In the Stefan–Maxwell approach, the basic scheme for the transport equations is

$$\text{driving force} = \text{friction coefficient} \times \text{relative velocity}.$$

In multicomponent systems the right-hand side of this equation contains a sum of terms.

It is important to note that both the diffusion coefficients and the friction coefficients depend on the composition of the multicomponent system (as well as on temperature and pressure). The strength of the Stefan–Maxwell approach is that it provides a particular functional dependence of the friction coefficients on the concentrations and proposes the use of another type of diffusion coefficient. Thus, when concentrated solutions are considered, the Fickian diffusion coefficients depend strongly on the concentrations, while the Stefan–Maxwell diffusion coefficients are roughly constant. This explains why the Stefan–Maxwell approach is often preferred for the description of transport processes in concentrated solutions.[16]

2.4.2 Diffusion of a neutral component

The Stefan–Maxwell approach can be introduced by considering a two-component solution. The solvent is denoted by the index $i = 0$ and the solute by $i = 1$. The driving force for the motion of the solute is the negative gradient of its chemical potential. During its motion, the solute can only experience friction with the solvent, and then the Stefan–Maxwell transport equation can be written as

$$c_1 \vec{\nabla} \mu_1 = K_{1,0}(\vec{v}_0 - \vec{v}_1), \tag{2.130}$$

where $K_{1,0}$ is the friction coefficient between solvent and solute. The solute molar flux density in the Hittorf reference frame is

$$\vec{j}_1^{\,\mathrm{H}} = c_1(\vec{v}_1 - \vec{v}_0) = -\frac{c_1^2}{K_{1,0}} \vec{\nabla} \mu_1. \tag{2.131}$$

Introducing the Stefan–Maxwell diffusion coefficient

$$\overline{D}_{1,0} \equiv RT \frac{c_1 x_0}{K_{1,0}}, \tag{2.132}$$

[16] Although it should be applied to dilute solutions only, the Nernst–Planck approach is, by far, the most widely used in electrochemistry because it provides a simple and satisfactory understanding of the transport processes.

where x_0 is the molar fraction of the solvent in the solution, the flux equation can also be written as

$$\vec{j}_1^{\mathrm{H}} = -\frac{\overline{D}_{1,0} c_1}{RT x_0} \vec{\nabla} \mu_1. \tag{2.133}$$

In dilute solutions, ideal behaviour can be assumed so that $\vec{\nabla}\mu_1 \approx RT\vec{\nabla} \ln c_1$ and $x_0 \approx 1$. Equation (2.133) then takes the form of Fick's first law, $\vec{j}_1^{\mathrm{H}} = -D_1 \vec{\nabla} c_1$, so that the Fick and Stefan–Maxwell diffusion coefficients, D_1 and $\overline{D}_{1,0}$, become equivalent. In concentrated solutions, however, they can differ significantly.

From Stokes' law for the frictional force of a viscous continuum on a spherical particle of radius R_1 [14], the friction coefficient can be estimated as $K_{1,0} \approx 6\pi R_1 \eta N_A c_1$, where N_A is Avogadro's constant, and eqn (2.132) then leads to

$$D_1 \approx RT \frac{c_1 x_0}{K_{1,0}} \approx \frac{RT}{6\pi R_1 \eta N_A} \tag{2.134}$$

which is known as the *Stokes–Einstein equation.*

In closing, it is interesting to observe that it is also possible to write an equation like eqn (2.130) for the solvent, that is,

$$c_0 \vec{\nabla} \mu_0 = K_{1,0}(\vec{v}_1 - \vec{v}_0). \tag{2.135}$$

This equation brings nothing new to the formulation of the problem because it contains just the same information as eqn (2.130). However, it serves to stress an important point. Equations (2.130) and (2.135) are only valid if the mechanical equilibrium condition

$$c_0 \vec{\nabla} \mu_0 + c_1 \vec{\nabla} \mu_1 = \vec{\nabla} p = \vec{0} \tag{2.136}$$

is fulfilled. This occurs in free solution, but not in membrane processes, where a pressure gradient is either imposed or develops as a consequence of the solute concentration gradient (i.e. due to the osmotic pressure gradient). In fact, when the Stefan–Maxwell approach is applied to transport across membranes, there is an additional friction force with the membrane that needs to be added to the right-hand side of these equations. This is due to the fact that the membrane is an additional component in the system, making it a multicomponent one. In Sections 2.4.3 and 2.4.4, transport across membranes is not considered and the condition $\vec{\nabla} p = \vec{0}$ is accepted implicitly.

2.4.3 Binary electrolyte solution

Consider now the transport processes in the solution of a binary electrolyte that dissociates completely into ν_1 ions of charge number z_1 and ν_2 ions of charge number z_2, such that $z_1\nu_1 + z_2\nu_2 = 0$. The Stefan–Maxwell transport equations are

$$c_1\vec{\nabla}\tilde{\mu}_1 = K_{1,0}(\vec{v}_0 - \vec{v}_1) + K_{1,2}(\vec{v}_2 - \vec{v}_1), \tag{2.137}$$

$$c_2\vec{\nabla}\tilde{\mu}_2 = K_{2,0}(\vec{v}_0 - \vec{v}_2) + K_{2,1}(\vec{v}_1 - \vec{v}_2), \tag{2.138}$$

where Onsager's reciprocal theorem

$$K_{i,j} = K_{j,i} \tag{2.139}$$

also applies to the friction coefficients between species i and j. We aim at determining the molar ionic flux densities in the Hittorf reference frame

$$\vec{j}_i^{\mathrm{H}} = c_i(\vec{v}_i - \vec{v}_0), \tag{2.140}$$

making use of the local electroneutrality condition

$$z_1c_1 + z_2c_2 = 0 \tag{2.141}$$

and the definition of the conduction current density

$$\vec{I} = F(z_1\vec{j}_1^{\mathrm{H}} + z_2\vec{j}_2^{\mathrm{H}}) = Fz_1c_1(\vec{v}_1 - \vec{v}_2). \tag{2.142}$$

In the absence of electric current, eqn (2.142) requires that $\vec{v}_1 = \vec{v}_2$, and eqns (2.137) and (2.138) simplify to

$$\begin{aligned}\vec{j}_i^{\mathrm{H}} = c_i(\vec{v}_i - \vec{v}_0) &= -\frac{\overline{D}_{i,0}c_i}{RTx_0}(\vec{\nabla}\mu_i + z_iF\vec{\nabla}\phi)\\ &= -\frac{\overline{D}_{i,0}c_i}{x_0}[\vec{\nabla}\ln(\gamma_ic_i) + z_if\,\vec{\nabla}\phi],\end{aligned} \tag{2.143}$$

where $\overline{D}_{i,0} \equiv RTc_ix_0/K_{i,0}$. Equation (2.143) resembles quite closely the Nernst–Planck equation, eqn (2.68).

In the case of pure electric conduction, that is, in the absence of concentration gradients, the Stefan–Maxwell transport equations reduce to

$$z_1c_1F\vec{\nabla}\phi = K_{1,0}(\vec{v}_0 - \vec{v}_1) + K_{1,2}(\vec{v}_2 - \vec{v}_1), \tag{2.144}$$

$$z_2c_2F\vec{\nabla}\phi = K_{2,0}(\vec{v}_0 - \vec{v}_2) + K_{2,1}(\vec{v}_1 - \vec{v}_2), \tag{2.145}$$

and lead to Ohm's law

$$\vec{I} = -\kappa\vec{\nabla}\phi_{\mathrm{ohm}} = -\kappa\vec{\nabla}\phi \quad (\vec{\nabla}c_i = \vec{0}), \tag{2.146}$$

where the electric conductivity is

$$\kappa \equiv \frac{z_1^2F^2c_1^2(K_{1,0} + K_{2,0})}{K_{1,0}K_{2,0} + K_{1,2}(K_{1,0} + K_{2,0})}. \tag{2.147}$$

It is interesting to note that the friction between the ions, which is described by the last terms in the right-hand sides of eqns (2.144) and (2.145), is only relevant in the presence of electric current because $\vec{I} = Fz_1c_1(\vec{v}_1 - \vec{v}_2)$. In relation to this it can be shown that the diffusion potential gradient is

$$\vec{\nabla}\phi_{\text{dif}} = -\frac{1}{F}\left(\frac{\bar{t}_1}{z_1}\vec{\nabla}\mu_1 + \frac{\bar{t}_2}{z_2}\vec{\nabla}\mu_2\right), \tag{2.148}$$

which does not depend on $K_{1,2}$ because the ionic transport numbers are defined as

$$\bar{t}_1 \equiv \frac{K_{2,0}}{K_{1,0} + K_{2,0}} = \frac{\nu_2\overline{D}_{1,0}}{\nu_2\overline{D}_{1,0} + \nu_1\overline{D}_{2,0}} = 1 - \bar{t}_2. \tag{2.149}$$

When concentration gradients and electric current are both present, the transport equations can be transformed to a diffusion–conduction form, quite similar to eqn (2.60). Introducing the gradient of the chemical potential of the electrolyte through

$$c_{12}\vec{\nabla}\mu_{12} = c_1\vec{\nabla}\tilde{\mu}_1 + c_2\vec{\nabla}\tilde{\mu}_2 = c_1\vec{\nabla}\mu_1 + c_2\vec{\nabla}\mu_2, \tag{2.150}$$

where we have made use of the relation $c_i = \nu_i c_{12}$, the above equations can be written in matrix form as

$$\begin{pmatrix} -K_{1,0} & -K_{2,0} \\ z_1c_1 & z_2c_2 \end{pmatrix} \begin{pmatrix} \vec{v}_1 - \vec{v}_0 \\ \vec{v}_2 - \vec{v}_0 \end{pmatrix} = \begin{pmatrix} c_{12}\vec{\nabla}\mu_{12} \\ \vec{I}/F \end{pmatrix}, \tag{2.151}$$

and inverted to give

$$\begin{pmatrix} \vec{j}_1^{\text{H}} \\ \vec{j}_2^{\text{H}} \end{pmatrix} = \begin{pmatrix} c_1(\vec{v}_1 - \vec{v}_0) \\ c_2(\vec{v}_2 - \vec{v}_0) \end{pmatrix} = \frac{1}{K_{1,0} + K_{2,0}} \begin{pmatrix} -c_1 & K_{2,0}/z_1 \\ -c_2 & K_{1,0}/z_2 \end{pmatrix} \begin{pmatrix} c_{12}\vec{\nabla}\mu_{12} \\ \vec{I}/F \end{pmatrix}. \tag{2.152}$$

In concentrated solutions, $\vec{\nabla}\mu_{12} = \nu_{12}RT\vec{\nabla}\ln(\gamma_{12}c_{12})$ and the diffusion–conduction equation becomes

$$\vec{j}_i^{\text{H}} = -\nu_i\frac{\overline{D}_{12}}{x_0}\left(1 + \frac{\mathrm{d}\ln\gamma_{12}}{\mathrm{d}\ln c_{12}}\right)\vec{\nabla}c_{12} + \frac{\bar{t}_i\vec{I}}{z_iF}, \tag{2.153}$$

where the diffusion coefficient of the electrolyte is

$$\overline{D}_{12} \equiv \frac{\nu_{12}c_{12}x_0RT}{K_{1,0} + K_{2,0}} = \frac{\nu_{12}\overline{D}_{1,0}\overline{D}_{2,0}}{\nu_2\overline{D}_{1,0} + \nu_1\overline{D}_{2,0}}. \tag{2.154}$$

In dilute solutions ($x_1, x_2 \ll x_0$), the friction coefficients can be approximated by $K_{i,0} = RTc_i/D_i$ and ideal behaviour can be assumed to evaluate

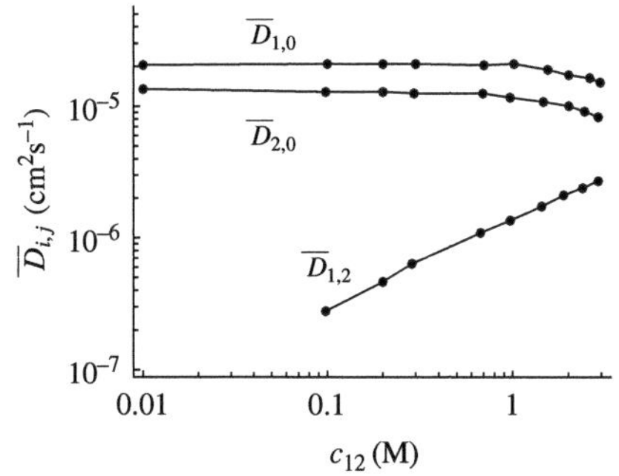

Fig. 2.2.
Concentration dependence of the Stefan–Maxwell diffusion coefficients in an aqueous solution of sodium chloride: $\overline{D}_{1,0}$ water-chloride ion, $\overline{D}_{2,0}$ water-sodium ion, $\overline{D}_{1,2}$ sodium ion-chloride ion. (Data taken from Ref. [7] with permission.)

the chemical potential gradient as $\vec{\nabla}\mu_{12} = \nu_{12}RT\vec{\nabla}\ln c_{12}$, so that the above equations reduce to

$$\vec{j}_i^{\mathrm{H}} = -\nu_i D_{12}\vec{\nabla}c_{12} + \frac{t_i\vec{I}}{z_i F}, \tag{2.155}$$

$$D_{12} = \frac{\nu_{12}c_{12}RT}{K_{1,0}+K_{2,0}} = \frac{\nu_{12}D_1D_2}{\nu_2D_1+\nu_1D_2}, \tag{2.156}$$

$$\kappa = \frac{z_1^2F^2c_1^2(K_{1,0}+K_{2,0})}{K_{1,0}K_{2,0}+K_{1,2}(K_{1,0}+K_{2,0})} = \frac{F^2(z_1^2D_1c_1+z_2^2D_2c_2)}{RT}, \tag{2.157}$$

$$t_1 = \frac{K_{2,0}}{K_{1,0}+K_{2,0}} = \frac{\nu_2D_1}{\nu_2D_1+\nu_1D_2} = 1-t_2. \tag{2.158}$$

Thus, once again, the Nernst–Planck and Stefan–Maxwell approaches are found to be equivalent in dilute solutions.

Note that $\overline{D}_{12}$, $\overline{t}_i$, and the mean activity coefficient γ_{12} can be measured at different concentrations [9]. Thus, the diffusion coefficients $\overline{D}_{i,0}$ (i = 1, 2) can be determined from eqns (2.149) and (2.154). Similarly, the cross-diffusion coefficient[17] $\overline{D}_{1,2}$ can be evaluated from electrical conductivity measurements. Figure 2.2 shows the Stefan–Maxwell diffusion coefficients in sodium chloride aqueous solutions. The cross-diffusion coefficient $\overline{D}_{1,2}$ increases with the electrolyte concentration, roughly as $(c_{12})^{0.5}$, but it is significantly smaller than the coefficients $\overline{D}_{i,0}$ (i = 1, 2) throughout the whole concentration range.

2.4.4 Multi-ionic systems

In a multi-ionic system, the Stefan–Maxwell equations are

$$c_i\vec{\nabla}\tilde{\mu}_i = \sum_j K_{i,j}(\vec{v}_j - \vec{v}_i), \tag{2.159}$$

[17] The cross-coefficient $\overline{D}_{1,2}$ should not be confused with the diffusion coefficient of the electrolyte $\overline{D}_{12}$.

where the sum also includes the solvent. The friction coefficients are written as

$$K_{i,j} = K_{j,i} = RTc_{\mathrm{T}} \frac{x_i x_j}{\overline{D}_{i,j}}, \tag{2.160}$$

where c_{T} is the total molar concentration (including the solvent) and the reciprocal relation $\overline{D}_{i,j} = \overline{D}_{j,i}$ also holds for the Stefan–Maxwell diffusion coefficients. Equation (2.160) can be written in terms of the flux densities as

$$\frac{c_i}{RT}\vec{\nabla}\widetilde{\mu}_i = \sum_j \frac{1}{\overline{D}_{i,j}}(x_i\vec{j}_j - x_j\vec{j}_i) = -\sum_j B_{i,j}\vec{j}_j, \tag{2.161}$$

where

$$B_{i,j} = -\frac{x_i}{\overline{D}_{i,j}} + \delta_{ij}\sum_{k=0} \frac{x_k}{\overline{D}_{i,k}}. \tag{2.162}$$

Equation (2.161) represents a linear system of equations that can be formulated in matrix form and solved for the flux densities. We do not pursue this approach any further here and the interested reader is referred to specialized books such as [7] and [15]. Newman's book [16] is also a recommended reference for electrochemical systems.

Note, finally, that the Stefan–Maxwell equations only involve relative velocities and they are therefore independent of the reference frame. At the same time, this implies that additional equations are needed to describe the absolute motion of the solution with respect to, e.g., the laboratory [7].

Exercises

2.1 The statement that every flux density $\vec{J}_i (i = 1, 2, \ldots)$ is a function of the generalized forces can be formally written as $\vec{J}_i = \vec{f}_i(\vec{X}_1, \vec{X}_2, \ldots)$ where $\vec{f}_i$ denotes a generic function. Derive the linear phenomenological equations by expanding $\vec{f}_i$ as a Taylor series around the equilibrium state, characterized by the condition $\vec{X}_j = \vec{0}$ for all j. What are the expressions for the phenomenological coefficients and the dissipation function? What condition do the phenomenological coefficients have to satisfy to ensure that the entropy production remains positive?

2.2 The phenomenological equations

$$J_1 = L_{1,1}X_1 + L_{1,2}X_2$$
$$J_2 = L_{2,1}X_1 + L_{2,2}X_2$$

can also be written in the form

$$X_1 = K_{1,1}J_1 + K_{1,2}J_2$$
$$X_2 = K_{2,1}J_1 + K_{2,2}J_2,$$

where the forces are functions of fluxes. This leads to a different kind of formalism, but the phenomenological coefficients can be related to each other. Derive the friction coefficients $K_{i,j}$ in terms of the phenomenological transport coefficients $L_{i,j}$. What condition must fulfil the coefficients $K_{i,j}$, and what does it mean in practice? Are Onsager's reciprocal relations still valid?

2.3 (a) Show that the invariance of the dissipation function under a transformation of fluxes and forces requires that the matrixes $\overleftrightarrow{\alpha}_J$ and $\overleftrightarrow{\alpha}_X$ that transform a set of fluxes $\vec{J}$ and forces $\vec{X}$ into another set $\vec{J}' = \overleftrightarrow{\alpha}_J \cdot \vec{J}$ and $\vec{X}' = \overleftrightarrow{\alpha}_X \cdot \vec{X}$, satisfy the relation $\overleftrightarrow{\alpha}_J^T = \overleftrightarrow{\alpha}_X^{-1}$, where the superscript T denote the transpose matrix.

2.4 Consider a binary electrolyte solution. Using the relation between the ionic and the component phenomenological coefficients

$$\begin{pmatrix} L_{12,12} & L_{12,\phi} \\ L_{12,\phi} & L_{\phi,\phi} \end{pmatrix} = \begin{pmatrix} t_2/\nu_1 & t_1/\nu_2 \\ z_1 F & z_2 F \end{pmatrix} \begin{pmatrix} l_{1,1} & l_{1,2} \\ l_{1,2} & l_{2,2} \end{pmatrix} \begin{pmatrix} t_2/\nu_1 & z_1 F \\ t_1/\nu_2 & z_2 F \end{pmatrix}$$

as well as the relations $z_1\nu_1 + z_2\nu_2 = 0$ and $t_1 + t_2 = 1$, show that

$$L_{12,\phi} = 0, L_{\phi,\phi} = \kappa, \text{ and } L_{12,12} = \frac{t_1 t_2 \kappa}{(F z_1 \nu_1)^2} + \frac{l_{1,2}}{\nu_1 \nu_2}.$$

2.5 Show that the ionic phenomenological coefficients $l_{i,j}(i,j = 1,2)$ can be written in terms of κ, t_i, and $L_{12,12}$ as

$$l_{i,j} = \frac{t_i t_j \kappa}{F^2 z_i z_j} + \nu_i \nu_j L_{12,12}.$$

2.6 Starting from the dissipation function

$$-\theta_{\text{ed}} = \vec{j}_1 \cdot \vec{\nabla}\tilde{\mu}_1 + \vec{j}_2 \cdot \vec{\nabla}\tilde{\mu}_2$$

(a) show that the transport equations only involve one transport coefficient when transport occurs subject to the constraint $j_2 = 0$ and find the relation between this coefficient and the ionic coefficients $l_{i,j}$ $(i,j = 1,2)$.

(b) Show that the transport equations only involve one transport coefficient when transport occurs subject to the constraint $I = 0$ and find the relation between this coefficient and the ionic coefficients $l_{i,j}$ $(i,j = 1,2)$.

Starting from the dissipation function

$$-\theta_{\text{ed}} = \vec{J}_{12} \cdot \vec{\nabla}\mu_{12} + \vec{I} \cdot \vec{\nabla}\phi_{\text{ohm}}$$

(c) show that the transport equations only involve one transport coefficient when transport occurs subject to the constraint $j_2 = 0$ and find the relation between this coefficient and the component coefficients $L_{i,j}(i,j = 12,\phi)$.

(d) Show that the transport equations only involve one transport coefficient when transport occurs subject to the constraint $I = 0$ and find the relation between this coefficient and the component coefficients $L_{i,j}(i,j = 12,\phi)$.

2.7 Check the invariance of the dissipation function under the transformation between ionic and component fluxes in a ternary system, i.e. prove that the following equality holds

$$\begin{aligned} -\theta_{\text{ed}} &= \vec{j}_1 \cdot \vec{\nabla}\tilde{\mu}_1 + \vec{j}_2 \cdot \vec{\nabla}\tilde{\mu}_2 + \vec{j}_3 \cdot \vec{\nabla}\tilde{\mu}_3 \\ &= \vec{J}_{13} \cdot \vec{\nabla}\mu_{13} + \vec{J}_{23} \cdot \vec{\nabla}\tilde{\mu}_{23} + \vec{I} \cdot \vec{\nabla}\phi_{\text{ohm}}. \end{aligned}$$

2.8 (a) Substituting the gradients

$$\begin{aligned}\vec{\nabla}\mu_{13} &= \nu_1\vec{\nabla}\mu_1 + \nu_{3,1}\vec{\nabla}\mu_3 = (\nu_1 + \nu_{3,1})RT\vec{\nabla}\ln c_{13} \\ &\quad + \nu_{3,1}RT\vec{\nabla}\ln(1 + \nu_{3,2}c_{23}/\nu_{3,1}c_{13}) \\ \vec{\nabla}\mu_{23} &= \nu_2\vec{\nabla}\mu_2 + \nu_{3,2}\vec{\nabla}\mu_3 = (\nu_2 + \nu_{3,2})RT\vec{\nabla}\ln c_{23} \\ &\quad + \nu_{3,2}RT\vec{\nabla}\ln(1 + \nu_{3,1}c_{13}/\nu_{3,2}c_{23})\end{aligned}$$

in

$$\begin{aligned}\begin{pmatrix}\vec{j}_1^{\mathrm{H}} \\ \vec{j}_2^{\mathrm{H}} \\ \vec{j}_3^{\mathrm{H}}\end{pmatrix} &= \begin{pmatrix}\nu_1 & 0 & t_1/z_1F \\ 0 & \nu_2 & t_2/z_2F \\ \nu_{3,1} & \nu_{3,2} & t_3/z_3F\end{pmatrix}\begin{pmatrix}\vec{j}_{13}^{\mathrm{H}} \\ \vec{j}_{23}^{\mathrm{H}} \\ \vec{I}\end{pmatrix} \\ &= -\begin{pmatrix}\nu_1 & 0 & t_1/z_1F \\ 0 & \nu_2 & t_2/z_2F \\ \nu_{3,1} & \nu_{3,2} & t_3/z_3F\end{pmatrix}\begin{pmatrix}L_{13,13} & L_{13,23} & 0 \\ L_{13,23} & L_{23,23} & 0 \\ 0 & 0 & \kappa\end{pmatrix}\begin{pmatrix}\vec{\nabla}\mu_{13} \\ \vec{\nabla}\mu_{23} \\ \vec{\nabla}\phi_{\mathrm{ohm}}\end{pmatrix}\end{aligned}$$

and comparing the resulting equations with

$$\begin{aligned}\vec{j}_1^{\mathrm{H}} &= -\nu_1(D_{13,13}\vec{\nabla}c_{13} + D_{13,23}\vec{\nabla}c_{23}) + \frac{t_1\vec{I}}{z_1F} \\ \vec{j}_2^{\mathrm{H}} &= -\nu_2(D_{23,13}\vec{\nabla}c_{13} + D_{23,23}\vec{\nabla}c_{23}) + \frac{t_2\vec{I}}{z_2F} \\ \vec{j}_3^{\mathrm{H}} &= -\nu_{3,1}(D_{13,13}\vec{\nabla}c_{13} + D_{13,23}\vec{\nabla}c_{23}) \\ &\quad -\nu_{3,2}(D_{23,13}\vec{\nabla}c_{13} + D_{23,23}\vec{\nabla}c_{23}) + \frac{t_3\vec{I}}{z_3F}\end{aligned}$$

find the relations between the component phenomenological $L_{I,K}$ and diffusion coefficients $D_{I,K}(I, K = 13, 23)$ in this ternary electrolyte solution.
(b) Substituting in these relations the expressions

$$\begin{aligned}L_{13,13} &= \frac{t_1(1-t_1)\kappa}{(Fz_1\nu_1)^2} \\ L_{13,23} = L_{23,13} &= -\frac{t_1t_2\kappa}{F^2z_1\nu_1z_2\nu_2} \\ L_{23,23} &= \frac{t_2(1-t_2)\kappa}{(Fz_2\nu_2)^2}\end{aligned}$$

determine the diffusion coefficients $D_{I,K}$ $(I, K = 13, 23)$.

2.9 Consider the description of transport in a binary electrolyte solution within the Nernst–Planck approximation.

(a) Starting from the diffusion–migration form of the ionic flux equation

$$\vec{j}_i = -D_i(\vec{\nabla} c_i + z_i c_i f\, \vec{\nabla}\phi)$$

prove that

$$\vec{\nabla}\phi = \vec{\nabla}\phi_{\text{dif}} + \vec{\nabla}\phi_{\text{ohm}} = -\frac{1}{f}\left(\frac{t_1}{z_1} + \frac{t_2}{z_2}\right)\vec{\nabla}\ln c_{12} - \frac{\vec{I}}{\kappa}$$

$$= -\frac{1}{f}\frac{D_1 - D_2}{z_1 D_1 - z_2 D_2}\vec{\nabla}\ln c_{12} - \frac{\vec{I}}{\kappa}.$$

(b) Starting from the chemical diffusion–ohmic conduction equation

$$\vec{j}_i = -\nu_i D_{12}\vec{\nabla} c_{12} + \frac{t_i \vec{I}}{z_i F} = -D_{12}\vec{\nabla} c_i + \frac{t_i \vec{I}}{z_i F},$$

and from

$$\vec{\nabla}\phi = \frac{\Gamma}{f}\vec{\nabla}\ln c_{12},$$

prove that $\vec{\nabla}\phi = \vec{\nabla}\phi_{\text{dif}} + \vec{\nabla}\phi_{\text{ohm}}$. Remember that

$$\Gamma = -\frac{1}{z_1 z_2}\frac{z_1 j_1/D_1 + z_2 j_2/D_2}{j_1/D_1 + j_2/D_2}.$$

(c) Would it be possible to make the derivation in case (b) without using the equation $\vec{\nabla}\phi = (\Gamma/f)\vec{\nabla}\ln c_{12}$?

2.10 Find the effective diffusion coefficient of an electrolyte that can form ion pairs according to the reaction

$$M^+ + A^- \rightleftarrows \text{IP},$$

where M^+, A^-, and IP represent the cation, the anion, and the ion pair, respectively. Assume chemical equilibrium and denote the association constant by K_{IP}.

2.11 Show that the effective diffusion coefficient of a symmetric, weak binary electrolyte

$$D_{12,\text{T}} \equiv \frac{\nu_{12}(1-\alpha)D_{12,\text{u}} + \alpha D_{12}}{\nu_{12}(1-\alpha) + \alpha}$$

can be written as

$$D_{12,\text{T}} \equiv D_{12,\text{u}} + (D_{12} - D_{12,\text{u}})\sqrt{\frac{K}{K + 4c_{12,\text{T}}}}.$$

2.12 Show that in the case of an arbitrary binary electrolyte, the convention

$$\sum_i \frac{c_i}{z_i}\ln\gamma_i = 0$$

leads to

$$\ln \gamma_{12} = \frac{\nu_1}{\nu_2} \ln \gamma_1 = \frac{\nu_2}{\nu_1} \ln \gamma_2.$$

2.13 Find the relation between the activity coefficient corrections

$$\beta_1 \equiv \left(1 + \frac{\mathrm{d}\ln \gamma_1}{\mathrm{d}\ln c_1}\right) \text{ and } \beta_2 \equiv \left(1 + \frac{\mathrm{d}\ln \gamma_2}{\mathrm{d}\ln c_2}\right)$$

in the case of an asymmetric 2:1 binary electrolyte.

2.14 Evaluate

$$\beta_1 \equiv 1 + \frac{\mathrm{d}\ln \gamma_1}{\mathrm{d}\ln c_1} = 1 + \frac{\mathrm{d}\ln \gamma_1}{\mathrm{d}\ln c_{12}}$$

at c_{12}= 0.01 M and 0.1 M using the Debye–Hückel limiting law

$$\ln \gamma_i \approx -\alpha z_i^2 \sqrt{I},$$

where $I = (z_1^2 c_1 + z_2^2 c_2)/2$ is the ionic strength and $\alpha = 1.1779\mathrm{M}^{-1/2}$ at 25°C. Discuss the importance of the activity corrections in 1:1 and 2:1 electrolytes.

2.15 (a) Within the Stefan–Maxwell approach, discuss whether friction effects between the ions act to decrease or to increase the solution electric conductivity

$$\kappa = \frac{z_1^2 F^2 c_1^2 (K_{1,0} + K_{2,0})}{K_{1,0}K_{2,0} + K_{1,2}(K_{1,0} + K_{2,0})}.$$

(b) Using the relation $K_{i,j} = RTc_\mathrm{T} x_i x_j / \overline{D}_{i,j}$, where c_T is the total molar concentration (including the solvent), show that the electrical conductivity can also be expressed as

$$\kappa = \frac{F^2}{RT} \frac{z_1^2 c_1 \overline{D}_{1,0} + z_2^2 c_2 \overline{D}_{2,0}}{x_0 + (x_1 \overline{D}_{2,0} + x_2 \overline{D}_{1,0})/\overline{D}_{1,2}}.$$

2.16 In order to determine the cross-diffusion coefficient $\overline{D}_{1,2}$ from the measured electrical conductivity, the equation given in the previous exercise must be used. Show that this equation can be solved for $\overline{D}_{1,2}$ to give

$$\overline{D}_{1,2} = \frac{\nu_1 \nu_2}{c_\mathrm{T}} \left[\frac{(F z_1 \nu_1 c_{12})^2}{\kappa} - \frac{\bar{t}_1 \bar{t}_2 \nu_{12} c_{12} x_0}{\overline{D}_{12}} \right]^{-1}.$$

2.17 (a) In the Stefan–Maxwell approach, show that the generalized Ohm's law

$$\vec{I} = -\kappa \vec{\nabla} \phi_\mathrm{ohm}$$

is valid even in the presence of concentration gradients, if we define

$$\vec{\nabla} \phi_\mathrm{ohm} \equiv \frac{1}{F} \left(\frac{\bar{t}_1}{z_1} \vec{\nabla} \tilde{\mu}_1 + \frac{\bar{t}_2}{z_2} \vec{\nabla} \tilde{\mu}_2 \right),$$

where the transport numbers are

$$\bar{t}_1 = \frac{K_{2,0}}{K_{1,0} + K_{2,0}} = \frac{\nu_2 D_1}{\nu_2 D_1 + \nu_1 D_2} = 1 - \bar{t}_2.$$

(b) Show that the electrolyte flux density

$$\vec{j}_{12}^{\,\mathrm{H}} = \frac{\bar{t}_2}{\nu_1}\vec{j}_1^{\,\mathrm{H}} + \frac{\bar{t}_1}{\nu_2}\vec{j}_2^{\,\mathrm{H}}$$

is given by

$$\vec{j}_{12}^{\,\mathrm{H}} = -L_{12,12}\vec{\nabla}\mu_{12} = \frac{\nu_{12}c_{12}^2}{K_{1,0} + K_{2,0}}\vec{\nabla}\mu_{12}$$

even in the presence of electric current.

(c) The Stefan–Maxwell equations

$$c_1\vec{\nabla}\widetilde{\mu}_1 = K_{1,0}(\vec{v}_0 - \vec{v}_1) + K_{1,2}(\vec{v}_2 - \vec{v}_1)$$
$$c_2\vec{\nabla}\widetilde{\mu}_2 = K_{2,0}(\vec{v}_0 - \vec{v}_2) + K_{2,1}(\vec{v}_1 - \vec{v}_2)$$

contain three transport coefficients. The equations $\vec{I} = -\kappa\vec{\nabla}\phi_{\mathrm{ohm}}$ and $\vec{j}_{12}^{\,\mathrm{H}} = -L_{12,12}\vec{\nabla}\mu_{12}$ can be derived from the former, but contain only two transport coefficients. How can you explain then that they are also of general validity?

(d) Check the consistency of the results of this exercise with those of Exercise 2.4.

2.18 (a) In order to guarantee that the two ions in a binary electrolyte solution move at the same velocity in the absence of electric current ($\vec{v}_1 = \vec{v}_2$ when $z_1\vec{j}_1^{\,\mathrm{H}} + z_2\vec{j}_2^{\,\mathrm{H}} = \vec{0}$) in spite of their different friction coefficients with the solvent ($K_{1,0} \neq K_{2,0}$ or $\overline{D}_{1,0} \neq \overline{D}_{2,0}$), an electric potential gradient must develop in the solution. This electric field is internally generated rather than externally imposed. From the flux equation

$$\vec{j}_i^{\,\mathrm{H}} = c_i(\vec{v}_i - \vec{v}_0) = -\frac{\overline{D}_{i,0}c_i}{RTx_0}(\vec{\nabla}\mu_i + z_iF\vec{\nabla}\phi)$$
$$= -\frac{\overline{D}_{i,0}c_i}{x_0}[\vec{\nabla}\ln(\gamma_i c_i) + z_i f\,\vec{\nabla}\phi],$$

and determining the activity coefficients according to the convention

$$\sum_i \frac{c_i}{z_i}\ln\gamma_i = 0,$$

show that the diffusion potential gradient in the Stefan–Maxwell approach is

$$\vec{\nabla}\phi_{\mathrm{dif}} = \frac{RT}{F}\frac{\overline{D}_{2,0} - \overline{D}_{1,0}}{z_1\overline{D}_{1,0} - z_2\overline{D}_{2,0}}\left[1 + \frac{\nu_1^2\overline{D}_{2,0} - \nu_2^2\overline{D}_{1,0}}{\nu_1\nu_2(\overline{D}_{2,0} - \overline{D}_{1,0})}\frac{\mathrm{d}\ln\gamma_{12}}{\mathrm{d}\ln c_{12}}\right]\vec{\nabla}\ln c_{12}.$$

(b) Explain why $\vec{\nabla}\phi_{\mathrm{dif}}$ does not depend on the cross-diffusion coefficient $\overline{D}_{1,2}$.

(c) Show that the diffusion potential gradient in the Stefan–Maxwell approach can also be written as

$$\vec{\nabla}\phi_{\text{dif}} = -\frac{1}{F}\left(\frac{\bar{t}_1}{z_1}\vec{\nabla}\mu_1 + \frac{\bar{t}_2}{z_2}\vec{\nabla}\mu_2\right) \quad .$$

2.19 In the Nernst–Planck approach the principle of independence allows us to write the flux density of species i as

$$\vec{j}_i = -\frac{t_i\kappa}{z_i^2F^2}\vec{\nabla}\mu_i.$$

In the Stefan–Maxwell approach the principle of independence does not apply and the flux density of species i depends also on the electrochemical potential gradients of other species. This means that, besides the term shown in the previous equation, additional terms must arise due to the friction between the ionic species

$$\vec{j}_i = -\frac{\bar{t}_i\kappa}{z_i^2F^2}\vec{\nabla}\mu_i + (\vec{j}_i)_{\text{cross effects}}.$$

Considering the case of a binary electrolyte solution, find the term due to cross-effects in the Stefan–Maxwell approach.

2.20 The following table shows the definitions of the electrical conductivity and the migrational transport numbers in the TIP, Nernst–Planck, and Stefan–Maxwell approaches. However, while those corresponding to the TIP and Nernst–Planck approaches are valid for multi-ionic systems, the Stefan–Maxwell ones are valid only for a binary system. Deduce the corresponding Stefan–Maxwell equations for a multi-ionic system.

	Electrical conductivity	Migrational transport number
TIP	$\kappa \equiv F^2\sum_i\sum_k z_iz_kl_{i,k}$	$t_i \equiv \frac{z_i\sum_k z_kl_{i,k}}{\sum_j\sum_k z_jz_kl_{j,k}}$
Nernst–Planck	$\kappa \equiv \frac{F^2}{RT}\sum_k z_k^2D_kc_k$	$t_i \equiv \frac{z_i^2D_ic_i}{\sum_k z_k^2D_kc_k}$
Stefan–Maxwell (binary)	$\kappa = \frac{F^2z_1^2c_1^2}{(K_{1,0}^{-1}+K_{2,0}^{-1})^{-1}+K_{1,2}}$	$\bar{t}_i = \frac{K_{i,0}^{-1}}{K_{1,0}^{-1}+K_{2,0}^{-1}}$

References

[1] A. Katchalsky and P.F. Curran, *Nonequilibrium Thermodynamics in Biophysics*, Harvard U.P., Cambridge, MA, 1965.

[2] S.R. de Groot and P. Mazur, *Non-equilibrium Thermodynamics*, North-Holland, Amsterdam, 1962.

[3] K. Kontturi, 'Countercurrent electrolysis in thin porous membranes', *Acta Polytech. Scand.*, 152 (1983) 1–40.

[4] J. Garrido and J.A. Manzanares, 'Observable electric potential and electrostatic potential in electrochemical systems', *J. Phys. Chem. B*, 104 (2000) 658–662.

[5] J. Maier, *Physical Chemistry of Ionic Materials*, Wiley, New York, 2004.

[6] A. Ekman, S. Liukkonen, and K. Kontturi, 'Diffusion and electric conduction in multicomponent electrolyte systems', *Electrochim. Acta,* 23 (1978) 243–250.

[7] J.A. Wesselingh and R. Krishna, *Mass Transfer in Multicomponent Mixtures*, Delft U.P., Delft, 2000.

[8] W.E. Morf, *The Principles of Ion-Selective Electrodes and of Membrane Transport*, Elsevier, Amsterdam, 1981.

[9] V.M.M. Lobo and J.L. Quaresma, *Electrolyte Solutions: Literature Data on Thermodynamic and Transport Properties*, Vols. I–II, Coimbra Editora, Coimbra 1981.

[10] R.P. Buck, 'Kinetics of bulk and interfacial ionic motion: microscopic bases and limits ofr the Nernst–Planck equation applied to membrane systems', *J. Membrane Sci.*, 17 (1984) 1–62.

[11] R. Schlögl, *Stofftransport durch Membranen*, D. Steinkopff, Darmstadt, 1964.

[12] O. Lamm, 'An analysis of the dynamical equations of three component diffusion for the determination of friction coefficients. I', *Ark. Kemi. Mineral. Geol. B*, 18 (1944) 1813.

[13] L. Onsager, 'Theories and problems of liquid diffusion', *Ann. N. Y. Acad. Sci.*, 46 (1945) 241–265.

[14] J.M.G. Barthel, H. Krienke, and W. Kunz, *Physical Chemistry of Electrolyte Solutions. Modern Aspects*, Steinkopff-Springer, Darmstadt, 1998.

[15] R. Taylor and R. Krishna, *Multicomponent Mass Transfer*, Wiley, New York, 1993.

[16] J.S. Newman, *Electrochemical Systems*, 3rd edn., Prentice Hall, Englewood Cliffs, N.J., 2004.

3 Transport at electrodes

3.1 Faraday's law

Mass transport at electrodes takes place due to electrode reactions, where the electrode is a surface source or a sink of the species involved in the reaction. As a consequence, concentration differences are created in the vicinity of the electrode, which gives rise to mass transport.

An electrode reaction can involve the following mechanisms:

1) a species is reduced (or oxidized) at the electrode and precipitates on the surface (e.g. $CuSO_4 + 2e^- \rightarrow Cu(s) + SO_4^{2-}$);
2) a species is reduced (or oxidized) at the electrode, remains in solution and is transported away from the electrode (e.g. $Fe^{2+} \rightarrow Fe^{3+} + e^-$);
3) a species reacts on the electrode and is transported into the electrode (e.g. Hg);
4) electrochemical (anodic) dissolution of an electrode (e.g. $Cu(s) \rightarrow Cu^{2+} + 2e^-$).

The rate of an electrochemical reaction is accurately known via Faraday's law. If the number of electrons exchanged in the electrode reaction is n and the electric current density is I, *Faraday's law* establishes that the reaction rate (density) is

$$r = \frac{I}{nF}. \tag{3.1}$$

This is an exceptional and valuable feature of electrochemistry: the reaction rate can be monitored or controlled *in situ*. Furthermore, with modern equipment very low currents $i = IA$, where A is the electrode area, can be measured. A current of the order of 10^{-9} A implies a reaction rate rA of the order of 10^{-14}mol s^{-1}, a current of 10^{-12} A implies a reaction rate of 10^{-17}mol s^{-1}, and a current of 10^{-15} A implies a reaction rate of the order of 10^{-20}mol s^{-1}. Thus, a femtoamp current measures an event where 6000 electrons in a second are transferred across the electrode solution/interface.

Our *sign convention* is such that the reaction rate r is always positive, and I and n have the same sign. This sign is positive in anodic oxidations (i.e. when the current flows from the electrode to the bulk solution) and negative in cathodic reductions. Similarly, the flux density of an *electroactive species* (i.e. one that participates in the electrode reaction) is positive when it moves from the electrode to the bulk solution and negative otherwise. In relation to the spatial position co-ordinate x, the electrode is located at $x = 0$ and the solution in the positive x region. The fluxes and the current density are considered positive in the positive x direction, and the potential drop $\Delta\phi$ in the solution adjacent to the electrode is defined as the potential in the bulk solution minus the potential at the electrode surface, i.e. with the same sign as the electric potential gradient $\mathrm{d}\phi/\mathrm{d}x$. Hence, for an oxidation reaction $\Delta\phi$ is negative, and for a reduction reaction it is positive. With this convention, the generalized Ohm's law is written as $I = -\kappa \mathrm{d}\phi_{\mathrm{ohm}}/\mathrm{d}x$, where $\kappa > 0$ is the electrical conductivity, and $\mathrm{d}\phi_{\mathrm{ohm}}/\mathrm{d}x$ (or $\Delta\phi_{\mathrm{ohm}}$) and I have opposite signs.

The general form of electrode reaction can be written as

$$\sum_i \nu_i \mathrm{B}_i^{z_i} + n\mathrm{e}^- = 0, \tag{3.2}$$

where charge conservation requires that $n = \sum_i z_i \nu_i$. In eqn (3.2), the stoichiometric number of species i, ν_i, and that of the electron, n, are positive for products and negative for reactants. For instance, in the oxidation $\mathrm{R} \rightarrow \mathrm{O} + n\mathrm{e}^-$, $\nu_{\mathrm{O}} = -\nu_{\mathrm{R}} = 1$ and $n > 0$, and in the reduction $\mathrm{O} + |n|\mathrm{e}^- \rightarrow \mathrm{R}$, $\nu_{\mathrm{R}} = -\nu_{\mathrm{O}} = 1$ and $n < 0$.

The reaction rate is related to transport through the mass balance. Faraday's law implies that the flux density at the electrode surface of an electroactive species i is

$$j_i = \frac{\nu_i I}{nF}. \tag{3.3}$$

Obviously, when there is only one electroactive ion in solution, i.e. when the electrode reaction involves only one ionic species in solution, eqn (3.3) becomes

$$j_i = \frac{I}{z_i F}. \tag{3.4}$$

For instance, in the cathodic deposition of copper, $\mathrm{Cu}^{2+}(\mathrm{aq.}) + 2\mathrm{e}^- \rightarrow \mathrm{Cu(s)}$, we have $n = -2$ and only one electroactive ion in the aqueous solution, the cupric ion, which has $\nu_{\mathrm{Cu}^{2+}} = -1$ and $z_{\mathrm{Cu}^{2+}} = 2$. Thus, $I < 0$ and $j_{\mathrm{Cu}^{2+}} = I/2F < 0$, which means that these ions are transported from the solution to the electrode surface (where they are deposited). On the contrary, in the redox electrode reaction $\mathrm{Cu}^{2+}(\mathrm{aq.}) + \mathrm{e}^- \rightarrow \mathrm{Cu}^+(\mathrm{aq.})$ there are two active ions in solution and eqn (3.3) can be applied to any of them with $I < 0$ and $n = -1$. Thus, the flux density of the cupric ion is $j_{\mathrm{Cu}^{2+}} < 0$ because it is consumed at the electrode, and that of the cuprous ion is $j_{\mathrm{Cu}^+} > 0$ because it is produced by the electrode reaction and transported to the solution.

In three-dimensional systems I and j_i in eqn (3.3) must be interpreted as the components of the corresponding vector quantities in the direction normal to the electrode surface (positive normal direction is taken from the electrode to the solution). Nevertheless, the electrode surface is usually equipotential and has uniform accessibility to the reactants, so that $\vec{I}$ and $\vec{j}_i$ are actually directed normal to the electrode surface.

3.2 Electrode processes in stationary state

3.2.1 Nernst diffusion boundary layer

In stationary transport processes, the continuity equation of a species i not involved in homogeneous chemical reactions reduces to $\vec{\nabla} \cdot \vec{j}_i = -\partial c_i/\partial t = 0$. Stirring is necessary to establish the stationary transport conditions because otherwise the situation becomes ambiguous due to natural convection, which tends to stir the solution in an uncontrolled fashion. An exception is an electrode that creates a spherical diffusion process; this case is discussed later in Section 3.2.7. When the solution is stirred, the velocity profile can be determined by solving the Navier–Stokes equation subject to the appropriate boundary conditions. There is an abundance of literature [1] concerning different hydrodynamic conditions, but here we present the general features taking the case of a planar electrode, where the solution is stirred with a propeller, as an example.

When species i reacts at the electrode surface, the situation under limiting current conditions can be depicted as shown in Fig. 3.1. The diffusion boundary layer can be some orders of magnitude thinner than the hydrodynamic boundary layer. The thickness of these layers cannot be defined unambiguously because the concentration and the velocity approach their bulk values asymptotically. As far as the diffusion boundary layer is concerned, the concept of Nernst or unstirred layer adjacent to the electrode surface is customarily introduced. In this layer, it is assumed that ionic transport takes place only by diffusion and migration, the convective contribution being negligible. Its thickness δ is defined as the distance through which the linear portion of the concentration profile of a reacting species (usually the one that is totally consumed at the surface under limiting conditions) must be extrapolated to reach its bulk concentration. This concept of an unstirred or stagnant layer does not apply as such in other geometries or stirring conditions and must be modified accordingly. It must also be emphasized that an experimental determination of the thickness of the diffusion boundary layer is rarely possible, and therefore, the problem is treated with the aid of the limiting current (density) as discussed below.

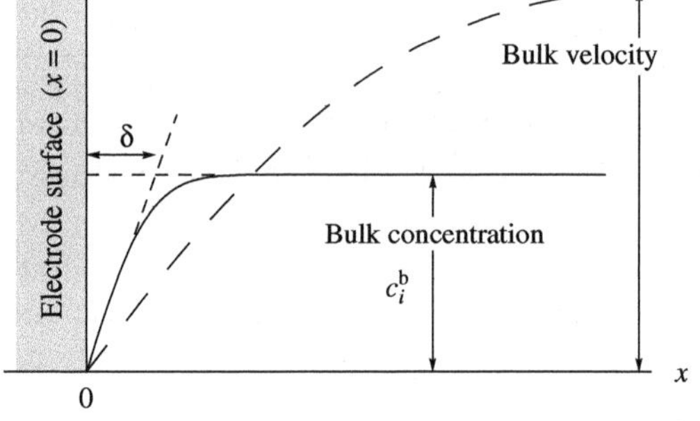

Fig. 3.1. Solution velocity (long dashed line) and concentration (solid line) profiles in the vicinity of an electrode. The diffusion boundary layer δ is defined by the crossing of the linearly extrapolated concentration profile at the electrode surface and the bulk concentration (short dashed lines).

3.2.2 Limiting current density

The transport to or from the electrode can take place by diffusion, migration and convection. At the electrode surface, however, the solution velocity usually vanishes due to the non-slip and non-penetrability conditions, and transport takes place by diffusion and migration only. When an electroactive species is consumed at the electrode surface, its surface concentration is lower than in the bulk solution. Since migration is proportional to the ionic concentration, the migrational contribution to the overall mass transport is often smaller than the diffusional one. In particular, when the surface concentration of reactant i vanishes, its flux density is purely diffusive close to the electrode surface,[1] and is given by $\vec{j}_i = -D_i\vec{\nabla}c_i$. In one-dimensional systems, the concentration gradient at the electrode surface located at $x = 0$ then takes its maximum value $(\mathrm{d}c_i/\mathrm{d}x)_{x=0} = c_i^{\mathrm{b}}/\delta$ (Fig. 3.1), and the system reaches the limiting current density

$$I_{\mathrm{L},i} \equiv \frac{nFj_{\mathrm{L},i}}{\nu_i} = -\frac{nFD_ic_i^{\mathrm{b}}}{\nu_i\delta}. \tag{3.5}$$

This is the *limiting diffusion current density of species i*. Note that δ can be expressed in terms of $I_{\mathrm{L},i}$, which is a measurable quantity.

By definition, the flux density under limiting conditions for species i ($c_i(0) = 0$) is negative, $j_{\mathrm{L},i} \equiv -D_ic_i^{\mathrm{b}}/\delta < 0$. This means that in order for species i to reach limiting conditions, it must be consumed at the electrode. Species i is then a reactant in the reaction and $\nu_i < 0$. The sign of $I_{L,i}$ in eqn (3.5) is then that of n. Thus, in the anodic oxidation $\mathrm{R} \to \mathrm{O} + ne^-$, the limiting diffusion current density of species R is $I_{\mathrm{L,R}} > 0$ because $n > 0$ and $\nu_{\mathrm{R}} = -1$. Similarly, in the cathodic reduction $\mathrm{O} + |n|\mathrm{e}^- \to \mathrm{R}$, the limiting diffusion current density of species O is $I_{\mathrm{L,O}} < 0$ because $n < 0$ and $\nu_{\mathrm{O}} = -1$. In principle, it is also possible to apply the definition of $I_{\mathrm{L},i}$ in eqn (3.5) to the products of the electrode reaction but $I_{\mathrm{L},i}$ then becomes just an auxiliary variable. For instance, in the anodic oxidation $\mathrm{R} \to \mathrm{O} + ne^-$, the limiting diffusion current density of species O is $I_{\mathrm{L,O}} < 0$ because $n > 0$ and $\nu_{\mathrm{O}} > 0$. But, since I is defined to be positive in oxidations, it is clear that the situation $I = I_{\mathrm{L,O}}$ can never be realized and the reaction product cannot reach limiting conditions.

When the electrode reaction involves several reactants, the limiting current density of the system is determined by the reactant that runs out first; in the following sections we chose subscript 1 for such species. In oxidations, the limiting current density of the system $I_{\mathrm{L}} > 0$ is equal to the lowest of their limiting diffusion current densities of the reactants. In reductions, when $I_{\mathrm{L}} < 0$ and the reactants have $I_{\mathrm{L},i} < 0$, I_{L} is equal to the $I_{\mathrm{L},i}$ that has the lowest absolute magnitude.

[1] In the absence of supporting electrolyte and under limiting conditions, this is not valid because the electric field becomes so intense at the electrode surface located at $x = 0$ that other transport mechanisms (such as electroconvection) come into play.

3.2.3 Transport equations in multi-ionic solutions

We compile here the main transport equations, in the Nernst–Planck approach, from Chapter 2. In a multi-ionic system with N species in solution, the number of unknown variables is $N + 1$: the N molar concentrations c_i and the electric potential ϕ. The $N + 1$ equations that allow for their determination are the N equations for the flux densities

$$-\vec{j}_i = D_i(\vec{\nabla} c_i + z_i c_i f\, \vec{\nabla}\phi), \tag{3.6}$$

and the local electroneutrality assumption[2]

$$\sum_i z_i c_i = 0. \tag{3.7}$$

In non-stationary or transient processes, the time dependence of these variables is determined by the continuity equation

$$\frac{\partial c_i}{\partial t} = -\vec{\nabla} \cdot \vec{j}_i + \pi_i, \tag{3.8}$$

where π_i is the net production rate of species i due to homogeneous chemical reactions. For stationary processes and absence of homogeneous reactions, the continuity equation reduces to

$$\vec{\nabla} \cdot \vec{j}_i = 0. \tag{3.9}$$

Moreover, the fluxes are coupled through the equation for the electric current[3]

$$\vec{I} = F \sum_i z_i \vec{j}_i. \tag{3.10}$$

The Nernst–Planck equation can be rewritten by making the ionic contribution to ohmic conduction explicit (i.e. the so-called chemical diffusion–ohmic conduction form of the transport equations). In the case of binary electrolyte solutions, this takes the simple expression

$$\vec{j}_i = -D_{12}\vec{\nabla} c_i + \frac{t_i \vec{I}}{z_i F}, \tag{3.11}$$

and its general form in multi-ionic solutions (see Section 2.3.3) is

$$-\vec{j}_i = \sum_k D_{i,k}\vec{\nabla} c_k - \frac{t_i \vec{I}}{z_i F}, \tag{3.12}$$

[2] When the current density approaches the limiting value in the absence of a supporting electrolyte, this approximation fails and electroconvection must be taken into account.

[3] In electrical relaxation transients, the displacement current also needs to be taken into account.

where the cross-diffusion coefficients are

$$D_{i,k} \equiv D_i \delta_{ik} + \frac{t_i}{z_i} z_k (D_i - D_k), \tag{3.13}$$

and δ_{ik} is the Kronecker delta ($\delta_{ik} = 1$ when $i = k$ and $\delta_{ik} = 0$ when $i \neq k$). Equation (3.12) shows explicitly that the ionic flux densities are coupled in a system where the ionic species have different mobilities, due to the local electroneutrality requirement.[4]

Alternatively, by elimination of the electric field, the Nernst–Planck equations can be cast in the form

$$\frac{\vec{j}_i}{D_i} = -\vec{\nabla} c_i + \frac{z_i c_i}{\sum_j z_j^2 c_j} \sum_k z_k \frac{\vec{j}_k}{D_k}, \tag{3.14}$$

and solved in terms of sums such as

$$\sum_i \frac{\vec{j}_i}{D_i} = -\vec{\nabla} \sum_i c_i, \tag{3.15}$$

which can be integrated straightforwardly (assuming that the diffusion coefficients are constant) because the flux densities are constant under steady-state conditions. Note that in the case of binary solutions, eqn (3.15) is simply Fick's first equation for the electrolyte, $\vec{J}_{12} = -D_{12}\vec{\nabla} c_{12}$. This procedure is explained in further detail in Section 3.2.6.

3.2.4 Trace ions

In electrochemical practice, supporting electrolytes are commonly used to increase the electric conductivity of the solutions. Those other ions that do not contribute significantly to the conductivity and have negligible transport numbers, $t_i \ll 1$, are known as *trace ions*. Since the second term in the right-hand side of eqn (3.13) contains a factor t_i, the cross-diffusion coefficients reduce to $D_{i,k} \approx D_i \delta_{i,k}$ for a trace ion. This means that its transport decouples from that of the other ionic species, and its flux density is given by

$$-\vec{j}_i \approx D_i \vec{\nabla} c_i - \frac{t_i \vec{I}}{z_i F} \approx D_i \vec{\nabla} c_i. \tag{3.16}$$

Hence, a trace ion is mainly transported by diffusion and its flux equation can be approximated by Fick's first law even though it is charged and an electric field might be present in the solution. The last approximation in eqn (3.16) comes from the fact that the contribution of electric conduction to the transport of a trace ion is negligible, regardless of the relative magnitudes of $|z_i F \vec{j}_i|$ and

[4] However, this equation is of limited utility in the solution of transport problems because the transport numbers t_i are functions of the local ionic concentrations. Rather often, eqn (3.12) cannot be integrated analytically in multicomponent solutions.

$|\vec{I}|$. In fact, a trace ion might carry all the electric current in the vicinity of the electrode surface, so that $\vec{I} = z_i F \vec{j}_i$ but still $|t_i \vec{I}| \ll |z_i F \vec{j}_i|$.

Under steady-state conditions and in the absence of homogeneous reactions, the continuity equation for a trace ion reduces to $\vec{\nabla} \cdot \vec{j}_i = 0$. Equation (3.16) then implies that the spatial distribution of its concentration is given by the Laplace equation

$$\nabla^2 c_i = 0. \tag{3.17}$$

Thus, in one-dimensional systems the concentration profile of a trace ion is linear

$$c_i(x) = c_i(0) + [c_i^{\mathrm{b}} - c_i(0)]\frac{x}{\delta}, \tag{3.18}$$

and its surface concentration is

$$c_i(0) = c_i^{\mathrm{b}}\left(1 - \frac{nFj_i}{\nu_i I_{\mathrm{L},i}}\right) = c_i^{\mathrm{b}}\left(1 - \frac{I}{I_{\mathrm{L},i}}\right), \tag{3.19}$$

where $I_{\mathrm{L},i}$ is its limiting diffusion current density.

Equations (3.16) and (3.17) are the most common transport equations in modern electrochemistry. Although quite a number of assumptions have been made to derive them, it has to be realized that they are valid with sufficient accuracy also in cases where the present assumptions are not strictly fulfilled. Finally, an interesting feature of eqn (3.16) is that it contains the ionic diffusion coefficient, not that of the component, which makes its experimental determination feasible.

3.2.5 Solutions with only one electroactive ion

The steady-state transport equations can be easily solved when the solution contains only one electroactive ion. In this section we show the solution procedure for one-dimensional systems (in the absence of homogeneous reactions) and illustrate it with an example. Since the electroactive ions are either consumed or produced at the electrode surface, their flux density is different from zero in the electrode vicinity. On the contrary, the electroinactive ions have zero flux density there. Under steady-state conditions, the continuity equation requires that the flux density of every ion is independent of position. Hence, throughout the diffusion boundary layer, the ionic flux densities are

$$j_1 = \frac{I}{z_1 F}, \tag{3.20}$$

$$j_i = 0, \quad i \neq 1, \tag{3.21}$$

where the index $i = 1$ denotes the electroactive ion. Moreover, when there is only one electroactive ion, its stoichiometric number in the electrode reaction must be equal to n/z_1, where n is the stoichiometric number of the electrons in the reaction. Thus, the limiting current density of the active species can also be written as

$$I_{L,1} = -\frac{z_1 F D_1 c_1^b}{\delta}. \tag{3.22}$$

In the case of binary electrolyte solutions, the diffusion–conduction form of the transport equations

$$j_1 = \frac{I}{z_1 F} = -D_{12}\frac{dc_1}{dx} + \frac{t_1 I}{z_1 F} = -\nu_1 D_{12}\frac{dc_{12}}{dx} + \frac{t_1 I}{z_1 F} \tag{3.23}$$

$$j_2 = 0 = -D_{12}\frac{dc_2}{dx} + \frac{t_2 I}{z_2 F} = -\nu_2 D_{12}\frac{dc_{12}}{dx} + \frac{t_2 I}{z_2 F} \tag{3.24}$$

allows us to solve easily for the concentration gradient. This gradient

$$\frac{dc_{12}}{dx} = -\frac{(1-t_1)I}{z_1\nu_1 F D_{12}} = \frac{t_2 I}{z_2\nu_2 F D_{12}} \tag{3.25}$$

can be integrated subject to the boundary condition $c_{12}(\delta) = c_{12}^b$ and leads to the concentration profile

$$c_{12}(x) = c_{12}^b + \frac{(1-t_1)I}{z_1\nu_1 F D_{12}}(\delta - x). \tag{3.26}$$

In the case of a symmetric electrolyte this further simplifies to

$$c_{12}(x) = c_{12}^b + \frac{I}{2z_1 F D_1}(\delta - x). \tag{3.27}$$

The limiting current density can then be evaluated from the condition of vanishing electrolyte concentration at the electrode surface, $c_{12}(0) = 0$, as

$$I_L = -\frac{2z_1 F D_1 c_{12}^b}{\delta} \tag{3.28}$$

where the resulting minus sign comes from our sign convention (i.e. $c_{12}(0) = 0$ can only occur when $j_1 < 0$). Note that this is double the limiting diffusion current of the active species, $I_{L,1} = -z_1 F D_1 c_1^b/\delta$. This feature is known as the *supporting electrolyte paradox*: adding an inert electrolyte to the solution, the conductivity of the solution increases but the limiting current density decreases. As shown above, this happens because a trace ion has no migrational contribution in the transport equation, namely $t_1 \approx 0$. In multi-ionic solutions a similar approach is not possible because the transport numbers t_i are functions of the (unknown) local ionic concentrations and, hence, eqn (3.23) cannot be integrated analytically. Fortunately, the diffusion–migration form of the transport equations can be easily integrated as we show next.

Consider first the case of a binary solution. The diffusion–migration form of the transport equations

$$j_1 = \frac{I}{z_1 F} = -D_1 \left(\frac{\mathrm{d}c_1}{\mathrm{d}x} + z_1 c_1 f \frac{\mathrm{d}\phi}{\mathrm{d}x} \right) \tag{3.29}$$

$$j_2 = 0 = -D_2 \left(\frac{\mathrm{d}c_2}{\mathrm{d}x} + z_2 c_2 f \frac{\mathrm{d}\phi}{\mathrm{d}x} \right) \tag{3.30}$$

allows us to solve for the electric potential gradient[5]

$$\frac{\mathrm{d}\phi}{\mathrm{d}x} = -\frac{1}{z_2 f} \frac{\mathrm{dln}c_2}{\mathrm{d}x} = -\frac{1}{z_2 f} \frac{\mathrm{dln}c_{12}}{\mathrm{d}x}. \tag{3.31}$$

Moreover, using the local electroneutrality assumption $z_1 c_1 + z_2 c_2 = 0$, the electric potential gradient can be eliminated from the transport equations and the concentration gradient is found as

$$\frac{\mathrm{d}c_{12}}{\mathrm{d}x} = -\frac{I}{z_1 \nu_{12} F D_1}, \tag{3.32}$$

which is the same as that shown in eqn (3.25); here, $\nu_{12} = \nu_1 + \nu_2$. An interesting feature of this procedure is that it provides directly the electric potential drop in the diffusion boundary layer in the form of the Nernstian equation

$$\Delta\phi \equiv \phi^{\mathrm{b}} - \phi(0) = \frac{1}{z_2 f} \ln \frac{c_{12}(0)}{c_{12}^{\mathrm{b}}} \tag{3.33}$$

as well as the current–voltage curve

$$I = I_{\mathrm{L}}(1 - \mathrm{e}^{z_2 f \Delta\phi}), \tag{3.34}$$

where we have used the following expression for the surface concentration

$$c_{12}(0) = c_{12}^{\mathrm{b}} + \frac{I\delta}{z_1 \nu_{12} F D_1} = c_{12}^{\mathrm{b}} \left(1 - \frac{I}{I_{\mathrm{L}}} \right). \tag{3.35}$$

In the case of multi-ionic solutions, we take advantage of the fact that the fluxes of electroinactive ions are zero. Hence, from their Nernst–Planck equations it is readily seen that they distribute in the boundary layer $0 < x < \delta$ according to the Boltzmann equation

$$c_i(x) = c_i^{\mathrm{b}} \mathrm{e}^{-z_i \varphi}, \quad i \neq 1, \tag{3.36}$$

[5] Note that when changing $\mathrm{dln}c_2$ to $\mathrm{dln}c_{12}$ the stoichiometric coefficient ν_2 cancels out.

where $\varphi(x) = f[\phi(x) - \phi^{\mathrm{b}}]$ is the dimensionless local electric potential relative to the bulk solution. From eqn (3.36) and the local electroneutrality assumption, the concentration of the electroactive ion is

$$c_1(x) = -\frac{1}{z_1}\sum_{i\neq 1} z_i c_i = -\frac{1}{z_1}\sum_{i\neq 1} z_i c_i^{\mathrm{b}} \mathrm{e}^{-z_i\varphi}, \tag{3.37}$$

and its flux density is

$$j_1 = \frac{I}{z_1 F} = -D_1\left(\frac{\mathrm{d}c_1}{\mathrm{d}x} - \sum_{i\neq 1} z_i c_i \frac{\mathrm{d}\varphi}{\mathrm{d}x}\right) = -D_1\left(\frac{\mathrm{d}c_1}{\mathrm{d}x} + \sum_{i\neq 1}\frac{\mathrm{d}c_i}{\mathrm{d}x}\right)$$
$$= -D_1\frac{\mathrm{d}c_{\mathrm{T}}}{\mathrm{d}x}, \tag{3.38}$$

where $c_{\mathrm{T}} \equiv \sum_i c_i$ is the total ionic concentration. The same result is obtained also by calculating the sum $\sum_i (j_i/D_i)$ and using the electroneutrality condition. Then, only $I/(z_1 F D_1)$ is left on the left-hand side, and the terms on the right-hand side containing $\mathrm{d}\varphi/\mathrm{d}x$ cancel out. As the flux density j_1 is independent of position, this can be integrated from $x = 0$ to $x = \delta$ to give

$$j_1 = \frac{D_1}{\delta}[c_{\mathrm{T}}(0) - c_{\mathrm{T}}^b] = \frac{D_1}{\delta}\sum_{i\neq 1}\left(1 - \frac{z_i}{z_1}\right) c_i^{\mathrm{b}}(\mathrm{e}^{z_i f\Delta\phi} - 1), \tag{3.39}$$

where $\Delta\phi \equiv \phi^{\mathrm{b}} - \phi(0) = -\varphi(0)/f$ and we have used eqns (3.36) and (3.37). Equations (3.36) and (3.39) allow us to calculate the variation of the surface concentrations with the current density and the current–voltage curve of the system.

With the help of eqn (3.37), eqn (3.38) can also be written as a generalized Ohm's law

$$I = -\kappa_{\mathrm{eff}}\frac{\mathrm{d}\phi}{\mathrm{d}x} \tag{3.40}$$

where the effective electrical conductivity $\kappa_{\mathrm{eff}} \equiv (F^2/RT) D_1 \sum_i z_i^2 c_i$ only involves D_1 because the electroactive species is the only one that moves in the solution.

Particularly interesting is the variation of the surface concentration of the electroactive ion with the current density. If it is a trace ion, its surface concentration can be evaluated as

$$c_1(0) = c_1^{\mathrm{b}} + \frac{j_1\delta}{D_1} = c_1^{\mathrm{b}}\left(1 - \frac{I}{I_{\mathrm{L},1}}\right). \tag{3.41}$$

It is worth remembering the sign convention. When the active species is consumed at the electrode, $c_1(0) < c_1^{\mathrm{b}}$, $j_1 < 0$, and $I/I_{\mathrm{L},1} > 0$, and when it is generated there (e.g. in a metal electrode dissolution), $c_1(0) > c_1^b, j_1 > 0$, and $I/I_{L,1} < 0$.

If the electroactive ion does not behave as a trace ion, analytical expressions for its surface concentration as a function of the current density, similarly to eqn (3.41), can only be obtained in special cases. We illustrate this with an example. In particular, we consider the stationary transport in a dilute aqueous solution of H_2SO_4 and K_2SO_4 associated to the cathodic reaction $2H^+ + 2e^- \rightarrow H_2$.

We denote by index $i = 1$ the electroactive ion H^+ ($z_1 = +1$), by $i = 2$ the electroinactive cation K^+ ($z_2 = +1$), and by $i = 3$ the common anion SO_4^{2-} ($z_3 = -2$). The electrolytes H_2SO_4 and K_2SO_4 are characterized by indexes 13 and 23, respectively, and their concentrations in the bulk solution are c_{13}^b and c_{23}^b. The bulk ionic concentrations are then $c_1^b = 2c_{13}^b$, $c_2^b = 2c_{23}^b$, and $c_3^b = c_{13}^b + c_{23}^b$. Bisulphate is thus assumed to be completely dissociated into sulphate in this example.

From eqn (3.39), the current–voltage curve can be written as

$$I = \frac{FD_1}{\delta} 3c_3^b(e^{-2f\Delta\phi} - 1) = I_{L0}(1 - e^{-2f\Delta\phi}), \qquad (3.42)$$

where $I_{L0} \equiv -3FD_1c_3^b/\delta = -FD_1c_T^b/\delta < 0$ is an auxiliary variable. The limiting current density I_L is defined from the condition that the concentration of the H^+ ion vanishes at the electrode surface, and consequently $c_2(0) = 2c_3(0)$. From eqn (3.36), the limiting value of the electric potential drop in the diffusion boundary layer ($\Delta\phi_L \equiv \Delta\phi$ when $I = I_L$) is

$$\Delta\phi_L = -\frac{1}{3f} \ln \frac{c_2^b}{2c_3^b} = \frac{1}{3f} \ln(1 + c_{13}^b/c_{23}^b) \geq 0. \qquad (3.43)$$

Substituting eqn (3.43) in eqn (3.42), the limiting current density is obtained as

$$I_L = I_{L0}(1 - e^{-2f\Delta\phi_L}) = I_{L0}[1 - (1 + c_{13}^b/c_{23}^b)^{-2/3}] < 0. \qquad (3.44)$$

Equation (3.44) has two interesting limiting cases (Fig. 3.2). First, when $c_{13}^b \ll c_{23}^b$ we recover the trace ion case, $I_L \approx I_{L,1} \ll I_{L0}$.[6] Also, the potential drop in the diffusion boundary layer $\Delta\phi_L$ vanishes. Second, when $c_{23}^b \ll c_{13}^b$, the system behaves as a binary electrolyte solution, and it is directly seen from eqn (3.43) that $\Delta\phi_L \rightarrow \infty$, and from eqn (3.44) that the limiting current density becomes $I_L \approx I_{L0} \approx (1 - z_1/z_3)I_{L,1}$.

[6] The limiting value of eqn (3.44) can be calculated with, e.g., the Taylor series as c_{13}^b approaches zero.

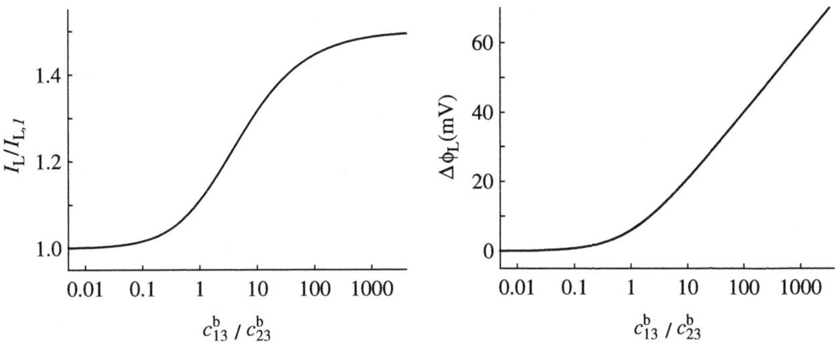

Fig. 3.2.
Limiting current density and potential drop in the boundary layer under limiting conditions as a function of the electrolyte concentration ratio c_{13}^b/c_{23}^b. When $c_{13}^b/c_{23}^b \ll 0.1$, K_2SO_4 plays the role of a supporting electrolyte and H^+ behaves as a trace ion, so that the potential drop vanishes, migration is negligible, and the limiting current is $I_{L,1}$. When $c_{13}^b/c_{23}^b \gg 100$, the reduction of migration by K_2SO_4 is negligible, and the limiting current increases by a factor $1 - z_1/z_3 = 1.5$.

The surface concentrations can be evaluated from eqns (3.36) and (3.42) as

$$\frac{c_3(0)}{c_3^b} = e^{-2f\Delta\phi} = 1 - \frac{I}{I_{L0}} = 1 - \frac{I}{I_L}[1 - (1 + c_{13}^b/c_{23}^b)^{-2/3}], \qquad (3.45)$$

$$\frac{c_2(0)}{c_2^b} = e^{f\Delta\phi} = \left[\frac{c_3(0)}{c_3^b}\right]^{-1/2}, \qquad (3.46)$$

$$c_1(0) = -c_2(0) + 2c_3(0) = 2c_3(0)\left[1 - \left(\frac{I_{L0} - I_L}{I_{L0} - I}\right)^{3/2}\right]. \qquad (3.47)$$

These concentrations are equal to the corresponding bulk values when the current density vanishes, and to $c_1(0) = 0$, $c_2(0) = 2c_3(0)$ and $c_3(0) = (c_{23}^b)^{2/3}(c_{13}^b + c_{23}^b)^{1/3}$ when the current density approaches the limiting value.

We can see from the current–voltage curves in Fig. 3.3 that the behaviour of the solution is practically ohmic in the presence of excess inert electrolyte. Indeed, it is easy to check that the current–voltage curve then reduces to $I = -\kappa_{\text{eff}}\Delta\phi/\delta$, where $\kappa_{\text{eff}} = (F^2/RT)D_1\sum_i z_i^2 c_i^b$, as expected from eqn (3.40).

3.2.6 Solutions with several electroactive species

We begin the study of these systems with an example. In particular, we consider the stationary transport in a dilute aqueous solution of $CuSO_4$ and H_2SO_4. These electrolytes are denoted by indexes 13 and 23, respectively, and their concentrations in the bulk solution are c_{13}^b and c_{23}^b. Depending on the concentrations, different species exist in the solution. We consider here only such concentrations that the free sulphate ion SO_4^{2-} does not exist, namely $c_{23}^b > 0.1$ M. The systems considered are[7]

[7] To the best of our knowledge, copper bisulphate is an imaginary species.

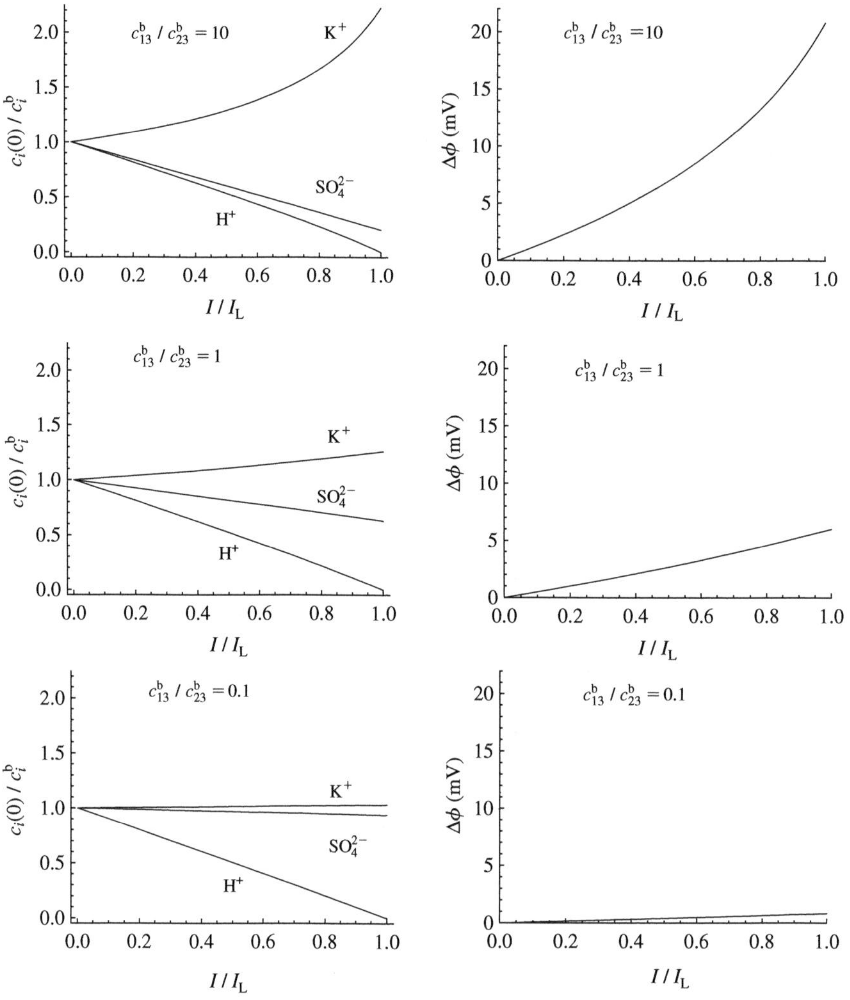

Fig. 3.3.
Surface concentrations and potential drop in the boundary layer as a function of the current density for the cases $c^b_{13}/c^b_{23} = 10$, 1, and 0.1.

$$\begin{aligned} i)&\ Cu^{2+} + H^+ + HSO_4^- \\ ii)&\ Cu(HSO_4)^+ + H^+ + HSO_4^- \\ iii)&\ Cu(HSO_4)_2 + H^+ + HSO_4^- \end{aligned}$$

or a mixture thereof and the electrode reactions are, respectively,

$$Cu^{2+} + 2e^- \rightarrow Cu(s) \tag{3.48}$$

$$Cu(HSO_4)^+ + 2e^- \rightarrow Cu(s) + HSO_4^- \tag{3.49}$$

$$Cu(HSO_4)_2 + 2e^- \rightarrow Cu(s) + 2HSO_4^-. \tag{3.50}$$

In the first and third cases there is only one electroactive ion in solution. The solution of the transport equations then follows the procedure explained in the previous section. We concentrate here on case *ii*).

We denote the species $Cu(HSO_4)^+$, H^+, and HSO_4^- by indexes $i = 1, 2$, and 3, respectively, and their bulk concentrations are $c_1^b = c_{13}^b$, $c_2^b = c_{23}^b - c_{13}^b$, and $c_3^b = c_{23}^b$. From the electrode reaction in eqn (3.49) and Faraday's law, the ionic flux densities are known as $j_1 = -j_3 = I/2F$ and $j_2 = 0$. The Nernst–Planck equations are then

$$-\frac{j_1}{D_1} = -\frac{I}{2FD_1} = \frac{dc_1}{dx} + c_1 f \frac{d\phi}{dx}, \tag{3.51}$$

$$-\frac{j_2}{D_2} = 0 = \frac{dc_2}{dx} + c_2 f \frac{d\phi}{dx}, \tag{3.52}$$

$$-\frac{j_3}{D_3} = \frac{I}{2FD_3} = \frac{dc_3}{dx} - c_3 f \frac{d\phi}{dx}, \tag{3.53}$$

and the local electroneutrality condition requires that $c_3 = c_1 + c_2$. These equations can be rearranged as

$$-\left(\frac{j_1}{D_1} + \frac{j_2}{D_2} + \frac{j_3}{D_3}\right) = -\frac{I}{2F}\left(\frac{1}{D_1} - \frac{1}{D_3}\right) = 2\frac{dc_3}{dx}, \tag{3.54}$$

$$-\left(\frac{j_1}{D_1} + \frac{j_2}{D_2} - \frac{j_3}{D_3}\right) = -\frac{I}{2F}\left(\frac{1}{D_1} + \frac{1}{D_3}\right) = 2c_3 f \frac{d\phi}{dx}, \tag{3.55}$$

$$\Gamma \equiv \frac{\frac{j_1}{D_1} + \frac{j_2}{D_2} - \frac{j_3}{D_3}}{\frac{j_1}{D_1} + \frac{j_2}{D_2} + \frac{j_3}{D_3}} = \frac{D_3 + D_1}{D_3 - D_1} = \frac{f\frac{d\phi}{dx}}{\frac{d\ln c_3}{dx}}. \tag{3.56}$$

The interest in this algebraic manipulation is that the Nernst–Planck equations form a system of coupled, non-linear differential equations, while eqns (3.54) and (3.56) are two uncoupled, linear equations. Moreover, their left-hand sides are known constants from Faraday's law.

The integration of eqns (3.54) and (3.56) yields

$$c_3(0) = c_3^b + \frac{I\delta}{2\Gamma F D_{13}} = c_3^b\left(1 - \frac{I}{I_{L0}}\right) = c_3^b e^{-f\Delta\phi/\Gamma}, \tag{3.57}$$

$$\Delta\phi \equiv \phi^b - \phi(0) = -\frac{\Gamma}{f}\ln\frac{c_3(0)}{c_3^b} = -\frac{\Gamma}{f}\ln\left(1 - \frac{I}{I_{L0}}\right), \tag{3.58}$$

where

$$D_{13} \equiv \frac{2D_1D_3}{D_1 + D_3}, \tag{3.59}$$

$$I_{L0} \equiv -\frac{2\Gamma F D_{13} c_3^b}{\delta}. \tag{3.60}$$

The surface concentration of the H^+ ion is

$$c_2(0) = c_2^{\mathrm{b}} \left[\frac{c_3(0)}{c_3^{\mathrm{b}}} \right]^{-\Gamma} = c_2^{\mathrm{b}} \left(1 - \frac{I}{I_{\mathrm{L0}}} \right)^{-\Gamma}, \tag{3.61}$$

which is obtained by eliminating the electric field from eqns (3.52) and (3.56).

The surface concentration of the $Cu(HSO_4)^+$ ion is determined from the local electroneutrality condition as $c_1(0) = c_3(0) - c_2(0)$. The limiting current density I_{L} is attained when $c_1(0) = 0$ (Fig. 3.4). Under these conditions, eqns (3.57) and (3.61) lead to

$$\Delta\phi_{\mathrm{L}} = -\frac{\Gamma}{f(1+\Gamma)} \ln \frac{c_2^{\mathrm{b}}}{c_3^{\mathrm{b}}} = -\frac{\Gamma}{f} \ln \left(1 - \frac{I_{\mathrm{L}}}{I_{\mathrm{L0}}} \right) \geq 0, \tag{3.62}$$

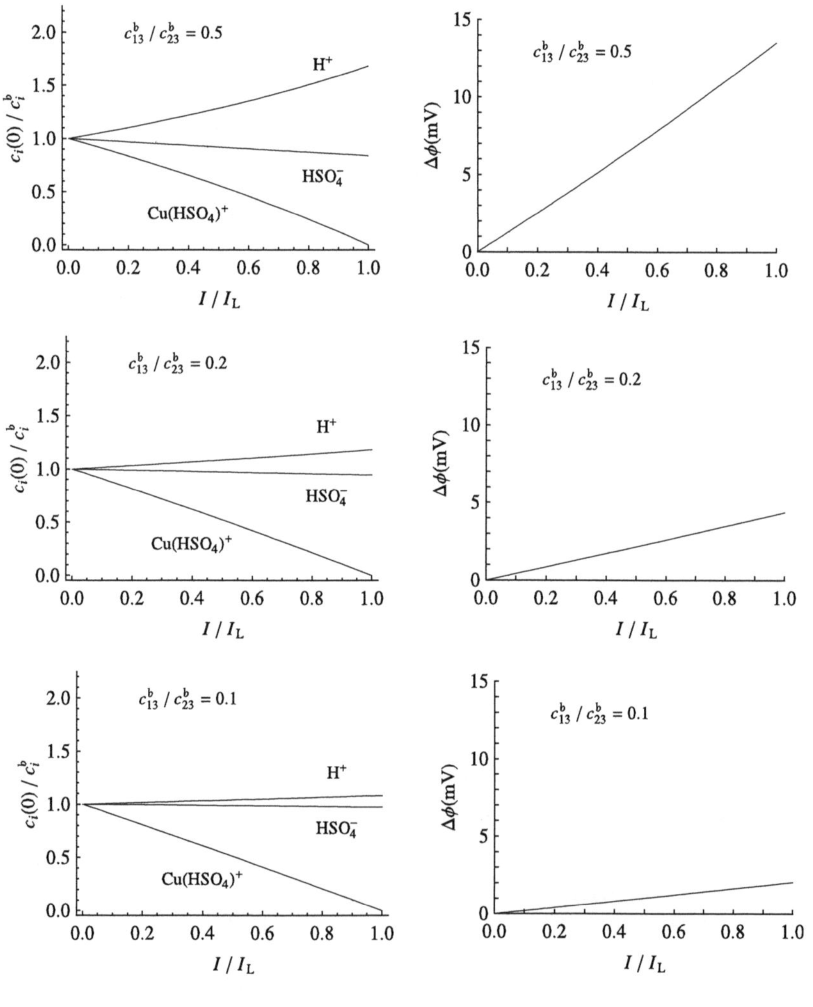

Fig. 3.4.
Surface concentrations and potential drop in the boundary layer as a function of the current density for the cases $c_{13}^{\mathrm{b}}/c_{23}^{\mathrm{b}} = 0.5$, 0.2, and 0.1.

and it becomes now clear that I_{L0} is the value of the limiting current density in the case $c_2^{\mathrm{b}} = 0$ (i.e. $c_{13}^{\mathrm{b}} = c_{23}^{\mathrm{b}}$), while in general $I_{\mathrm{L}} = I_{\mathrm{L0}}(1 - \mathrm{e}^{-f\,\Delta\phi_{\mathrm{L}}/\Gamma})$.

Once this example has been explained, we aim next at finding a general solution procedure for multi-ionic solutions with several electroactive species. Such a method is necessarily complicated if we consider any charge numbers, but relatively simple if we restrict discussion to a situation of symmetric electrolytes in which all ions have either charge number z or $-z$. Similarly to eqns (3.54)–(3.58) we perform the following algebraic manipulation of the Nernst–Planck equations

$$-G_0 \equiv -\sum_i \frac{j_i}{D_i} = \frac{\mathrm{d}}{\mathrm{d}x}\sum_i c_i \equiv \frac{\mathrm{d}c_{\mathrm{T}}}{\mathrm{d}x}, \tag{3.63}$$

$$-G_1 \equiv -\sum_i \frac{z_i j_i}{D_i} = \sum_i z_i^2 c_i f \frac{\mathrm{d}\phi}{\mathrm{d}x} = z^2 c_{\mathrm{T}} f \frac{\mathrm{d}\phi}{\mathrm{d}x}, \tag{3.64}$$

$$\Gamma \equiv \frac{G_1}{z^2 G_0} = f \frac{\mathrm{d}\phi}{\mathrm{d}\ln c_{\mathrm{T}}}, \tag{3.65}$$

where the local electroneutrality condition $\sum_i z_i c_i = 0$ has been used. Since Γ is a constant that can be evaluated from Faraday's law as

$$\Gamma \equiv \frac{G_1}{z^2 G_0} = \frac{\sum_i z_i j_i/D_i}{z^2 \sum_k j_k/D_k} = \frac{\sum_i z_i \nu_i/D_i}{z^2 \sum_k \nu_k/D_k}, \tag{3.66}$$

where ν_i is the stoichiometric coefficient of species i in the electrode reaction, the integration of these equations yields

$$c_{\mathrm{T}}(0) = c_{\mathrm{T}}^{\mathrm{b}} + G_0\delta = c_{\mathrm{T}}^{\mathrm{b}}\left(1 - \frac{I}{I_{\mathrm{L0}}}\right) = c_{\mathrm{T}}^{\mathrm{b}}\mathrm{e}^{-f\,\Delta\phi/\Gamma}, \tag{3.67}$$

$$\Delta\phi \equiv \phi^{\mathrm{b}} - \phi(0) = -\frac{\Gamma}{f}\ln\frac{c_{\mathrm{T}}(0)}{c_{\mathrm{T}}^{\mathrm{b}}} = -\frac{\Gamma}{f}\ln\left(1 - \frac{I}{I_{\mathrm{L0}}}\right), \tag{3.68}$$

$$\phi(x) - \phi^{\mathrm{b}} = \frac{\Gamma}{f}\ln\frac{c_{\mathrm{T}}(x)}{c_{\mathrm{T}}^{\mathrm{b}}}, \tag{3.69}$$

where

$$I_{\mathrm{L0}} \equiv -\frac{c_{\mathrm{T}}^{\mathrm{b}} I}{\delta G_0} = -\frac{nFc_{\mathrm{T}}^{\mathrm{b}}}{\delta \sum_i \nu_i/D_i}. \tag{3.70}$$

Finally, the surface concentrations can be calculated as

$$c_i(0) = \left[c_i^{\mathrm{b}} + \frac{j_i}{D_i}\int_0^{\delta} \mathrm{e}^{z_i f\phi(x)}\mathrm{d}x\right]\mathrm{e}^{-z_i f\phi(0)}$$
$$= c_i^{\mathrm{b}}\left[1 - \frac{j_i}{D_i c_i^{\mathrm{b}}}\frac{\int_0^{\delta} c_T^{z_i\Gamma}\mathrm{d}c_T}{G_0 (c_{\mathrm{T}}^{\mathrm{b}})^{z_i\Gamma}}\right]\left[\frac{c_{\mathrm{T}}(0)}{c_{\mathrm{T}}^{\mathrm{b}}}\right]^{-z_i\Gamma}$$

$$= c_i^{\mathrm{b}} \left\{ 1 - \frac{j_i \delta}{D_i c_i^{\mathrm{b}}} \frac{I_{\mathrm{L0}}}{(1+z_i\Gamma)I} \left[\left(1 - \frac{I}{I_{\mathrm{L0}}}\right)^{1+z_i\Gamma} - 1 \right] \right\} \left(1 - \frac{I}{I_{\mathrm{L0}}}\right)^{-z_i\Gamma}$$

$$= c_i^{\mathrm{b}} \left\{ 1 + \frac{c_{\mathrm{T}}^{\mathrm{b}} \nu_i / D_i}{(1+z_i\Gamma) c_i^{\mathrm{b}} \sum_j \nu_j / D_j} \left[\left(1 - \frac{I}{I_{\mathrm{L0}}}\right)^{1+z_i\Gamma} - 1 \right] \right\} \left(1 - \frac{I}{I_{\mathrm{L0}}}\right)^{-z_i\Gamma} \tag{3.71}$$

where the integral has been calculated with the help of eqns (3.63) and (3.65). Note that I_{L0} is not the limiting current density but a convenient auxiliary variable, and that eqn (3.71) is only valid if $z_i\Gamma \neq -1$. If there is only one active species, then $\Gamma = 1/z_1$ and eqn (3.71) reduces to

$$c_1(0) = \frac{c_{\mathrm{T}}^{\mathrm{b}}}{2} \left(1 - \frac{I}{I_{\mathrm{L0}}}\right) - \frac{c_{\mathrm{T}}^{\mathrm{b}} - 2c_1^{\mathrm{b}}}{2} \left(1 - \frac{I}{I_{\mathrm{L0}}}\right)^{-1}, \tag{3.72}$$

$$c_i(0) = c_i^{\mathrm{b}} \left(1 - \frac{I}{I_{\mathrm{L0}}}\right)^{-z_i/z_1}, \quad i \neq 1. \tag{3.73}$$

If we use the subscript $i = 1$ for that reactant whose surface concentration vanishes first when increasing the current density, i.e. for the one that satisfies

$$\left[1 - \frac{(1+z_1\Gamma)I_{\mathrm{L},1}}{I_{\mathrm{L0}}}\right]^{1/(1+z_1\Gamma)} \geq \left[1 - \frac{(1+z_i\Gamma)I_{\mathrm{L},i}}{I_{\mathrm{L0}}}\right]^{1/(1+z_i\Gamma)} \quad \text{for all } i, \tag{3.74}$$

where $I_{\mathrm{L},i} \equiv -nFD_i c_i^{\mathrm{b}}/(\nu_i\delta)$ is the limiting diffusion current density of species i, then eqn (3.71) allows us to determine the limiting current density as

$$I_{\mathrm{L}} = I_{\mathrm{L0}}(1 - \mathrm{e}^{-f\,\Delta\phi_{\mathrm{L}}}) = I_{\mathrm{L0}} \left\{ 1 - \left[1 - \frac{(1+z_1\Gamma)I_{\mathrm{L},1}}{I_{\mathrm{L0}}}\right]^{1/(1+z_1\Gamma)} \right\}$$

$$= I_{\mathrm{L0}} \left\{ 1 - \left[1 - \frac{(1+z_1\Gamma)c_1^{\mathrm{b}} \sum_j \nu_j/D_j}{c_{\mathrm{T}}^{\mathrm{b}} \nu_1/D_1}\right]^{1/(1+z_1\Gamma)} \right\}, \tag{3.75}$$

which reduces to

$$I_{\mathrm{L}} = I_{\mathrm{L0}} \left(1 - \sqrt{1 - \frac{2c_1^{\mathrm{b}}}{c_{\mathrm{T}}^{\mathrm{b}}}}\right), \tag{3.76}$$

if species 1 is the only active species in solution. Note also that when all the electroactive ions behave as trace ions, $\Gamma = 0$, and eqn (3.74) reduces to $I_{\mathrm{L},1}/I_{\mathrm{L0}} \leq I_{\mathrm{L},i}/I_{\mathrm{L0}}$, as expected if species 1 determines the limiting current.

It should be stressed that we have been able to find an analytical expression for the surface concentrations, eqn (3.71), because we have restricted discussion to the case of two valency classes (i.e. charge numbers z and $-z$).

Although the steady-state transport equations for a general multi-ionic system can also be solved [2], an approximate solution procedure proves to be rather accurate and much simpler in those systems. We refer to the Goldman constant-field approximation, which leads to the following equation for the surface concentrations

$$c_i(0) \approx c_i^{\mathrm{b}} \mathrm{e}^{z_i f \Delta\phi} + \frac{\mathrm{e}^{z_i f \Delta\phi} - 1}{z_i f \Delta\phi} \frac{j_i \delta}{D_i}, \tag{3.77}$$

where the electric potential gradient has been written as $\mathrm{d}\phi/\mathrm{d}x = \Delta\phi/\delta = [\phi^{\mathrm{b}} - \phi(0)]/\delta$. In eqn (3.77) the ionic flux densities are known from Faraday's law and the potential drop is chosen so that the electroneutrality condition $\sum_i z_i c_i(0) = 0$ is satisfied.

Table 3.1 summarizes the expressions derived in Sections 3.2.4–3.2.6.

3.2.7 Transport in spherical geometry

Stationary transport conditions usually require vigorous stirring of the electrolyte solutions. An exception to this rule is the transport in spherical geometry, where the system need not be stirred in order to achieve the steady state. Consider that the transport of an electroactive species i takes place along the radial direction in a semi-infinite medium limited by a hemispherical electrode of radius a. In the presence of a supporting electrolyte this species behaves as a trace ion and its flux density is

$$j_i \approx -D_i \frac{\mathrm{d}c_i}{\mathrm{d}r}, \tag{3.78}$$

where r is the distance from the electrode centre. The steady-state continuity equation, $\mathrm{d}j_i/\mathrm{d}r = 0$, then implies that the spatial distribution of its molar concentration is given by the Laplace equation

$$\frac{\mathrm{d}^2 c_i}{\mathrm{d}r^2} + \frac{2}{r}\frac{\mathrm{d}c_i}{\mathrm{d}r} = \frac{1}{r^2}\frac{\mathrm{d}}{\mathrm{d}r}\left(r^2 \frac{\mathrm{d}c_i}{\mathrm{d}r}\right) = 0, \tag{3.79}$$

and the boundary conditions

$$\left(\frac{\mathrm{d}c_i}{\mathrm{d}r}\right)_{r=a} = -\frac{I}{z_i F D_i}, \tag{3.80}$$

$$c_i(r \to \infty) = c_i^{\mathrm{b}}, \tag{3.81}$$

where the first one comes from Faraday's law (in the case of an electrode reaction with a single electroactive species i) and the convention of defining the current density as positive in the direction from the electrode to the solution (i.e. in the positive radial direction).

The solution of eqn (3.79) is

$$c_i(r) = c_i^{\mathrm{b}} + [c_i(a) - c_i^{\mathrm{b}}]\frac{a}{r}, \tag{3.82}$$

Table 3.1. Limiting current density, potential drop in the boundary layer under limiting conditions, current–voltage relation, and surface concentrations for different types of solutions

Binary solution, symmetric electrolyte $I_L = -2z_1FD_1c_{12}^b/\delta$ $z_1\Delta\phi_L \to \infty$ $I = I_L(1 - e^{z_2f\Delta\phi})$ $c_{12}(0) = c_{12}^b(1 - I/I_L)$	Binary solution, asymmetric electrolyte $I_L = -(1 - z_1/z_2)z_1FD_1c_1^b/\delta$ $z_1\Delta\phi_L \to \infty$ $I = I_L(1 - e^{z_2f\Delta\phi})$ $c_{12}(0) = c_{12}^b(1 - I/I_L)$
Multi-ionic solution, only one active species and it is a trace ion $I_L = I_{L,1} \equiv -z_1FD_1c_1^b/\delta$ $\Delta\phi_L \approx 0$ $\Delta\phi \approx 0$ $c_1(0) = c_1^b(1 - I/I_L)$	Multi-ionic solution, all active species are trace ions, species 1 determines I_L $I_L = I_{L,1} \equiv -nFD_1c_1^b/\nu_1\delta$ $\Delta\phi_L \approx 0$ $\Delta\phi \approx 0$ $c_i(0) = c_i^b(1 - z_i\nu_iI/nI_{L,i})$
Multi-ionic solution, one active species, symmetric electrolytes (charge numbers z and $-z$) $I_L = I_{L0}(1 - \sqrt{1 - 2c_1^b/c_T^b})$ $2z_1f\Delta\phi_L = -\ln(1 - 2c_1^b/c_T^b)$ $I = I_{L0}(1 - e^{-z_1f\Delta\phi}),\ I_{L0} \equiv -z_1FD_1c_T^b/\delta$ $c_1(0) = \frac{c_T^b}{2}(1 - I/I_{L0}) - \frac{c_T^b - 2c_1^b}{2}(1 - I/I_{L0})^{-1}$ $c_i(0) = c_i^b(1 - I/I_{L0})^{-z_i/z_1}, i \neq 1$ $c_T(0) = c_T^b(1 - I/I_{L0})$	Multi-ionic solution, one active species, arbitrary electrolytes $I_L = (z_1FD_1/\delta)\sum_{i\neq1}(1 - z_i/z_1)\,c_i^b(e^{z_if\Delta\phi_L} - 1)$ $\sum_{i\neq1} z_ic_i^be^{z_if\Delta\phi_L} = 0$ $I = (z_1FD_1/\delta)\sum_{i\neq1}(1 - z_i/z_1)\,c_i^b(e^{z_if\Delta\phi} - 1)$ $c_1(0) = -(1/z_1)\sum_{i\neq1} z_ic_i(0)$ $c_i(0) = c_i^b\,e^{z_if\Delta\phi}, i \neq 1$
Multi-ionic solution, several active species, species 1 determines I_L, symmetric electrolytes (charge numbers z and $-z$) $I_L = I_{L0}\{1 - [1 - (1 + z_1\Gamma)/A_1]^{1/(1+z_1\Gamma)}\},\ I_{L0} \equiv -\frac{nFc_T^b/\delta}{\sum_j \nu_j/D_j}, A_i \equiv \frac{c_T^b\nu_i/D_i}{c_i^b\sum_j \nu_j/D_j}$ $(1 + z_1\Gamma)f\Delta\phi_L = -\Gamma\ln[1 - (1 + z_1\Gamma)/A_1]$ $I = I_{L0}(1 - e^{-f\Delta\phi/\Gamma})$ $c_i(0) = c_i^b(1 - I/I_{L0})^{-z_i\Gamma}\{1 + A_i[(1 - I/I_{L0})^{1+z_i\Gamma} - 1]/(1 + z_i\Gamma)\},\ z_i \neq -1/\Gamma$ $c_i(0) = c_i^b(1 - I/I_{L0})[1 + A_i\ln(1 - I/I_{L0})],\ z_i = -1/\Gamma$ $c_T(0) = c_T^b(1 - I/I_{L0})$	

and the relation between the surface concentration $c_i(a)$ and the current density I is

$$c_i(a) = c_i^b\left(1 - \frac{I}{I_L}\right), \tag{3.83}$$

where

$$I_L = -\frac{z_iFD_ic_i^b}{a} \tag{3.84}$$

is the limiting current density. From eqn (3.84) it can be seen that the mass-transfer rate is enhanced when the electrode radius a decreases. This observation has led to the development of ultramicroelectrodes, whose radii are of the order

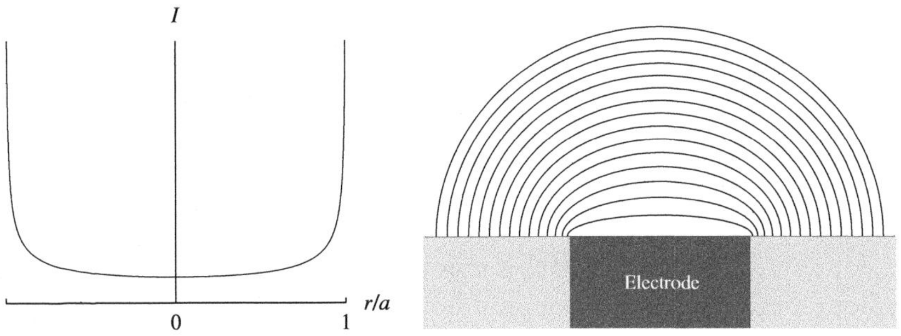

Fig. 3.5.
Radial distribution of the current density and equiconcentration lines on a disc electrode (arbitrary units).

of micrometers or less. When the mass-transfer rate is so enhanced, diffusion is no longer the rate-determining step of the reaction and kinetic measurements become more feasible.

Let us consider, for example, a first-order reaction taking place on the surface of a small spherical particle. The reaction rate is $r = kc_i(a)$, where k is the rate constant (in s^{-1} units). At steady state, the diffusion flux towards the surface equals to the reaction rate

$$kc_i(a) = D_i \left(\frac{\partial c_i}{\partial r} \right)_{r=a} = \frac{D_i}{a}[c_i^{\rm b} - c_i(a)], \tag{3.85}$$

from which the surface concentration $c_i(a)$ can be solved as

$$c_i(a) = \frac{c_i^{\rm b}}{1 + ka/D_i}. \tag{3.86}$$

The reaction rate thus becomes

$$r = \frac{kc_i^{\rm b}}{1 + ka/D_i}. \tag{3.87}$$

When[8] $\mathrm{Da} \equiv ka/D_i \ll 1$ the reaction is under kinetic control and the rate is $r \approx kc_i^{\rm b}$. When $\mathrm{Da} \gg 1$, the reaction is under diffusion control and $r \approx D_i c_i^{\rm b}/a$. Hence, by reducing the particle size a, reactions with higher values of the rate constant k can be monitored.

Unlike hemispherical electrodes, which are difficult to prepare, microdisc electrodes are commonly used because their diffusion field also has a quasi-spherical symmetry, as shown in Fig. 3.5. Without going into the details of the description of mass transport to a microdisc electrode[9] [4], we mention here some relevant characteristics. First, the limiting current at a microdisc electrode of radius a is

$$i_{\rm L} = \pi a^2 I_{\rm L} = -4z_i F D_i c_i^{\rm b} a, \tag{3.88}$$

[8] Da is known as the second Damköhler number [3].

[9] The solution is obtained in cylindrical co-ordinates in terms of modified Bessel functions.

to be compared with that at a hemispherical electrode $i_L = 2\pi a^2 I_L = -2\pi z_i F D_i c_i^b a$. It is observed that the limiting current is lower by a factor $2/\pi$ at the microdisc electrode. Second, the current distribution at the microdisc electrode is not uniform, that is, the current density varies with the radial position variable. It can be shown [5] that the functional dependence of the current distribution is

$$I(r) \propto \frac{1}{\sqrt{1-(r/a)^2}}, \tag{3.89}$$

which diverges at the disc edge $r = a$, as shown in Fig. 3.5. Yet, the integral over the disc is finite. In practice, the current density is limited at the disc edge by the kinetics of the electrode reaction.

Ultramicroelectrodes also have the advantage that the current flowing in the system is very low, and therefore the ohmic loss becomes negligible. Ultramicroelectrodes of various geometries are widely used in electroanalytical applications, and the interested reader is directed to, e.g., Ref. [6].

3.3 Hydrodynamic electrodes

Stirring of the electrolyte solution is required to establish a diffusion boundary layer of well-defined thickness, δ. If the solution is not stirred, this layer expands from the electrode towards the bulk of the solution as a function of time, until natural convection begins to mix the solution due to density differences created by the electrode process. Increasing the stirring rate, decreases δ and, consequently, enhances the mass transport rate. Because the mass transfer frequently is the rate-determining step of the overall electrode process, its enhancement makes feasible the determination of higher values of kinetic parameters. Different hydrodynamic methods have different characteristics of mass transfer, and in the following sections the most usual ones are briefly presented.

3.3.1 Rotating-disc electrode

Perhaps the most common hydrodynamic electrochemical method is the rotating-disc electrode (RDE). When a disc electrode is rotated, a well-defined velocity profile is developed at the electrode. This velocity profile is sketched in Fig. 3.6. A modification of the RDE is the rotating ring-disc electrode (RRDE) where there is an additional ring-shaped electrode around the disc. The potential of the ring can be varied independently, so that a species generated in the disc reaction is collected at the ring. Details of the RRDE can be found in standard electrochemistry textbooks [1, 7, 8].

a) The convective electrodiffusion equation

When describing mass transport, the flux density is decomposed into electrodiffusive and convective contributions, and the steady-state continuity equation for species i is written as

$$0 = \vec{\nabla} \cdot \vec{j}_i = \vec{\nabla} \cdot (\vec{j}_i^m + c_i \vec{v}) = \vec{\nabla} \cdot \vec{j}_i^m + \vec{v} \cdot \vec{\nabla} c_i, \tag{3.90}$$

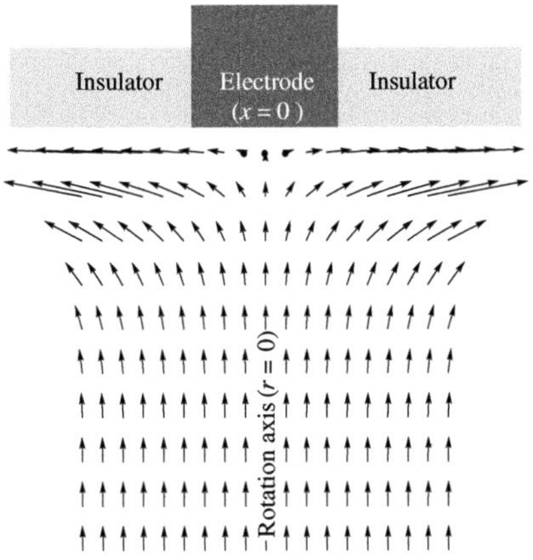

Fig. 3.6.
Fluid velocity field in a rotating-disc electrode.

where the continuity equation for the total mass, $\vec{\nabla} \cdot \vec{v} = 0$, has been used. Since the properties of the disc surface are uniform with respect to the electrode reaction, the flux density $\vec{j}_i^m$ and the concentration gradient $\vec{\nabla} c_i$ can be assumed to take place along the normal direction only. Thus, even though the solution velocity pattern is clearly three-dimensional, eqn (3.90) reduces to

$$0 = \frac{\mathrm{d}j_i^m}{\mathrm{d}x} + v_x \frac{\mathrm{d}c_i}{\mathrm{d}x}, \tag{3.91}$$

where x is the distance to the electrode surface, and the mass transport is considered as one-dimensional. Equation (3.91) describes the stationary convective electrodiffusion and must be solved under the boundary conditions $c_i(x \to \infty) = c_i^{\mathrm{b}}$ and $j_i^m(x = 0) = \nu_i I/nF$ [see eqn (3.3)].

The convective velocity is determined from the solution of the Navier–Stokes equation. In the case of a RDE, this solution can be obtained in the form of a series expansion in the variable $x(\omega/\nu)^{1/2}$, where ω is the electrode angular rotation frequency and ν is the kinematic solution viscosity. The first term of the series expansion of the normal component of the velocity is

$$v_x \approx -Bx^2, \tag{3.92}$$

where $B = -0.510\omega^{3/2}\nu^{-1/2}$ and the minus sign indicates that the fluid moves towards the electrode. Since higher-order terms have been neglected, eqn (3.92) is only valid when $x(\omega/\nu)^{1/2} \ll 1$, i.e. when x is much smaller than an upper bound $(\nu/\omega)^{1/2}$ that is of the order of 100 μm for a rotation frequency $\omega/2\pi =$ 100 Hz.

b) Trace ion

Consider first the case of a single electroactive ion (denoted by index 1) that behaves as a trace ion. Since $j_1^m(x) = -D_1 \mathrm{d}c_1/\mathrm{d}x$, eqn (3.91) reduces to

$$\frac{\mathrm{d}j_1^m}{\mathrm{d}x} = \frac{v_x}{D_1} j_1^m, \tag{3.93}$$

and can be solved as follows

$$j_1^m(x) = j_1^m(0)\exp\left[-(B/D_1)\int_0^x u^2 \mathrm{d}u\right] = j_1(0)\mathrm{e}^{-Bx^3/3D_1}, \tag{3.94}$$

where $j_1^m(0) = j_1(0)$ because $v_x(0) = 0$, and u is a dummy variable. As expected, $j_1^m(x)$ rapidly decreases with the distance to the electrode and becomes negligible when $x \gg (D_1/B)^{1/3} \approx \mathrm{Sc}^{1/3}(\nu/\omega)^{1/2}$, where $\mathrm{Sc} \equiv \nu/D_1 \approx 10^3$ is the Schmidt number.

As depicted in Fig. 3.1, the common procedure in electrochemistry is to describe the concentration profile established due to the combined action of electrodiffusion and convection by a linear profile that neglects convection

$$c_1(x) \approx c_1(0) + \left(\frac{\mathrm{d}c_1}{\mathrm{d}x}\right)_{x=0} x = c_1(0) + \frac{c_1^\mathrm{b} - c_1(0)}{\delta}x. \tag{3.95}$$

The effective thickness of the diffusion boundary layer is defined as

$$\delta \equiv \frac{c_1^\mathrm{b} - c_1(0)}{(\mathrm{d}c_1/\mathrm{d}x)_{x=0}} = \frac{\int_0^\infty (\mathrm{d}c_1/\mathrm{d}x)\mathrm{d}x}{(\mathrm{d}c_1/\mathrm{d}x)_{x=0}} = \frac{1}{j_1(0)}\int_0^\infty j_1^m(x)\mathrm{d}x, \tag{3.96}$$

and integration of eqn (3.94) leads to the result

$$\delta = \int_0^\infty \mathrm{e}^{-Bx^3/3D_1}\mathrm{d}x = \left(\frac{3D_1}{B}\right)^{1/3}\Gamma(4/3) = 1.61 D_1^{1/3}\omega^{-1/2}\nu^{1/6}, \tag{3.97}$$

where we have introduced the gamma function $\Gamma(n) \equiv \int_0^\infty \mathrm{e}^{-t}t^{n-1}\mathrm{d}t$ and used its value $\Gamma(4/3) = 0.893$. There is an apparent dilemma in the derivation of eqn (3.97): the integration is carried out to infinity, while the parabolic approximation in eqn (3.92) is valid only at short distances. The error made is, however, negligible because $\exp(-Bx^3/3D_1)$ goes to zero very fast as x increases.

When the surface concentration $c_1(0)$ vanishes, the electric current density $I = z_1 F j_1(0)$ takes its maximum value

$$I_\mathrm{L} = I_{\mathrm{L},1} = -\frac{z_1 F D_1 c_1^\mathrm{b}}{\delta} = -0.620 z_1 F D_1^{2/3} c_1^\mathrm{b} \omega^{1/2}\nu^{-1/6}. \tag{3.98}$$

This is known as the *Levich equation* and evidences the characteristic feature of the RDE: the limiting current density is proportional to the square root of the rotation frequency.

The actual concentration profile can be written after some cumbersome algebra in terms of the incomplete gamma function $\Gamma(n,x) \equiv \int_x^\infty \mathrm{e}^{-t}t^{n-1}\mathrm{d}t$

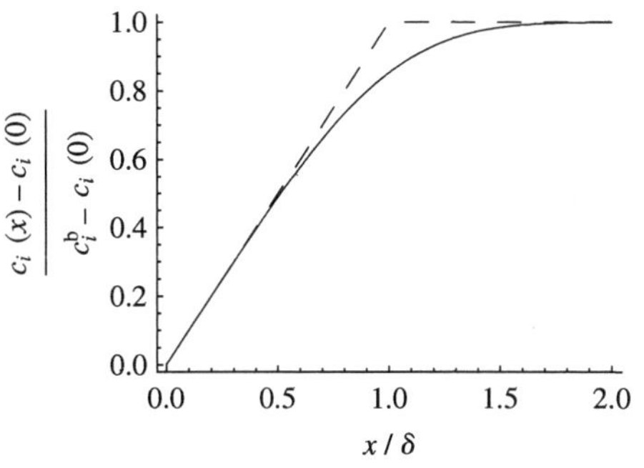

Fig. 3.7.
Concentration profile at the rotating-disc electrode (solid line) as described by eqn (3.99) and its linear approximation (dashed line).

[9] as[10]

$$c_1(x) = c_1(0) + \int_0^x \frac{dc_1}{du}\,du = c_1(0) - \frac{1}{D_1}\int_0^x j_1^m(u)\,du = c_1^b\left[1 - \frac{I}{I_L}\frac{\Gamma(1/3,\xi)}{\Gamma(1/3)}\right], \tag{3.99}$$

where

$$\xi = \frac{Bx^3}{3D_1} = \left[\Gamma(4/3)\frac{x}{\delta}\right]^3. \tag{3.100}$$

Figure 3.7 shows a representation of this profile and its comparison with eqn (3.95). Note that eqn (3.99) reduces to $c_1(0) = c_1^b(1 - I/I_L)$ at the electrode surface.

In relation to eqn (3.94) we have mentioned that the diffusive contribution to the flux density $j_1^m(x)$ decreases (in magnitude) with increasing distance to the electrode surface. The convective contribution $c_1 v_x$, on the contrary, is zero at the interface and increases with this distance because both c_1 and v_x also increase. The total flux density in the laboratory reference frame $j_1 = j_1^m + c_1 v_x$ also increases with increasing distance to the electrode, and it can be proved from eqns (3.90) and (3.91) that $dj_1/dx = c_1 dv_x/dx$. Thus, solute transport is predominantly diffusive in the vicinity of the electrode and convective in the outer region of the diffusion boundary layer (Fig. 3.8).

The fact that the flux density is the lowest at the electrode surface means that the species accessing the electrode and reacting there have flowed through a smaller cross-sectional area than that of the electrode. This can be seen also in Fig. 3.6 where the streamlines are radially separated from each other close to the electrode surface, indicating a decreased flux density normal to the electrode surface. Yet, at steady state, the flux remains constant throughout the whole system. It must be emphasized that the parabolic velocity profile, eqn (3.92), cannot be extended over the distance δ, i.e. the increase of the flux density is terminated at $x = \delta$.

[10] Note that $\int_0^x j_1^m(u)du = \int_0^\infty j_1^m(u)du - \int_x^\infty j_1^m(u)du$, which leads to a factor $\Gamma(1/3) - \Gamma(1/3,\xi)$.

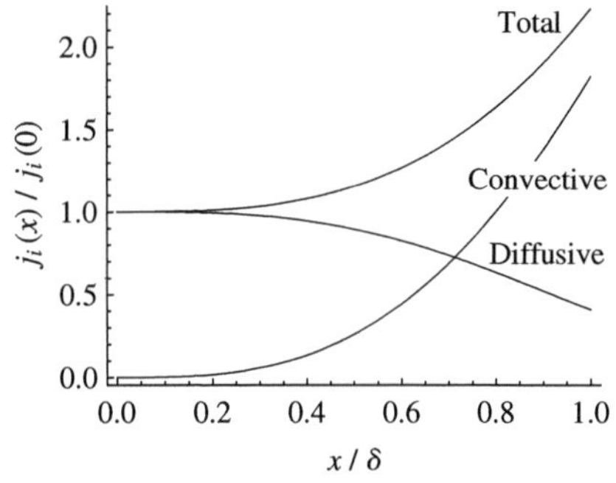

Fig. 3.8.
Contributions to the solute flux density at the rotating-disc electrode under limiting current conditions: diffusive j_1^m, convective $c_1 v_x$, and total j_1.

c) Binary electrolyte solution

Although most experimental situations involve the use of supporting electrolyte and this allows us to treat the electroactive ion as a trace ion, we describe now the transport at a RDE in the absence of supporting electrolyte [10, 11]. This is a more complicated situation in which migration is not negligible. Consider that the solute is a binary electrolyte. By making use of the local electroneutrality assumption, $z_1c_1 + z_2c_2 = 0$, the electric field can be eliminated from the two convective electrodiffusion equations[11]

$$\frac{d^2c_i}{dx^2} + z_i f \frac{d}{dx}\left(c_i \frac{d\phi}{dx}\right) = \frac{v_x}{D_i}\frac{dc_i}{dx}, \qquad (3.101)$$

and the convective diffusion equation for the electrolyte is then obtained as

$$\frac{d^2c_i}{dx^2} = \frac{v_x}{D_{12}}\frac{dc_i}{dx}, \qquad (3.102)$$

where D_{12} is the Nernst–Hartley electrolyte diffusion coefficient. This equation can be solved following the same procedure explained above and leads to

$$\delta = 1.61 D_{12}^{1/3}\omega^{-1/2}\nu^{1/6}. \qquad (3.103)$$

It is interesting to note that the thickness of the diffusion boundary layer is determined by D_{12}. This might seem surprising, e.g. in case where only species 1 is electroactive, because the boundary conditions at the electrode surface

$$(j_1)_{x=0} = \frac{I}{z_1F} = -D_1\left[\left(\frac{dc_1}{dx}\right)_{x=0} + z_1c_1(0)f\left(\frac{d\phi}{dx}\right)_{x=0}\right] \qquad (3.104)$$

$$0 = \left(\frac{dc_2}{dx}\right)_{x=0} + z_2c_2(0)f\left(\frac{d\phi}{dx}\right)_{x=0} \qquad (3.105)$$

do not involve D_2 and hence it could be expected that only D_1 determines δ. However, this is not the case because the ionic flux densities are position

[11] The electric field varies with position and hence these convective electrodiffusion equations cannot be solved with the help of the Goldman constant-field assumption.

dependent and D_2 is involved in the transport equations at all positions except for the electrode surface.

The limiting current density is defined in this system as

$$I_\mathrm{L} = \left(1 - \frac{z_1}{z_2}\right) I_{\mathrm{L},1} = -\left(1 - \frac{z_1}{z_2}\right) \frac{z_1 F D_1 c_1^\mathrm{b}}{\delta}, \tag{3.106}$$

and the surface concentration of the electroactive species is $c_1(0) = c_1^\mathrm{b}(1 - I/I_\mathrm{L})$. Finally, the electric field at the electrode surface is given from eqns (3.96) and (3.105) as

$$z_2 f \delta \left(\frac{\mathrm{d}\phi}{\mathrm{d}x}\right)_{x=0} = \left(1 - \frac{I_\mathrm{L}}{I}\right)^{-1}. \tag{3.107}$$

This field increases (in magnitude) with increasing current density and diverges under limiting conditions, which evidences deviations from the local electroneutrality assumption [1].

d) Ternary electrolyte solution

The last case that we consider is that of a ternary solution formed by the mixture of two strong binary electrolytes AC and DC. The ions A^z, D^z, and C^{-z} are denoted by the indexes 1, 2, and 3, respectively. Only the ion A^z is electroactive, and DC is an inert electrolyte. The concentration of inert electrolyte is such that the migrational contribution to the transport of A^z is small but not negligible, and we aim to describe the effect of the electric field on the limiting current density.

Due to the presence of excess inert electrolyte, we can assume that the electric field is constant and small (compared to $RT/F\delta$), so that eqn (3.101)

$$\frac{\mathrm{d}^2 c_1}{\mathrm{d}x^2} = \left(\frac{v_x}{D_1} - zf\frac{\mathrm{d}\phi}{\mathrm{d}x}\right)\frac{\mathrm{d}c_1}{\mathrm{d}x} = \left(-\frac{Bx^2}{D_1} - zf\frac{\mathrm{d}\phi}{\mathrm{d}x}\right)\frac{\mathrm{d}c_1}{\mathrm{d}x} \tag{3.108}$$

can be integrated to

$$\frac{\mathrm{d}c_1}{\mathrm{d}x} = C\mathrm{e}^{-\xi}\exp\left(-zf\frac{\mathrm{d}\phi}{\mathrm{d}x}x\right) \approx C\mathrm{e}^{-\xi}\left(1 - zf\frac{\mathrm{d}\phi}{\mathrm{d}x}x\right), \tag{3.109}$$

where $\xi = Bx^3/3D_1 = [(x/\delta)\Gamma(4/3)]^3$. Since the convective velocity is zero at the electrode surface, the boundary condition for the electroactive ion is

$$(j_1)_{x=0} = \frac{I}{zF} = -D_1\left[\left(\frac{\mathrm{d}c_1}{\mathrm{d}x}\right)_{x=0} + zc_1(0)f\frac{\mathrm{d}\phi}{\mathrm{d}x}\right] < 0, \tag{3.110}$$

and the integration constant C can be determined as

$$C = \left(\frac{\mathrm{d}c_1}{\mathrm{d}x}\right)_{x=0} = -\frac{I}{zFD_1} - zc_1(0)f\frac{\mathrm{d}\phi}{\mathrm{d}x}. \tag{3.111}$$

Integration of eqn (3.109) over the diffusion boundary layer yields

$$c_1^{\mathrm{b}} - c_1(0) \approx C\delta \left[1 - 0.566 z f \frac{\mathrm{d}\phi}{\mathrm{d}x}\delta\right], \tag{3.112}$$

where we have used eqn (3.97) and

$$\int_0^\infty \mathrm{e}^{-Bx^3/3D_1} x\,\mathrm{d}x = \left(\frac{3D_1}{B}\right)^{2/3} \frac{\Gamma(2/3)}{3} = \frac{\Gamma(2/3)}{\Gamma(1/3)\Gamma(4/3)}\delta^2 \approx 0.566\,\delta^2. \tag{3.113}$$

Finally, combining eqns (3.111) and (3.112), the surface concentration is

$$c_1(0) \approx c_1^{\mathrm{b}} \frac{1 - \frac{I}{I_{\mathrm{L},1}}\left[1 - 0.566 z f \frac{\mathrm{d}\phi}{\mathrm{d}x}\delta\right]}{1 - z f \frac{\mathrm{d}\phi}{\mathrm{d}x}\delta}, \tag{3.114}$$

where $I_{\mathrm{L},1} = -zFD_1c_1^{\mathrm{b}}/\delta$ and we have neglected second-order terms in the electric field, for the sake of consistency with the linear approximation introduced in eqn (3.109). Equation (3.114) describes the effect of the electric field on the transport of the electroactive ion toward the RDE.

If we further assume that the electric field is ohmic, the potential drop in the boundary layer is $\Delta\phi_{\mathrm{ohm}} = (\mathrm{d}\phi/\mathrm{d}x)\delta = -IR$, where $R = \delta/\kappa$ is the electrical resistance[12] of the solution in this layer, and the limiting current density can be written as

$$\begin{aligned} I_{\mathrm{L}} &= -\frac{zFD_1c_1^{\mathrm{b}}}{\delta}\left[1 - 0.566 z f \frac{\mathrm{d}\phi}{\mathrm{d}x}\delta\right]^{-1} \\ &\approx -\frac{zFD_1c_1^{\mathrm{b}}}{\delta}\left[1 + 0.566 z f\,\Delta\phi_{\mathrm{ohm}}\right]. \end{aligned} \tag{3.115}$$

Since the ions A^z (no matter whether they are cations or anions, i.e. regardless of the sign of z) are consumed at the electrode surface, we have that $(j_1)_{x=0} = I/zF < 0$ and $z\Delta\phi_{\mathrm{ohm}} = z[\phi(\delta) - \phi(0)] = -zIR > 0$. Therefore, we conclude that the effect of an ohmic electric field is to increase the magnitude of the limiting current density. In relation to this, it should be remembered that eqn (3.106) also described an increase in the magnitude of the limiting current density, in a factor $1 - z_1/z_2 > 1$, due to the effect of the electric field in a binary electrolyte solution, although in that case the field had both ohmic and diffusional contributions. The effect of the electric field on the limiting current density in a ternary electrolyte solution, taking into account both ohmic and diffusion contributions, is described, e.g., in Ref. [12].

[12] Strictly speaking, R is the product of the resistance and the area through which current is transported. The ohmic potential drop, IR, where I is the current density, has units of electric potential.

3.3.2 Channel-flow electrode

The channel-flow electrode (CFE) is frequently used as a detector in liquid chromatography, capillary electrophoresis, and flow-injection analysis [13]. In a typical experimental set-up (Fig. 3.9), a channel electrode is located downstream after the separation unit and is biased at a sufficiently positive (or negative) potential such that the analyte is readily oxidized (reduced) at the electrode. Thus, a CFE is operated under limiting current conditions. The reference electrode is usually situated upstream of the working electrode, preferably via a liquid junction, while the counter electrode is placed downstream to prevent contamination of the working electrode due to the electrolysis products formed at the counter electrode.

Another large application area of CFEs is in the study of homogeneous chemical reactions (C) coupled to electrochemical reactions (E). Since hydrodynamics has a strong effect on the concentration profiles of the species reacting in the bulk of the solution, while the electrochemical step takes place at the electrode only, it is possible to distinguish between, say, CE and EC or ECE reaction mechanisms.

The mass-transfer problem at a CFE can be solved in closed form, because the convection velocity can be determined from the solution of the Navier–Stokes equation. The solution flow takes place in the x direction, thus the velocity components v_y and v_z are zero; direction y is normal to the electrode surface and the channel width is defined in the z direction (Fig. 3.9). The diffusion equation to be solved is eqn (3.90), and in the trace ion case it becomes

$$D_1\left(\frac{\partial^2 c_1}{\partial x^2}+\frac{\partial^2 c_1}{\partial y^2}+\frac{\partial^2 c_1}{\partial z^2}\right)-v_x\frac{\partial c_1}{\partial x}=0. \tag{3.116}$$

The usual approximation made is that the diffusion in the z direction is negligible if the electrode width w is smaller than the channel width d [1]. Also, since convection is usually fast compared to diffusion, it appears that $(\partial^2 c_1/\partial x^2) \ll (\partial^2 c_1/\partial y^2)$.

In a fully developed Poiseuille flow the velocity profile has a parabolic form

$$v_x(y) = v_{\max}\frac{y}{h}\left(2-\frac{y}{h}\right), \tag{3.117}$$

with its maximum value $v_{\max}$ at the pore centre $y = h$ and becoming zero at the pore walls $y = 0$ and $y = 2h$ due to the non-slip boundary condition. The

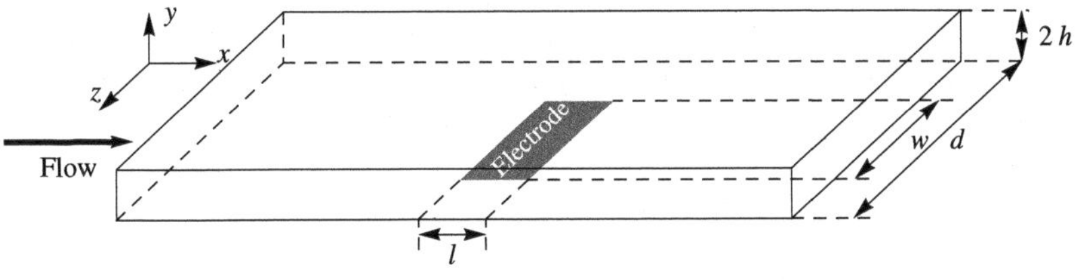

Fig. 3.9.
Sketch of the channel flow electrode. The solution flows along direction x.

maximum velocity is usually evaluated from the volume flow $\dot{V}$ as[13]

$$v_{\max} = \frac{3}{4}\frac{\dot{V}}{hd}. \tag{3.118}$$

Changes in the concentration take place in the close vicinity of the electrode and, therefore, Lévêque's approximation

$$v_x(y) \approx 2v_{\max}\frac{y}{h} \tag{3.119}$$

can be used. Hence, eqn (3.116) is reduced to

$$\left(\frac{\partial^2 c_1}{\partial y^2}\right)_x - \frac{2v_{\max}}{D_1 h} y \left(\frac{\partial c_1}{\partial x}\right)_y = 0. \tag{3.120}$$

Assuming that $c_1(x, y)$ depends on the position variables only through the combination

$$\xi \equiv \frac{2}{9}\frac{v_{\max}}{D_1 h}\frac{y^3}{x} \tag{3.121}$$

eqn (3.120) transforms to the following linear ordinary differential equation for $c_1(\xi)$

$$\frac{\mathrm{d}^2 c_1}{\mathrm{d}\xi^2} + \left(1 + \frac{2}{3\xi}\right)\frac{\mathrm{d}c_1}{\mathrm{d}\xi} = 0. \tag{3.122}$$

Under limiting current conditions, the boundary conditions for this equation are $c_1(0) = 0$ and $c_1(\infty) = c_1^{\mathrm{b}}$, and its integration yields

$$c_1(\xi) = c_1^{\mathrm{b}} \frac{\int_0^{\xi} \mathrm{e}^{-\zeta}\zeta^{-2/3}\mathrm{d}\zeta}{\int_0^{\infty} \mathrm{e}^{-\xi}\xi^{-2/3}\mathrm{d}\xi} = c_1^{b}\left[1 - \frac{\Gamma(1/3, \xi)}{\Gamma(1/3)}\right], \quad (I = I_{\mathrm{L}}). \tag{3.123}$$

The limiting current flowing across the electrode is then

$$\begin{aligned} i_{\mathrm{L}} &= -z_1 F D_1 w \int_0^l \left(\frac{\partial c_1}{\partial y}\right)_{y=0} \mathrm{d}x \\ &= -z_1 F D_1 c_1^{\mathrm{b}} w \left(\frac{v_{\max}}{D_1 h}\right)^{1/3} \frac{6^{1/3}}{\Gamma(1/3)} \int_0^l x^{-1/3}\mathrm{d}x, \end{aligned} \tag{3.124}$$

[13] This equation neglects the edge effects on the velocity profile caused by channel walls at $z = 0$ and $z = d$ but this is a good approximation because $h \ll d$.

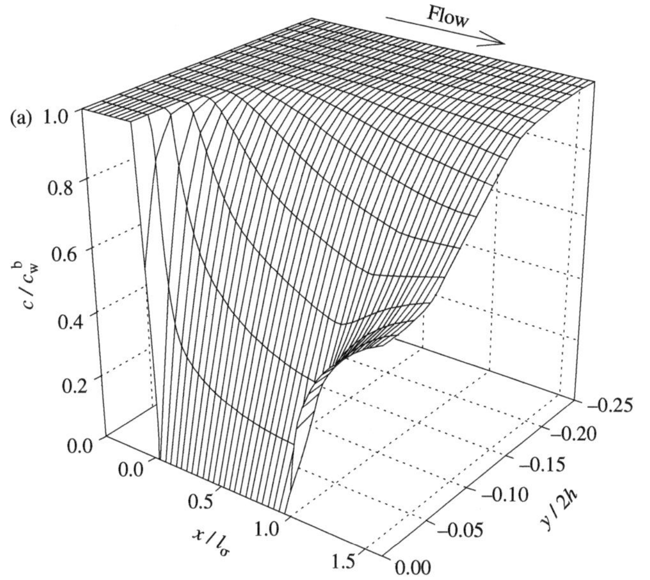

Fig. 3.10.
Simulated equiconcentration surface at the channel flow electrode. (Reproduced from Ref. [18] with permission.)

where l is the length of the electrode (in the flow direction), and using eqn (3.118) and the value $6^{-1/3}\Gamma(1/3) \approx 1.4743$, it can be transformed to the *Levich equation*

$$i_L = -0.9244 z_1 F D_1^{2/3} c_1^b w \left(\frac{l^2 \dot{V}}{h^2 d}\right)^{1/3}. \tag{3.125}$$

Sometimes, the numerical coefficient is replaced by 0.835 depending on the simplifying assumptions made in the derivation of the velocity profile [14]. The essential finding is, however, that the limiting current is proportional to the cube root of the volume flow rate. In any case, a detector is always calibrated with known solutions prior to use.

Complete current–voltage characteristics taking electrode kinetics into account have been provided by Matsuda [15]. Various reaction mechanisms have been treated in non-stationary cases by Compton *et al.* [13, 16]. The analysis of the effect of the channel geometry is also omitted here, as the simulation of the transport problems in arbitrary geometry is feasible nowadays with numerical software packages [17] (Fig. 3.10).

3.3.3 Wall-jet electrode

An impinging jet electrode is an interesting modification of a RDE. Instead of rotating the disc, which induces a convective flow towards the electrode, a jet stream is directed towards the electrode (Fig. 3.11), thus creating a well-defined flow profile [20] as depicted in Figure 3.12.

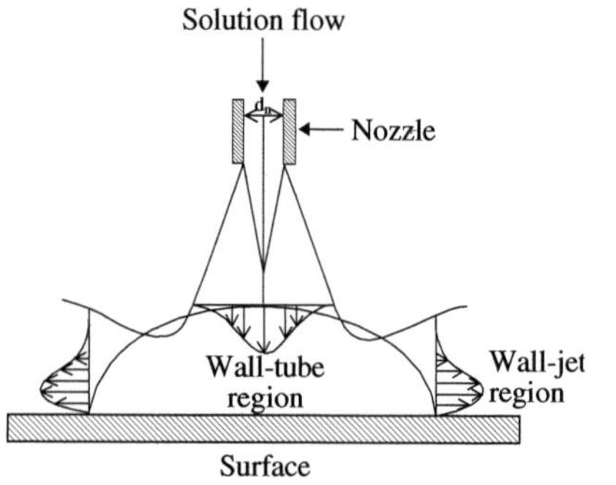

Fig. 3.11.
Schematic picture of a wall-jet electrode. At the wall-tube region the axial and at the wall-jet region the radial velocity is governing. The lengths of the arrows indicate their relative magnitude. (Reproduced from Ref. [19] with permission.)

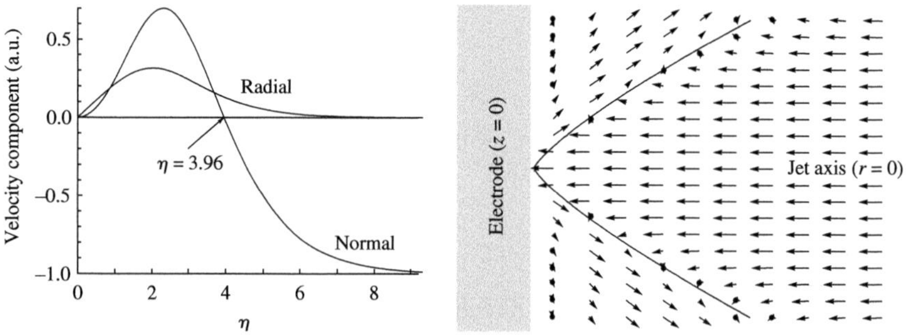

Fig. 3.12.
(Left) Dimensionless radial (v_r) and normal (v_n) velocity components as a function of the dimensionless distance η from the electrode. (Right) Fluid velocity field in the (r, z) plane also showing the curve $\eta = 3.96$ where the normal component reverses direction.

The dimensionless distance from the electrode is defined as [21]

$$\eta = \left(\frac{135M}{32\nu^3}\right)^{1/4} \frac{z}{r^{5/4}}, \tag{3.126}$$

where r and z are the radial and axial co-ordinates, respectively, $M \equiv k^4\dot{V}^3/(2\pi^3 d_n^2)$, ν is the kinematic viscosity of the solution, $\dot{V}$ is the volume flow rate of the jet, d_n is its diameter, and $k = 0.86$ is an experimental constant.

As can be seen in Fig. 3.12, the normal velocity vanishes when $\eta = 3.96$. At distances shorter than that the flow is directed towards the electrode and at larger distances away from the electrode. This velocity pattern means that the electrode only sees fresh solution passing through the jet nozzle. In the electrode vicinity, $\eta \ll 1$, the velocity components can be approximated as

$$v_r \approx \frac{2}{9}\left(\frac{15M}{2\nu r^3}\right)^{1/2} \eta \tag{3.127}$$

$$v_n \approx \frac{7}{36}\left(\frac{40M\,\nu}{3r^5}\right)^{1/4} \eta^2. \tag{3.128}$$

Neglecting the radial diffusion term, $(\partial^2 c_1/\partial r^2)$, the convective diffusion equation to be solved is

$$D_1 \frac{\partial^2 c_1}{\partial z^2} - v_r \frac{\partial c_1}{\partial r} - v_n \frac{\partial c_1}{\partial z} = 0. \qquad (3.129)$$

Transforming from variables (r, z) to variables (ρ, ξ) defined as $\rho \equiv (r/a)^{9/8}$ and

$$\xi \equiv M^{1/4} \left(\frac{9}{8D_1}\right)^{1/3} \left(\frac{125}{216\nu^5}\right)^{1/12} (a^3 r^7)^{-1/8} z, \qquad (3.130)$$

where a is the electrode radius, eqn (3.129) is converted to

$$\frac{\partial^2 c_1}{\partial \xi^2} = \xi \frac{\partial c_1}{\partial \rho}. \qquad (3.131)$$

The boundary conditions on $c_1(\rho, \xi)$ under limiting conditions are $c_1(\rho, 0) = 0$, $c_1(\rho, \infty) = c_1^{\mathrm{b}}$, and $c_1(0, \xi) = c_1^{\mathrm{b}}$. This equation can be solved by Laplace transformation with respect to variable ρ. Denoting by $\tilde{c}_1(s, \xi) = \mathcal{L}[c_1(\rho,\xi)]$ the transformed concentration, eqn (3.131) becomes

$$\frac{\mathrm{d}^2 \tilde{c}_1}{\mathrm{d}\xi^2} = \xi (s\tilde{c}_1 - c_1^b), \qquad (3.132)$$

and its solution is given in terms of the Airy function as

$$\tilde{c}_1 = \frac{c_1^b}{s}\left[1 - \frac{\mathrm{Ai}(s^{1/3}\xi)}{\mathrm{Ai}(0)}\right], \quad (I = I_{\mathrm{L}}). \qquad (3.133)$$

The concentration gradient at the electrode surface can be evaluated using the inverse Laplace transform $\mathcal{L}^{-1}$ as

$$\begin{aligned}
\left(\frac{\partial c_1}{\partial z}\right)_{z=0} &= \left(\frac{\partial \xi}{\partial z}\right)_{z=0} \mathcal{L}^{-1}\left(\frac{\mathrm{d}\tilde{c}_1}{\mathrm{d}\xi}\right)_{\xi=0} = -c_1^{\mathrm{b}} \frac{\xi}{z} \frac{\mathrm{Ai}'(0)}{\mathrm{Ai}(0)} \mathcal{L}^{-1}(s^{-2/3}) \\
&= c_1^{\mathrm{b}} \frac{\xi}{z} \frac{3^{2/3}\Gamma(2/3)}{3^{1/3}\Gamma(1/3)} \frac{\rho^{-1/3}}{\Gamma(2/3)} = c_1^{\mathrm{b}} \frac{\xi}{z} \frac{3^{1/3}}{\Gamma(1/3)} \rho^{-1/3} \\
&= c_1^{\mathrm{b}} M^{1/4} \left(\frac{9}{8D_1}\right)^{1/3} \left(\frac{125}{216\nu^5}\right)^{1/12} \frac{3^{1/3}}{\Gamma(1/3)} r^{-5/4}, \qquad (3.134)
\end{aligned}$$

where we have used the properties of the Airy function $\mathrm{Ai}(0) = [3^{2/3}\Gamma(2/3)]^{-1}$ and $\mathrm{Ai}'(0) = -[3^{1/3}\Gamma(1/3)]^{-1}$. The $r^{-5/4}$ dependence of the concentration gradient implies that the electrode is non-uniformly accessible because the current density is proportional to this gradient. Finally, the limiting current is obtained as

$$\begin{aligned}
i_{\mathrm{L}} &= -2\pi z_1 F D_1 \int_0^a \left(\frac{\partial c_1}{\partial z}\right)_{z=0} r \mathrm{d}r \\
&= -2\pi z_1 F D_1 c_1^{\mathrm{b}} M^{1/4} \left(\frac{9}{8D_1}\right)^{1/3} \left(\frac{125}{216\nu^5}\right)^{1/12} \frac{3^{1/3}}{\Gamma(1/3)} \int_0^a r^{-1/4} \mathrm{d}r \\
&= -1.5971 k z_1 F D_1^{2/3} c_1^{\mathrm{b}} \frac{\dot{V}^{3/4} a^{3/4}}{d_n^{1/2} \nu^{5/12}}.
\end{aligned} \tag{3.135}$$

The characteristic features of the wall-jet electrode are that the limiting current density varies with the volume flow rate to the 3/4 power, and that the diffusion boundary layer thickness is not uniform along the electrode surface.

The wall-jet electrode has the advantage that very high convection rates are possible, which makes the study of fast heterogeneous kinetics feasible. Compared with the RDE, the mass-transfer rate at the wall-jet electrode can correspond to the rotation frequency of 500 000 Hz [19]. In a channel flow, for example, the convection rate is limited by the onset of turbulence in the cell, when the Reynolds number exceeds the value of ca. 2000.

3.4 Non-stationary or transient electrode processes

3.4.1 Introduction

The key equation to describe non-stationary transport processes in the absence of homogeneous chemical reactions is the continuity equation

$$\frac{\partial c_i}{\partial t} = -\vec{\nabla} \cdot \vec{j}_i. \tag{3.136}$$

In one-dimensional problems with no convection, this equation leads to

$$\frac{\partial c_i}{\partial t} = D_i \left[\frac{\partial^2 c_i}{\partial x^2} + z_i f \frac{\partial}{\partial x}\left(c_i \frac{\partial \phi}{\partial x}\right)\right], \tag{3.137}$$

where the Nernst–Planck equation has been used. If migration is negligible due to the presence of an excess of supporting electrolyte, this reduces to the *diffusion equation*

$$\frac{\partial c_i}{\partial t} \approx D_i \frac{\partial^2 c_i}{\partial x^2}. \tag{3.138}$$

In the case of neutral solutes, including, e.g., the electrolyte in a binary solution, eqn (3.138) is exact and receives the name of *Fick's second law*.

In the following sections we consider that there is only one electroactive ion in solution ($i = 1$) and it behaves as a trace ion. Equation (3.138) then has to be solved under the appropriate boundary conditions. The concentration $c_1(0, t)$ at the electrode surface is specified in chronoamperometric techniques, while the concentration gradient

$$\left(\frac{\partial c_1}{\partial x}\right)_{x=0} = -\frac{I(t)}{z_1 F D_1} \tag{3.139}$$

is known in chronopotentiometric techniques. Note that the sign convention is such that $I > 0$ if the electroactive ion is anodically dissolved, and $I < 0$ if it is reduced at the cathode. The boundary condition that specifies the bulk solution concentration

$$c_1(x \to \infty, t) = c_1^{\mathrm{b}} \tag{3.140}$$

can be replaced by

$$c_1(\delta, t) = c_1^{\mathrm{b}} \tag{3.141}$$

when the solution is mixed and the diffusion boundary layer has a finite thickness δ. Finally, the initial condition is

$$c_1(x, 0) = c_1^{\mathrm{b}}. \tag{3.142}$$

Equation (3.138) can be solved by the method of Laplace transformation [22]. It is then converted to the linear ordinary differential equation

$$s\tilde{c}_1 - c_1^b = D_1 \frac{\mathrm{d}^2 \tilde{c}_1}{\mathrm{d}x^2}, \tag{3.143}$$

where

$$\tilde{c}_1(x, s) \equiv \int_0^\infty c_1(x, t)\,\mathrm{e}^{-st}\mathrm{d}t \tag{3.144}$$

is the Laplace transformed concentration. The general solution of eqn (3.143) has the form

$$\tilde{c}_1 = \frac{c_1^{\mathrm{b}}}{s} + A\,\mathrm{e}^{qx} + B\,\mathrm{e}^{-qx} = \frac{c_1^{\mathrm{b}}}{s} + C\,\sinh qx + E\,\cosh qx, \tag{3.145}$$

where $q \equiv \sqrt{s/D_1}$. In the absence of mixing, the solution based on the exponentials is preferred because eqn (3.140) imposes that $A = 0$. In the presence of mixing, the solution based on the hyperbolic trigonometric functions must be used and eqn (3.141) requires that

$$E = -C\tanh q\delta. \tag{3.146}$$

In the following sections we determine the coefficients B or C for some of the most common transient electrochemical techniques.

3.4.2 Current step in the absence of mixing

In chronopotentiometric techniques, the electric current density is known as a function of time and eqn (3.139) can be used to determine the coefficient B in eqn (3.145) as

$$B = \frac{\tilde{I}(s)}{z_1 F \sqrt{s D_1}} = \frac{\tilde{I}(s)}{z_1 F D_1} \frac{1}{q}, \tag{3.147}$$

where $\tilde{I}(s)$ is the Laplace transform of the current density. The solution of eqn (3.138) in the time domain is formally given by

$$c_1(x,t) = c_1^{\mathrm{b}} + \mathcal{L}^{-1}\left(\frac{\tilde{I}}{z_1 F D_1} \frac{\mathrm{e}^{-qx}}{q}\right), \tag{3.148}$$

where $\mathcal{L}^{-1}$ denotes the inverse Laplace transformation.

In a current step (galvanostatic method) the function $I(t)$ is zero for $t < 0$ and takes the constant value I_0 for $t > 0$. Its Laplace transform is $\tilde{I}(s) = I_0/s$ and the inverse Laplace transform in eqn (3.148) can then be found in Laplace tables [9, 22–24] as

$$c_1(x,t) = c_1^b + \frac{2I_0}{z_1 F \sqrt{D_1}} \sqrt{t}\, [\pi^{-1/2} \mathrm{e}^{-\zeta^2} - \zeta \operatorname{erfc}(\zeta)], \tag{3.149}$$

where $\zeta \equiv x/(2\sqrt{D_1 t})$ is the *Boltzmann variable*, and erfc is the complementary error function. The function inside the brackets in eqn (3.149) is the first integral of the complementary error function [9]

$$\operatorname{ierfc}(\zeta) \equiv \pi^{-1/2} \mathrm{e}^{-\zeta^2} - \zeta \operatorname{erfc}(\zeta). \tag{3.150}$$

The concentration profile in eqn (3.149) is represented in Fig. 3.13.

At the electrode surface, the concentration takes the value

$$c_1(0,t) = c_1^{\mathrm{b}} + \frac{2I_0}{z_1 F \sqrt{\pi D_1}} \sqrt{t} = c_1^{\mathrm{b}} \left[1 - \sqrt{\frac{t}{\tau}}\right]. \tag{3.151}$$

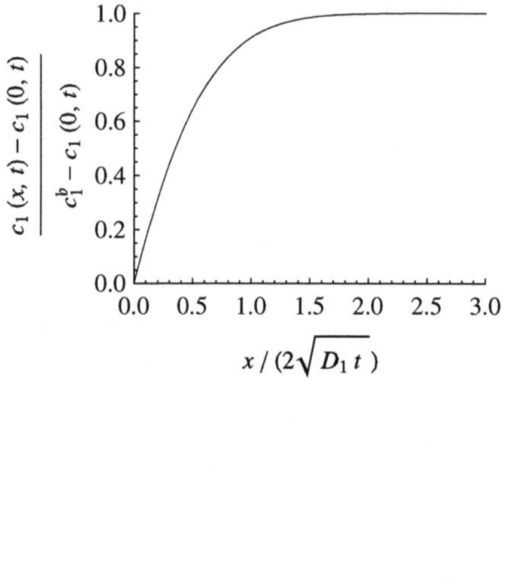

Fig. 3.13.
Concentration profile in a current step.

If the electroactive species is reduced at the cathode, then $I_0 < 0$ and the surface concentration vanishes (Fig. 3.14) after a transition time[14] that is given by the *Sand equation*

$$\tau = \pi D_1 \left(\frac{z_1 F c_1^{\mathrm{b}}}{2 I_0} \right)^2 . \qquad (3.152)$$

Using the values $I_0 = -0.1\,\mathrm{mA\,cm^{-2}}$, $D_1 = 10^{-5}\mathrm{cm^2\,s^{-1}}$, $z_1 = 1$, and $c_1^{\mathrm{b}} = 10^{-6}\mathrm{mol\,cm^{-3}}$, the transition time can be estimated as

$$\tau = \pi \ 10^{-5} \left(\frac{96\,500 \times 10^{-6}}{2 \times 10^{-4}} \right)^2 \mathrm{s} \approx 7\,\mathrm{s}.$$

Obviously, for smaller current densities, the transition time would be larger. But it must be observed that the experimental times cannot be very much larger because convection might then play a role in the mass transport, and the solution obtained would no longer be valid.

The electrode potential E, i.e. the potential at the electrode with respect to the solution, is given by the Nernst equation, which in this case takes the form

$$E(t) = E^{\circ\prime} + \frac{RT}{z_1 F} \ln c_1(0,t) = E(0) + \frac{RT}{z_1 F} \ln \left[1 - \sqrt{\frac{t}{\tau}} \right], \qquad (3.153)$$

where $E^{\circ\prime}$ is the formal standard potential and $E(0) = E^{\circ\prime} + (RT/z_1 F) \ln c_1^{\mathrm{b}}$ is the initial electrode potential; the concentrations in these equations must be expressed in M units. It is clearly seen in eqn (3.153) that E diverges when the transition time is approached (Fig. 3.15). In experimental practice, this means that the electrode potential changes so much that new electrode processes take place (before an actual divergence can occur).

The above equations provide an accurate description of the concentration changes during a current-step experiment. When the current is established at

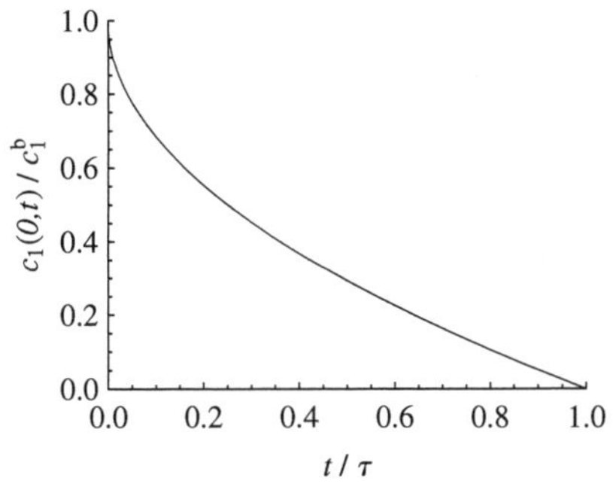

Fig. 3.14.
Time variation of the surface concentration in a current step.

[14] When this transition time is determined experimentally, the current density I_0 is chosen so that natural convection and double-layer charging do not interfere with the measurement.

time $t = 0$, the concentration of the electroactive species in the vicinity of the electrode changes in such a way that its gradient at the surface is determined by the current. The concentration changes propagate towards the bulk solution and the surface concentration decreases with increasing time. Eventually, the surface concentration vanishes at the transition time τ. The diffusion front has then covered a distance of ca. $4\sqrt{D_1\tau}$ (Fig. 3.16).

The transient transport processes can be better understood with the help of the concept of *diffusion length*, $L_d(t) \equiv 2\sqrt{D_1 t/\pi}$. This is the distance across which concentration changes propagate by diffusion in a time t, and the Boltzmann variable can be rewritten as $\zeta = x/\sqrt{\pi}L_d(t)$. When $\zeta \gg 1$ the concentration changes caused by the electrode reaction have not reached position x yet and the concentration $c_1(x,t)$ is still equal to the initial value c_1^b. On the contrary, the concentration is significantly different from the initial value at those positions x in which $\zeta < 1$. Interestingly, Figs. 3.13 and 3.16 show that at time t the concentration changes are confined within a region of thickness ca. $3L_d$, which is thus the maximum distance at which the digital simulation of this electrochemical process needs to be extended; this conclusion also holds for other techniques, such as the current scan and the voltage step.

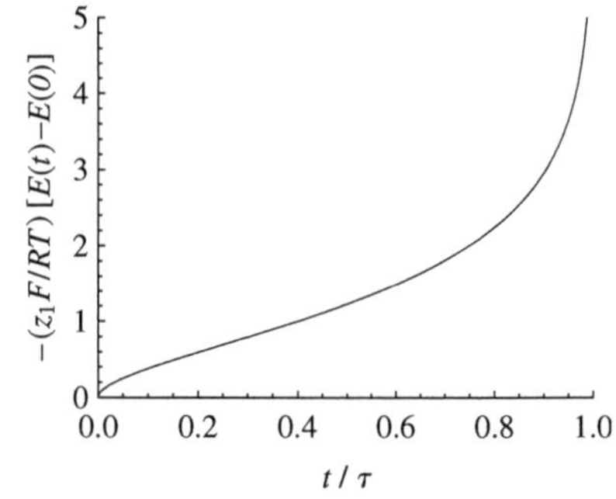

Fig. 3.15.
Time variation of the electrode potential in a current step.

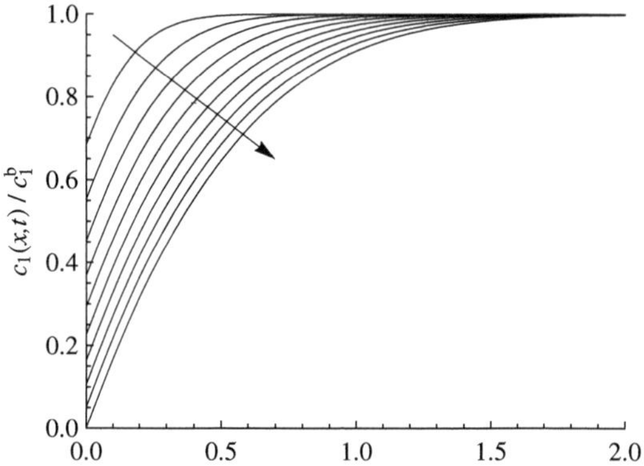

Fig. 3.16.
Concentration profile during a current step experiment at times $t/\tau = 0.1, 0.2, \ldots, 0.9$, and 1.0 (increasing in the arrow direction). Note that the concentration gradient at the electrode surface is independent of time because it is fixed by the constant current.

3.4.3 Current step with mixing

When the solution is mixed during the current-step experiment, the thickness δ of the diffusion boundary layer remains constant and the solution of the transport equations differs from that in eqn (3.149). Following the discussion at the end of the previous section, the situation is now different because, in addition to x and $L_d(t)$, δ is another characteristic length of the problem. Accordingly, there is an additional characteristic time of the system, $\tau_\delta \equiv \pi\delta^2/4D_1$ which is the time required by diffusion to reach the outer end of the boundary layer, δ.

Regarding the relative values of the current densities, we can distinguish two possible regimes: $I_0/I_{L,1} > 1$ and $I_0/I_{L,1} \leq 1$, where $I_{L,1} \equiv -z_1FD_1c_1^b/\delta$ is the limiting diffusion current density of the electroactive species. When $I_0/I_{L,1} > 1$ the surface concentration vanishes before the concentration changes reach the outer end of the boundary layer. That is, the transition time τ defined by Sand's equation is smaller than τ_δ. Hence, the influence of mixing cannot be noticed and this situation does not differ practically from that considered in the previous section. Thus, we restrict the present study to the case $I_0/I_{L,1} \leq 1$ and $\tau \geq \tau_\delta$.

Due to the mixing, the concentration changes cannot propagate by diffusion beyond $x = \delta$. At short times ($t \ll \tau_\delta$ and $L_d \ll \delta$), δ is not a relevant variable and the concentration c_1 must be a function of ζ (and t) practically identical to that in eqn (3.149). At large times ($t \gg \tau_\delta$ and $L_d \gg \delta$), the diffusion length L_d is not a relevant variable, and the concentration c_1 must be a function of x/δ. In particular, for the current-step technique under consideration, a steady state is reached at large times, and the diffusion equation reduces to

$$\frac{d^2c_1}{dx^2} = 0, \quad t \gg \tau_\delta. \tag{3.154}$$

The solution of this equation is

$$c_1(x) = c_1^b + [c_1(0) - c_1^b]\left(1 - \frac{x}{\delta}\right), \quad t \gg \tau_\delta, \tag{3.155}$$

and the surface concentration can be calculated from eqn (3.139) as

$$c_1(0) = c_1^b\left(1 - \frac{I_0}{I_{L,1}}\right), \quad t \gg \tau_\delta. \tag{3.156}$$

At intermediate times, the concentration $c_1(x,t)$ must be determined from the solution of the diffusion equation, eqn (3.138). Using the last expression in eqn (3.145) and eqn (3.146), the transformed concentration can be written as

$$\tilde{c}_1 = \frac{c_1^b}{s} - C\frac{\sinh q(\delta - x)}{\cosh q\delta}, \tag{3.157}$$

where $q \equiv \sqrt{s/D_1}$. Since the transformed current density is $\tilde{I}(s) = I_0/s$ in a current step, eqn (3.139) can be used to determine the coefficient C as

$$C = -\frac{\tilde{I}(s)}{z_1F\sqrt{sD_1}} = -\frac{I_0}{z_1F\sqrt{s^3D_1}}. \tag{3.158}$$

The concentration is then

$$c_1(x,t) = c_1^{\mathrm{b}} - \frac{I_0}{z_1 F D_1}\,\mathcal{L}^{-1}\left[\frac{\sinh q(\delta - x)}{\cosh q\delta}\,\frac{1}{qs}\right] \tag{3.159}$$

but this inverse transform cannot be found in the tables. Instead, we use the series expansion

$$\begin{aligned}\frac{1}{\cosh q\delta} &= 2\mathrm{e}^{-q\delta}(1+\mathrm{e}^{-2q\delta})^{-1} = 2\mathrm{e}^{-q\delta}\sum_{n=0}^{\infty}(-1)^n\mathrm{e}^{-2nq\delta} \\ &= 2\sum_{n=0}^{\infty}(-1)^n\mathrm{e}^{-(2n+1)q\delta}.\end{aligned} \tag{3.160}$$

The concentration can now be calculated using the same inverse Laplace transform as in eqn (3.149) as

$$\begin{aligned}c_1(x,t) &= c_1^{\mathrm{b}} + \frac{I_0}{z_1 F D_1}\,\mathcal{L}^{-1}\left[\sum_{n=0}^{\infty}(-1)^n\frac{\mathrm{e}^{-q(2n\delta+x)}}{qs} + \sum_{n=1}^{\infty}(-1)^n\frac{\mathrm{e}^{-q(2n\delta-x)}}{qs}\right] \\ &= c_1^{\mathrm{b}} + \frac{2I_0\sqrt{t}}{z_1 F\sqrt{D_1}}\left[\sum_{n=0}^{\infty}(-1)^n\,\mathrm{ierfc}\left(\frac{2n\delta+x}{2\sqrt{D_1 t}}\right)\right. \\ &\quad \left. + \sum_{n=1}^{\infty}(-1)^n\,\mathrm{ierfc}\left(\frac{2n\delta-x}{2\sqrt{D_1 t}}\right)\right].\end{aligned} \tag{3.161}$$

This concentration profile has been represented in Figs. 3.17 and 3.18. Equation (3.161) has the correct limiting behaviour. In the absence of mixing the diffusion boundary layer thickness goes to infinity (i.e. $\delta \gg L_\mathrm{d}$ at all times), and the functions ierfc tend to zero except for the one corresponding to $n = 0$. Therefore, eqn (3.161) becomes equal to eqn (3.149), as expected. In fact, regardless of the value of δ, eqn (3.161) becomes equal to eqn (3.149) at times short enough that $t \ll \tau_\delta$ and $\delta \gg L_\mathrm{d}$.

From eqns (3.152) and (3.161) the surface concentration is

$$c_1(0,t) = c_1^{\mathrm{b}}\left\{1 - \sqrt{\frac{t}{\tau}}\left[1 + 2\sqrt{\pi}\sum_{n=1}^{\infty}(-1)^n\,\mathrm{ierfc}\left(\frac{n\delta}{\sqrt{D_1 t}}\right)\right]\right\}, \tag{3.162}$$

which tends asymptotically towards zero. Since ierfc is a monotonous and rapidly decreasing function, we conclude that the mixing increases the surface concentration (Fig. 3.19), that is, $c_1(0,t) \geq c_1^{\mathrm{b}}(1 - \sqrt{t/\tau})$.

In the time range $t \gg \tau_\delta$, the series in eqns (3.161) and (3.162) converge to the values given in eqns (3.155) and (3.156), respectively, as can be seen in Figs. 3.17 and 3.18. However, the convergence is rather slow and it is difficult to

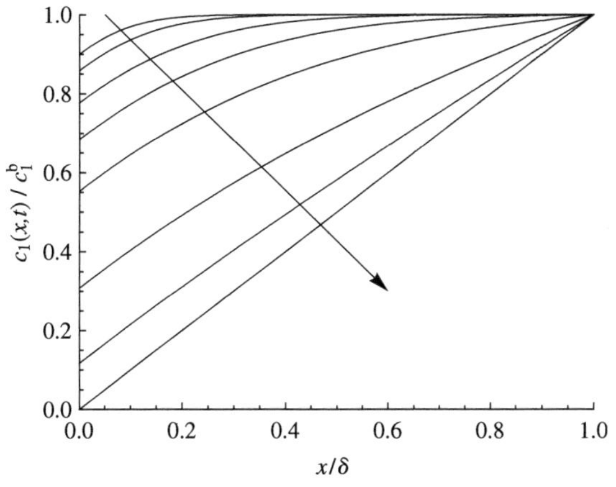

Fig. 3.17.
Concentration profile during a current step with mixing for $I_0 = I_{L,1}$ and times $t/\tau_\delta = 0.01, 0.02, 0.05, 0.1, 0.2, 0.5, 1.0$, and 5.0 (increasing in the arrow direction). The concentration gradient at the electrode surface is independent of time (and fixed by the constant current) only at short times, while at large times it is determined by the boundary layer thickness.

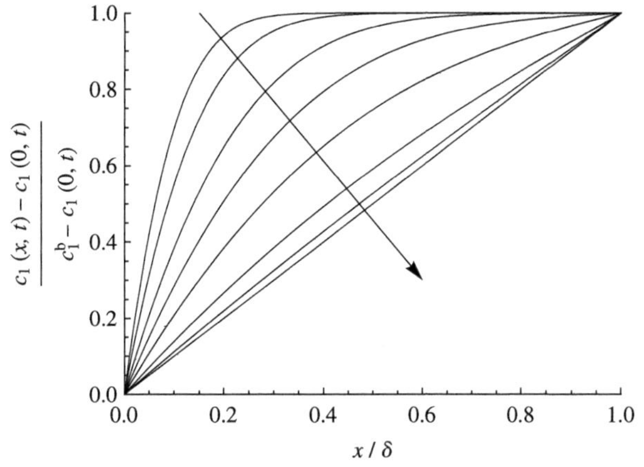

Fig. 3.18.
Concentration profile in a current step with mixing at times $t/\tau_\delta = 0.01, 0.02, 0.05, 0.1, 0.2, 0.5, 1.0$, and 5.0 (increasing in the arrow direction). This plot is valid for any value of the current density.

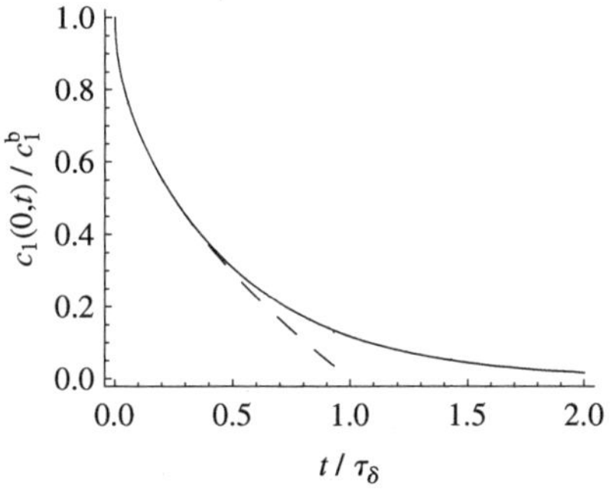

Fig. 3.19.
Time variation of the surface concentration in a current step with mixing for $I_0 = I_{L,1}$ (solid line) and the comparison with the variation in the absence of mixing (dashed line). Note that $I_0 = I_{L,1}$ implies $\tau = \tau_\delta$.

check this limiting behaviour analytically from eqn (3.161). Fortunately, with the help of the approximation

$$\begin{aligned}
\text{ierfc}(z) &= \sum_{n=1}^{\infty} [\text{ierfc}(z+ny-y) - \text{ierfc}(z+ny)] \\
&\approx y \sum_{n=1}^{\infty} \text{erfc}\left(z+ny-\frac{y}{2}\right) \approx 2y \sum_{n=1}^{\infty} \text{erfc}(z+2ny-y) \\
&\approx 2 \sum_{n=1}^{\infty} \left[\text{ierfc}\left(z+(2n-1)y-\frac{y}{2}\right) - \text{ierfc}\left(z+2ny-\frac{y}{2}\right)\right] \\
&= -2 \sum_{n=1}^{\infty} (-1)^n \text{ierfc}\left(z+ny-\frac{y}{2}\right),
\end{aligned} \tag{3.163}$$

where we have treated both y and n as dummy variables with the only restriction that $y \ll z$, eqn (3.161) reduces to

$$\begin{aligned}
c_1(x,t) \approx c_1^{\text{b}} \Bigg\{ 1 - \frac{1}{2}\sqrt{\frac{\pi t}{\tau}} \Bigg[2\,\text{ierfc}\left(\frac{x}{2\sqrt{D_1 t}}\right) - \text{ierfc}\left(\frac{\delta + x}{2\sqrt{D_1 t}}\right) \\
- \text{ierfc}\left(\frac{\delta - x}{2\sqrt{D_1 t}}\right)\Bigg]\Bigg\},
\end{aligned} \tag{3.164}$$

for $t \gg \tau_\delta$ and $L_\text{d} \gg \delta$. When the arguments of the ierfc functions are very small, we can further use that

$$\text{ierfc}(z) \approx \frac{1}{\sqrt{\pi}} - z \quad \text{when } z \ll 1, \tag{3.165}$$

and eqn (3.164) then becomes approximately equal to eqn (3.155); see also Fig. 3.17.

3.4.4 Current scan

The time dependence of a linear current scan from zero at time $t = 0$ is

$$I(t) = I_0 a t, \tag{3.166}$$

where $a > 0$ is a constant with dimensions of inverse of time. Its Laplace transform is

$$\tilde{I}(s) = \frac{I_0 a}{s^2}, \tag{3.167}$$

and eqn (3.166) can then be used to determine the coefficient B in eqn (3.145) as

$$B = \frac{\tilde{I}(s)}{z_1 F \sqrt{s D_1}} = \frac{I_0 a}{z_1 F D_1} \frac{1}{q s^2}, \tag{3.168}$$

where $q \equiv \sqrt{s/D_1}$ and we have considered that the solution is not mixed.

The solution of eqn (3.138) in the time domain is now formally given by

$$c_1(x,t) = c_1^{\mathrm{b}} + \mathcal{L}^{-1}\left(\frac{I_0 a}{z_1 F D_1}\frac{\mathrm{e}^{-qx}}{qs^2}\right). \tag{3.169}$$

This inverse Laplace transform can be found in Laplace tables [9, 22–24] and leads to

$$c_1(x,t) = c_1^{\mathrm{b}} + \frac{I_0 a}{z_1 F\sqrt{D_1}}(4t)^{3/2}\mathrm{i}^3\mathrm{erfc}(\zeta), \tag{3.170}$$

where $\zeta \equiv x/(2\sqrt{D_1 t})$ is the previously defined Boltzmann variable and

$$\mathrm{i}^3\mathrm{erfc}(\zeta) = \frac{\pi^{-1/2}}{6}(\zeta^2+1)\mathrm{e}^{-\zeta^2} - \frac{\zeta}{12}(2\zeta^2+3)\mathrm{erfc}(\zeta) \tag{3.171}$$

is the third integral of the complementary error function [9]. This concentration profile has been represented in Fig. 3.20.

The surface concentration

$$c_1(0,t) = c_1^{\mathrm{b}} + \frac{4}{3}\frac{I_0 a}{z_1 F\sqrt{\pi D_1}}t^{3/2} = c_1^{\mathrm{b}}\left[1 - \left(\frac{t}{\tau}\right)^{3/2}\right] \tag{3.172}$$

vanishes (for cathodic reductions where $I_0 < 0$) at the transition time

$$\tau \equiv \left(-\frac{3z_1 F c_1^{\mathrm{b}}\sqrt{\pi D_1}}{4I_0 a}\right)^{2/3} \equiv \pi D_1\left(\frac{3z_1 F c_1^{\mathrm{b}}}{4I_0 a\tau}\right)^2, \tag{3.173}$$

which varies with the scan rate a (Fig. 3.21).

If we compare this transition time with that observed using the current-step technique we have, from eqns (3.152) and (3.173), that

$$\left(\frac{\tau_{\mathrm{scan}}}{\tau_{\mathrm{step}}}\right)^{1/2} = \frac{3}{2}\frac{I_{\mathrm{step}}}{I(\tau_{\mathrm{scan}})}, \tag{3.174}$$

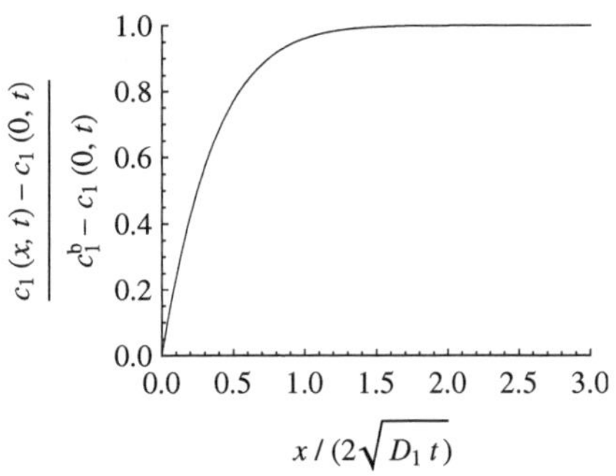

Fig. 3.20.
Concentration profile in a current scan (without mixing).

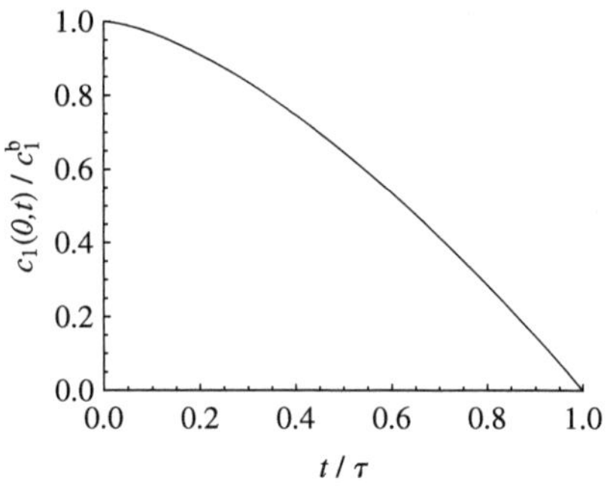

Fig. 3.21.
Time variation of the surface concentration in a current scan.

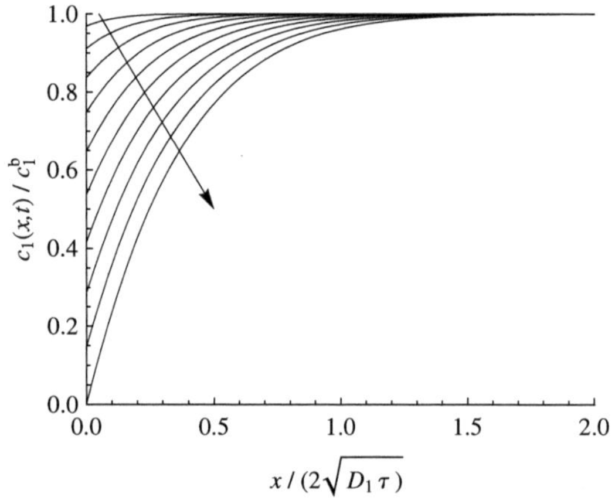

Fig. 3.22.
Concentration profile during a current scan at times $t/\tau = 0.1, 0.2, \ldots, 0.9$, and 1.0 (increasing in the direction of the arrow). The concentration gradient at the electrode surface is proportional to time because it is determined by the scanned current.

where $I(\tau_{\text{scan}}) = I_0 a\tau$; note that the second equality in eqn (3.173) is very helpful to verify eqn (3.174). If we choose the scan rate a so that the transition time is the same in these two techniques, then the scanned current at the transition time is 3/2 times higher than in the step technique.

The concentration changes during a current-scan experiment can be better understood with the help of Fig. 3.22. As time progresses, the concentration changes propagate towards the bulk solution and the surface concentration decreases in such a way that the gradient at the surface is determined by the current, and hence it is proportional to time. Eventually, a transition time τ is reached when the surface concentration vanishes. The diffusion front has then covered a distance of ca. $3\sqrt{D_1\tau}$.

3.4.5 Voltage step

In chronoamperometric techniques, the electrode potential and hence the surface concentration of the electroactive species is known as a function of time.

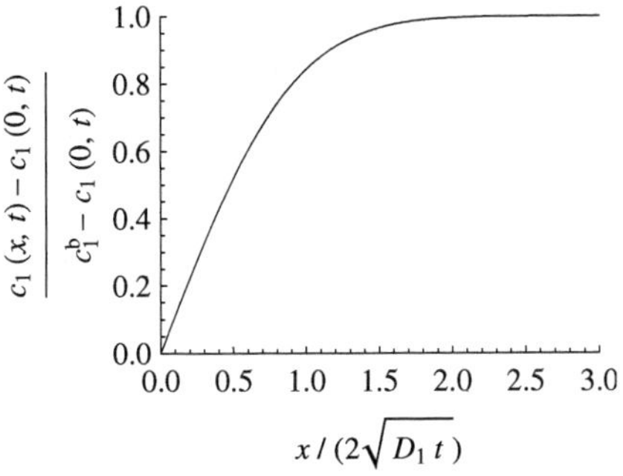

Fig. 3.23.
Concentration profile in a voltage step.

The coefficient B in eqn (3.145) can now be determined as

$$B = \tilde{c}_1(0, s) - \frac{c_1^b}{s}, \tag{3.175}$$

where $\tilde{c}_1(0, s)$ is the Laplace transform of the surface concentration $c_1(0, t)$.

In a voltage step (potentiostatic method) the electrode potential is biased to a constant negative (or positive) value at $t = 0$ so that the electroactive species is immediately reduced (or oxidized). The Nernst equation for the equilibrium electrode potential implies that the surface concentration of the electroactive species is fixed after the voltage step (i.e. for $t > 0$) to the value

$$c_1(0, t) = c_1^b e^{z_1 f \Delta E}, \tag{3.176}$$

where ΔE is the voltage step. This surface concentration is practically zero when the step is very large (i.e. $-z_1 f \Delta E \gg 1$). Since this concentration is independent of time, its Laplace transform is $\tilde{c}_1(0, s) = c_1^b e^{z_1 f \Delta E}/s$ and the solution of Fick's second law, eqn (3.138), in the time domain is formally given by

$$c_1(x, t) = c_1^b \left[1 + (e^{z_1 f \Delta E} - 1)\mathcal{L}^{-1}\left(\frac{e^{-qx}}{s}\right)\right]. \tag{3.177}$$

Looking for this inverse transform in the Laplace tables [9, 22–24] we get

$$c_1(x, t) = c_1^b[1 + (e^{z_1 f \Delta E} - 1)\,\mathrm{erfc}(\zeta)], \tag{3.178}$$

where $\zeta \equiv x/2\sqrt{D_1 t}$ is the Boltzmann variable. This concentration profile is represented in Fig. 3.23 in a dimensionless form that is valid for all values of ΔE. It can be seen from eqn (3.178) that the function represented in this figure is $\mathrm{erf}(\zeta)$. In fact, when $-z_1 f \Delta E \gg 1$, the concentration profile reduces to $c_1(x, t) \approx c_1^b \mathrm{erf}(\zeta)$. Finally, it is interesting to observe once again that the diffusion front proceeds no further than ca. $6\sqrt{D_1 t}$. Thus, digital simulations of diffusion in an electrochemical process do not need to extend beyond this distance.

The electric current density is then

$$I(t) = -z_1 F D_1 \left(\frac{\partial c_1}{\partial x}\right)_{x=0} = z_1 F c_1^{\mathrm{b}} (\mathrm{e}^{z_1 f \Delta E} - 1)\sqrt{\frac{D_1}{\pi t}}, \tag{3.179}$$

which is known as the *Cottrell equation.*

Exercises

3.1 In relation to the example of a ternary electrolyte worked out in Section 3.2.5, show that the system behaves as a binary electrolyte solution when $c_{23}^{\mathrm{b}} \ll c_{13}^{\mathrm{b}}$ and the ionic concentrations at the electrode surface are then

$$c_i(0) = c_i^{\mathrm{b}}\left(1 - \frac{I}{I_L}\right) \qquad i = 1, 3.$$

3.2 Consider the cathodic reduction of a univalent metal cation M^+ in an aqueous solution containing the salt M^+A^- and the acid H^+A^-. Derive the steady-state concentration profiles in the following cases:
(a) the salt M^+A^- is completely dissociated, and
(b) the salt M^+A^- is completely associated.
How does the diffusion coefficient of A^- affect the limiting current in both cases?

3.3 Consider the stationary ionic transport in an aqueous solution of $Cu(HSO_4)^+$, H^+, and HSO_4^- in the boundary layer close to an electrode where the reaction

$$Cu(HSO_4)^+ + 2e^- \rightarrow Cu(s) + HSO_4^-$$

takes place, i.e. case *ii*) in Section 3.2.6. Find the ionic concentrations at the electrode surface as particular cases of the general expression, eqn (3.71)

$$c_i(0) = c_i^{\mathrm{b}}\left\{1 - \frac{j_i \delta}{D_i c_i^{\mathrm{b}}}\frac{I_{\mathrm{L0}}}{(1 + z_i \Gamma) I}\left[\left(1 - \frac{I}{I_{\mathrm{L0}}}\right)^{1+z_i\Gamma} - 1\right]\right\}\left(1 - \frac{I}{I_{\mathrm{L0}}}\right)^{-z_i\Gamma}$$

worked out at the end of that section. Check also that $c_1(0) = c_3(0) - c_2(0)$.

3.4 Consider the stationary ionic transport in an aqueous solution of Cu^{+2}, H^+, and HSO_4^- in the boundary layer close to a cathode where the reaction

$$Cu^{+2} + 2e^- \rightarrow Cu(s)$$

takes place, i.e. case *i*) in Section 3.2.6. Determine the limiting current density and the potential drop in the diffusion boundary layer as a function of the bulk concentrations of $Cu(HSO_4)_2$ and H_2SO_4. Study also the limiting cases in which either of these concentrations vanishes.

3.5 A metal electrode is placed in an aqueous solution of a binary electrolyte A^+B^-. The initial electrolyte concentration is c^{b}. By imposing a constant current density I, the reduction of the cation takes place until the stationary state is achieved. Calculate the final concentration profile and the limiting current density in the following cases:
(a) The solution is well stirred and the thickness of the Nernst layer close to the electrode is δ.

(b) From the beginning, an ideally selective cation-exchange membrane is placed at a distance δ. From the cathode, and the electrolyte solution is stagnant in between them.

3.6 Describe the time-dependent concentration and electric potential distribution in the two cases of the previous exercise.

3.7 An electrode is equilibrated with a solution that contains a redox couple A^{z_1}/A^{z_2}. From time $t = 0$, a current step of amplitude I_0 is applied and the electrode reaction

$$A^{z_1} \rightleftarrows A^{z_2} + ne^-$$

proceeds under quasi-equilibrium conditions until species A^{z_1} runs out at the electrode surface. Using the Nernst equation

$$E(t) = E^{\circ\prime} + \frac{RT}{nF}\ln\frac{c_2(0,t)}{c_1(0,t)},$$

where $E^{\circ\prime}$ is the formal standard potential, and assuming that $\sqrt{D_1}c_1^{\mathrm{b}} \approx \sqrt{D_2}c_2^{\mathrm{b}}$, show that the electrode potential can be expressed as

$$E(t) = E(0) + \frac{2RT}{nF}\operatorname{arctanh}\sqrt{\frac{t}{\tau}},$$

where τ is the transition time.

3.8 Solve the diffusion equation for a current step with mixing using the method of separating the variables, i.e. introducing the transformation $c_1 = c_1^{\mathrm{b}} + X(x)T(t)$ and writing the concentration of the electroactive ion as a Fourier series.

3.9 The voltage step can be (mathematically) described as a chronopotentiometric technique in which the current is proportional to $t^{-1/2}$. Solve the diffusion equation imposing that the current is $I = \alpha t^{-1/2}$, and evaluate the surface concentration. Then find the value of α by comparing your result with the equation $c_1(0,t) = c_1^{\mathrm{b}} e^{z_1 f \Delta E}$.

3.10 Consider the irreversible reaction of species 1 at the electrode forming species 2, which undergoes a homogeneous, irreversible reaction forming species 3 (i.e. the so-called EC mechanism). By solving the equations

$$\frac{\partial c_1}{\partial t} = D\frac{\partial^2 c_1}{\partial x^2},$$

$$0 = D\frac{d^2 c_2}{dx^2} - kc_2^m,$$

where $m \geq 1$ is an integer, obtain the concentration profiles $c_1(x,t)$ and $c_2(x)$ with appropriate initial and boundary conditions. Note that the transport of species 2 is assumed to take place under steady state.

3.11 Consider the irreversible reaction of species 1 at the electrode forming species 2, which undergoes a homogeneous, irreversible reaction forming species 3 (i.e. the so-called EC mechanism). By solving the equations

$$\frac{\partial c_1}{\partial t} = D\frac{\partial^2 c_1}{\partial x^2}$$

$$\frac{\partial c_2}{\partial t} = D\frac{\partial^2 c_2}{\partial x^2} - kc_2$$

obtain the concentration profiles $c_1(x,t)$ and $c_2(x,t)$ with appropriate initial and boundary conditions. Compare your solution with that of the previous exercise for $m = 1$.

3.12 Find the Cottrell equation corresponding to a large voltage step in a hemispherical microelectrode of radius a by solving the diffusion equation in spherical geometry

$$\frac{\partial c}{\partial t} = D\frac{1}{r^2}\frac{\partial}{\partial r}\left(r^2\frac{\partial c}{\partial r}\right) = D\left(\frac{\partial^2 c}{\partial r^2} + \frac{2}{r}\frac{\partial c}{\partial r}\right)$$

under the initial and boundary conditions $c(r,0) = c^{\mathrm{b}}$ and $c(a,t) = 0$.
(Hint: Use the transformation $c(r,t) = c^{\mathrm{b}} + u(r,t)/r$.)

3.13 Describe the time evolution of the concentration profile during a current step in a hemispherical microelectrode of radius a by solving the diffusion equation in spherical geometry

$$\frac{\partial c}{\partial t} = D\frac{1}{r^2}\frac{\partial}{\partial r}\left(r^2\frac{\partial c}{\partial r}\right) = D\left(\frac{\partial^2 c}{\partial r^2} + \frac{2}{r}\frac{\partial c}{\partial r}\right)$$

under the initial and boundary conditions $c(r,0) = c^{\mathrm{b}}$ and $(\partial c/\partial r)_{r=a} = -I/zFD$. Is there a transition time for any value of the current density as in the case of a planar electrode?
(Hint: Use the transformation $c(r,t) = c^{\mathrm{b}} + u(r,t)/r$.)

3.14 A simplified conductance cell is formed by two parallel plates of the same metal M that have a geometrical surface area A and are separated by a distance d. For the sake of simplicity, consider that the cell has uniform cross-section A. The cell is filled with an aqueous solution of concentration c^{b} of a strong electrolyte M^+A^-, and the electrode reactions are $M^+ + e^- \rightleftarrows M$.

(a) Find the dependence of the electric current $i = AI$ through the cell with the applied potential difference between the metal plates, $\Delta_\alpha^\beta\phi \equiv \phi^\beta - \phi^\alpha$, where β and α denote the metal plates. In the solution phase use a position co-ordinate x ranging from 0 at the plate α to d at the plate β, and define the current density I as positive from α to β.

(b) From the total power $-i\Delta_\alpha^\beta\phi$ consumed by the cell during operation, evaluate the contribution from ohmic (or Joule) dissipation and the contribution from electrodiffusion.

(c) Describe the relation between $\Delta_\alpha^\beta\phi \equiv \phi^\beta - \phi^\alpha$ and I when $I \to 0$, and discuss whether diffusion (or concentration polarization) effects can then be neglected.

References

[1] See, e.g., V.G. Levich, *Physicochemical Hydrodynamics*, Prentice-Hall, Englewood Cliffs, N.J., 1962.

[2] J.A. Manzanares and K. Kontturi, 'Diffusion and migration' in E.J. Calvo (ed.), *Encyclopedia of Electrochemistry, vol.* 2, *Interfacial Kinetics and Mass Transport*, Wiley-VCH, Weinheim, 2003.

[3] E.L. Cussler, *Diffusion. Mass Transfer in Fluid Systems*, 2nd edn, Cambridge University Press, Cambridge, 1997.

[4] M. Fleischmann, S. Pons, D.R. Rolison, and P.P. Schmidt (ed.), *Ultramicroelectrodes*, Datatech Systems, Morganton, N.C., 1987.

[5] This can be shown with conformal mapping from the spherical to planar geometry (Schwarz–Christoffel transformation). See, e.g. F.B. Hildebrand, *Advanced Calculus for Applications*, 2nd edn, Prentice-Hall, Englewood Cliffs, N.J., 1976, Ch. 11.

[6] M.I. Montenegro, M.A. Queirós, and J.L. Daschbach (ed.), *Microelectrodes: Theory and Applications*, NATO ASI Series E: Applied Sciences, vol. 197, Kluwer, Dordrecht, 1991.

[7] Y.V. Pleskov and V.Y. Filinovskii, *The Rotating Disc Electrode*, Consultants Bureau, New York, 1976.

[8] A.J. Bard and L.R. Faulkner, *Electrochemical Methods*, John Wiley & Sons, 2nd edn, New York, 2001.

[9] M. Abramowitz and I.A. Stegun, *Handbook of Mathematical Functions*, Dover, New York, 1965.

[10] K.B. Oldham, 'Steady-state voltammetry at a rotating disk electrode in the absence of supporting electrolyte', *J. Phys. Chem. B*, 104 (2000) 4703–4706.

[11] N.P.C. Stevens, M.B. Rooney, A.M. Bond, and S.W. Feldberg, 'A comparison of simulated and experimental voltammograms obtained for the $[Fe(CN)_6]^{3-/4-}$ couple in the absence of added supporting electrolyte at a rotating disk electrode', *J. Phys. Chem. A*, 105 (2001) 9085–9093.

[12] W.J. Albery, 'Effect of the electric field on investigations using the rotating disk electrode', *Trans. Faraday. Soc.*, 61 (1965) 2063–2077.

[13] J.A. Cooper and R.G. Compton, 'Channel electrodes – A review', *Electroanalysis*, 10 (1998) 141–155; and references therein.

[14] T. Singh and J. Dutt, 'Cyclic voltammetry at the tubular graphite electrode. Reversible processes (theory)', *J. Electroanal. Chem.*, 190 (1985) 65–73.

[15] H. Matsuda, 'Zur Theorie der stationären Strom-Spannungs-Kurven von Redox-Elektrodenreaktionen in hydrodynamischer Voltammetrie. II. Laminare Rohr und Kanalströmungen', *J. Electroanal. Chem.*, 15 (1967) 325–336.

[16] R.G. Compton, M.B.G. Pilkington, G.M. Stearn, and P.R. Unwin, 'Mass transport to channel and tubular electrodes: The 'Singh and Dutt approximation'', *J. Electroanal. Chem.*, 238 (1987) 43–66; P.D. Morland and R.G. Compton, 'Heterogeneous and homogeneous EC and ECE processes at channel electrodes: analytical wave shape theory', *J. Phys. Chem. B*, 103 (1999) 8951–8959.

[17] For example: http://www.comsol.com/products/multiphysics/.

[18] P. Liljeroth, C. Johans, K. Kontturi, and J.A. Manzanares, 'Channel flow at an immobilised liquid|liquid interface', *J. Electroanal. Chem.*, 483 (2000) 37–46.

[19] J.V. MacPherson, N. Simjee, and P.R. Unwin, 'Hydrodynamic ultramicroelectrodes: kinetic and analytical applications', *Electrochim. Acta*, 47 (2001) 29–45.

[20] M.B. Glauert, 'The wall jet', *J. Fluid Mech.*, 1 (1956) 625–643.

[21] W.J. Albery and C.M.A. Brett, 'The wall-jet ring-disc electrode. Part I. Theory', *J. Electroanal. Chem.*, 148 (1983) 201–210.

[22] H.S. Carslaw and J.C. Jaeger, *Conduction of Heat in Solids*, Clarendon Press, Oxford, 1959.

[23] V.S. Arpaci, *Conduction Heat Transfer*, Addison-Wesley, Reading, MA, 1966.

[24] F. Oberhettinger and L. Badii, *Tables of Laplace Transforms*, Springer-Verlag, Berlin, 1993.

4 Transport in membranes

4.1 Transport across neutral porous membranes

In this section we describe different aspects of the one-dimensional transport processes (in the absence of homogeneous chemical reactions) across a chemically inert, porous membrane, which has the only function of creating a well-defined unmixed space between two compartments α and β. The compartments are ideally mixed, so that they have homogeneous solute concentrations c^{α} and c^{β} at all times. These concentrations may be constant if the compartment solutions are circulated, or vary with time as a result of the transport across the membrane if they are not. If the first case, a true steady state can be reached. In the second one, the transport process is time dependent because so they are the compartment concentrations. However, when the compartment volumes V^{α} and V^{β} are large (compared to the membrane volume), the solute flow leads to very slow time variations of $c^{\alpha}(t)$ and $c^{\beta}(t)$, and these can be considered constant as far as the transport across the membrane is concerned. This type of transient transport processes are known as *quasi-steady processes*. In Section 4.1.1 we consider the transport of a neutral solute under these conditions and then in Section 4.1.2 we analyse the validity of the quasi-steady-state assumption.

4.1.1 Quasi-steady diffusive transport between two closed compartments

Consider the transport of a neutral solute in the experimental set-up depicted in Fig. 4.1. Initially, the solute is present only in compartment α, where its concentration is c_0^{α}. The time variation of the concentration in compartment α is given by the mass balance

$$j = -\frac{V^{\alpha}}{A}\frac{\mathrm{d}c^{\alpha}}{\mathrm{d}t}, \tag{4.1}$$

where j is the solute flux density across the membrane and A is the membrane area. The solute concentration in compartment β can be obtained from the total mass balance in the cell

$$V^{\alpha}c_0^{\alpha} = V^{\alpha}c^{\alpha}(t) + V^{\beta}c^{\beta}(t) \tag{4.2}$$

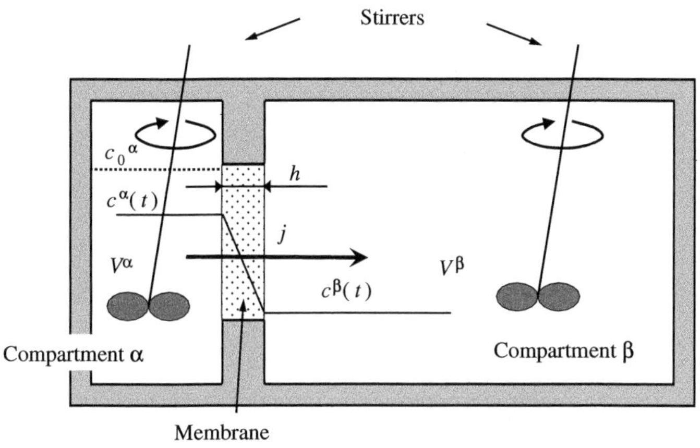

Fig. 4.1.
A cell divided into two compartments by a porous membrane. The compartments are well stirred so that concentration gradients only appear inside the membrane. The solute concentration in the compartments can vary with time due to the mass transport across the membrane.

as

$$c^\beta(t) = \frac{V^\alpha}{V^\beta}[c_0^\alpha - c^\alpha(t)]. \tag{4.3}$$

The solute concentration inside the membrane is denoted as $c(x,t)$ and its flux density j is given by Fick's first equation

$$j = -D\frac{\partial c}{\partial x}. \tag{4.4}$$

Under (quasi)steady-state conditions, it can be assumed that j is independent of position and eqn (4.4) can then be easily integrated over the membrane, extending from $x = 0$ where the solute concentration is $c(0,t) = c^\alpha(t)$ to $x = h$, where it is $c(h,t) = c^\beta(t)$, to give

$$j = -D\frac{c^\beta - c^\alpha}{h}. \tag{4.5}$$

The concentrations c^α and c^β, and the flux density j, however, are slowly varying functions of time. In particular, since the solute is initially present in compartment α only, the flux density varies from an initial maximum value Dc_0^α/h to zero at large times when the two compartments have the same concentration

$$c_\infty^\alpha = c_\infty^\beta = c_0^\alpha \frac{V^\alpha}{V^\alpha + V^\beta}. \tag{4.6}$$

Inserting eqns (4.5) and (4.3) into eqn (4.1), the solute concentration in compartment α is found to be given by the linear ordinary differential equation

$$\frac{\mathrm{d}c^\alpha}{\mathrm{d}t} + \left(\frac{1}{\tau^\alpha} + \frac{1}{\tau^\beta}\right) c^\alpha = \frac{c_0^\alpha}{\tau^\beta}, \tag{4.7}$$

where the coefficients

$$\tau^\alpha \equiv \frac{V^\alpha h}{AD} \tag{4.8}$$

$$\tau^\beta \equiv \frac{V^\beta h}{AD} \tag{4.9}$$

are the characteristic times required for diffusion to change significantly the concentrations of compartments α and β, respectively. The solution of eqn (4.7) with the initial condition $c^\alpha(0) = c_0^\alpha$ is

$$c^\alpha(t) = c_0^\alpha \frac{\tau^\alpha + \tau^\beta \,\mathrm{e}^{-t/\tau}}{\tau^\alpha + \tau^\beta} = c_0^\alpha \mathrm{e}^{-t/\tau} + c_\infty^\alpha (1 - \mathrm{e}^{-t/\tau}), \tag{4.10}$$

and, therefore,

$$c^\beta(t) = c_\infty^\beta (1 - \mathrm{e}^{-t/\tau}) \tag{4.11}$$

$$j = -\frac{V^\alpha}{A}\frac{\mathrm{d}c^\alpha}{\mathrm{d}t} = \frac{V^\alpha c_0^\alpha}{A\tau^\alpha}\mathrm{e}^{-t/\tau} = \frac{Dc_0^\alpha}{h}\mathrm{e}^{-t/\tau}, \tag{4.12}$$

where $1/\tau \equiv 1/\tau^\alpha + 1/\tau^\beta$. The fact that the relaxation time of the system towards equilibrium, τ, is a harmonic mean of τ^α and τ^β implies that the smaller compartment determines the system response. Thus, for instance, if compartment β is much larger in volume than compartment α, eqn (4.10) reduces to $c^\alpha(t) = c_0^\alpha \,\mathrm{e}^{-t/\tau^\alpha}$. Figure 4.2 shows the representation of eqns (4.10) and (4.11) for different values of the ratio $\tau^\beta/\tau^\alpha = V^\beta/V^\alpha$. Note that at large times the concentrations are given by eqn (4.6) and decrease with increasing V^β/V^α.

In Fig. 4.2 we have not used the values of the membrane thickness h and area A because of the use of dimensionless variables. In practice, however, the so-called membrane constant A/h is not known and has to be determined from the experimental data of $c^\beta(t)$. In particular, eqn (4.11) shows that a plot of

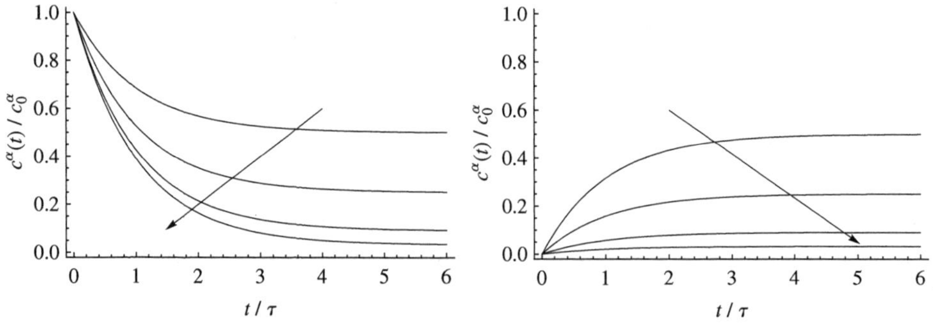

Fig. 4.2.
Time variation of the solute concentration in compartments α and β for $\tau^\beta/\tau^\alpha = 1, 3, 10$, and 30 (increasing in the arrow direction).

$\ln[1 - c^{\beta}(t)/c^{\beta}_{\infty}]$ vs. t must yield a straight line of negative slope $1/\tau \propto A/h$ from which A/h can be determined. Alternatively, the analysis of the initial behaviour leads to

$$V^{\beta} c^{\beta}(t) \approx (A/h) D c_0^{\alpha} t \quad \text{when } t \ll \tau, \tag{4.13}$$

from which A/h could also be determined if V^{β} and D were known.

4.1.2 Lag time in diffusion

The quasi-steady-state assumption used in Section 4.1.1 implies that the concentration profile inside the membrane is linear and can be described by the expression

$$\begin{aligned} c(x,t) &= c^{\alpha}(t)\left(1 - \frac{x}{h}\right) + c^{\beta}(t)\frac{x}{h} \\ &= c_0^{\alpha} e^{-t/\tau}\left(1 - \frac{x}{h}\right) + c_{\infty}^{\alpha}(1 - e^{-t/\tau}), \end{aligned} \tag{4.14}$$

where we have used that $c_{\infty}^{\alpha} = c_{\infty}^{\beta}$. This assumption might seem dubious because the linearity of the profile stems from the absence of time dependence of the diffusion process and, at the same time, the solute concentration in the compartments varies with time. The validity argument that was given in Section 4.1.1 is that the solute concentration in the compartments varies slowly with time when their volumes V^{α} and V^{β} are large compared to the membrane volume $V^{\mathrm{M}} = Ah$. We can now provide additional reasons. The time required for the solute to cross the membrane is the diffusional time $\tau^{\mathrm{M}} \equiv h^2/D = V^{\mathrm{M}} h/AD$. Since the flux density at short times $t \ll \tau$ is $j \approx D c_0^{\alpha}/h$ [see eqn (4.12)], the amount of solute that transfers to compartment β in a time τ^{M} is

$$\Delta n \approx A j \tau^{\mathrm{M}} \approx V^{\mathrm{M}} c_0^{\alpha}. \tag{4.15}$$

When this is compared with the amount of solute initially present in compartment α, $n_0^{\alpha} = V^{\alpha} c_0^{\alpha}$, it becomes clear that the condition $V^{\alpha} \gg V^{\mathrm{M}}$ implies that the changes in the solute concentration in the compartments are negligible for processes taking place in times of the order of τ^{M}, and hence the quasi-steady transport assumption seems reasonable. In fact, the weakest point in the use of this assumption is that the flux density in eqn (4.5) is considered to be established immediately, even though the membrane does not actually have a linear concentration profile at $t = 0$. In any case, a more complete analysis of the validity of the quasi-steady-state assumption seems to be convenient.

The actual time and spatial variation of the solute concentration inside the membrane should be obtained from the combination of Fick's first law and the continuity equation, that is, from Fick's second equation

$$\frac{\partial c}{\partial t} + \frac{\partial j}{\partial x} = \frac{\partial c}{\partial t} - D\frac{\partial^2 c}{\partial x^2} = 0. \tag{4.16}$$

We solve this equation next for a situation in which the concentration in compartments α and β are kept constant (by external circulation) at the values c^b and 0, respectively. At the beginning of the experiment, the solute concentration inside the membrane is $c(x,0) = 0$ and the boundary conditions for eqn (4.16) are then $c(0,t) = c^{\mathrm{b}}$ and $c(h,t) = 0$. Since the boundary conditions establish that the changes in concentration must take place over a spatial scale fixed by the membrane thickness h, a dimensional analysis of eqn (4.16) shows that the time scale for concentration changes is $\tau^{\mathrm{M}} \equiv h^2/D$. That is, the solution of eqn (4.16) must approach the steady-state behaviour

$$c(x,\infty) = c^{\mathrm{b}}\left(1 - \frac{x}{h}\right) \tag{4.17}$$

in a time of the order of τ^{M}.

Equation (4.16) does not have an analytical solution in closed form under these boundary conditions but can be solved by the method of separation of variables,[1] i.e. by writing the solution in the form

$$c(x,t) = c(x,\infty) + X(x)T(t). \tag{4.18}$$

Inserting this ansatz into eqn (4.16) we obtain

$$\frac{1}{DT}\frac{\mathrm{d}T}{\mathrm{d}t} = \frac{1}{X}\frac{\mathrm{d}^2X}{\mathrm{d}x^2}. \tag{4.19}$$

Since the right-hand side of eqn (4.19) is only a function of position and the left-hand side is only a function of time, we conclude that this can only be true if both sides are equal to a constant that we write as $-\lambda^2$. Thus, we can split eqn (4.19) into two ordinary differential equations

$$\frac{1}{DT}\frac{\mathrm{d}T}{\mathrm{d}t} = -\lambda^2, \tag{4.20}$$

$$\frac{1}{X}\frac{\mathrm{d}^2X}{\mathrm{d}x^2} = -\lambda^2, \tag{4.21}$$

whose solutions are

$$T = \mathrm{e}^{-\lambda^2 Dt}, \tag{4.22}$$

$$X = A\sin(\lambda x) + B\cos(\lambda x), \tag{4.23}$$

where we have imposed, without loss of generality, that $T(0) = 1$. The boundary condition $c(0,t) = c^{\mathrm{b}}$ requires that the coefficient B must vanish, and the

[1] Actually, the name of this method is superposition of separated solutions because the solution is not finally written in the form of eqn (4.18) but as a sum of terms with separated dependence in x and t.

boundary condition $c(h, t) = 0$ requires that $\sin(\lambda h) = 0$, which can be satisfied if λ is of the form

$$\lambda_n = \frac{n\pi}{h}, \quad n \text{ integer}. \tag{4.24}$$

Therefore, the complete solution must be

$$c(x,t) = c^{\mathrm{b}}\left(1 - \frac{x}{h}\right) + \sum_{n=1}^{\infty} A_n \sin\frac{n\pi x}{h} \exp\left(-\frac{n^2\pi^2 Dt}{h^2}\right). \tag{4.25}$$

Finally, the orthogonality condition $\int_0^h \sin(n\pi x/h)\sin(m\pi x/h)\mathrm{d}x = (h/2)\delta_{nm}$ and the initial condition $c(x, 0) = 0$ can be used to determine the coefficients A_n as

$$A_n = -\frac{2c^{\mathrm{b}}}{n\pi}. \tag{4.26}$$

The final solution is then

$$c(x,t) = c^{\mathrm{b}}\left(1 - \frac{x}{h}\right) - \frac{2c^{\mathrm{b}}}{\pi} \sum_{n=1}^{\infty} \frac{1}{n} \sin\frac{n\pi x}{h} \exp\left(-\frac{n^2\pi^2 Dt}{h^2}\right), \tag{4.27}$$

and its graphical representation appears in Fig. 4.3.

Figure 4.3 confirms that the time required to establish the linear concentration profile inside the membrane is of the order of τ^{M}. Therefore, it is only after this time that the flux density that enters the compartment β

$$\begin{aligned} j(h,t) &= -D\left(\frac{\partial c}{\partial x}\right)_{x=h} \\ &= \frac{Dc^{\mathrm{b}}}{h}\left[1 + 2\sum_{n=1}^{\infty}(-1)^n \exp\left(-\frac{n^2\pi^2 Dt}{h^2}\right)\right] \end{aligned} \tag{4.28}$$

reaches the steady-state value Dc^{b}/h (Fig. 4.4).

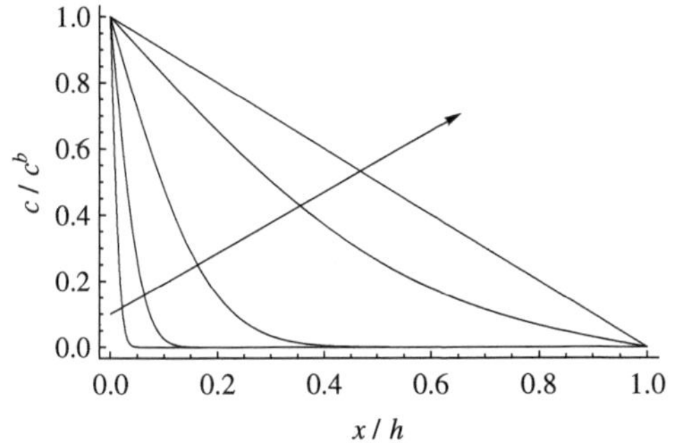

Fig. 4.3.
Concentration profiles inside the membrane at times $t/\tau^M = Dt/h^2 = 10^{-4}, 10^{-3}, 10^{-2}, 10^{-1}$, and 1 (increasing in the arrow direction); the first 100 terms have been computed in the series of eqn (4.27).

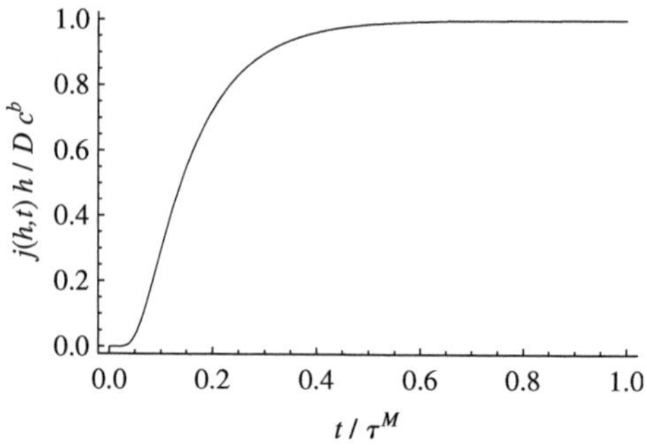

Fig. 4.4.
Time variation of the dimensionless flux density that reaches the compartment β according to eqn (4.28); the first 100 terms have been computed in the series.

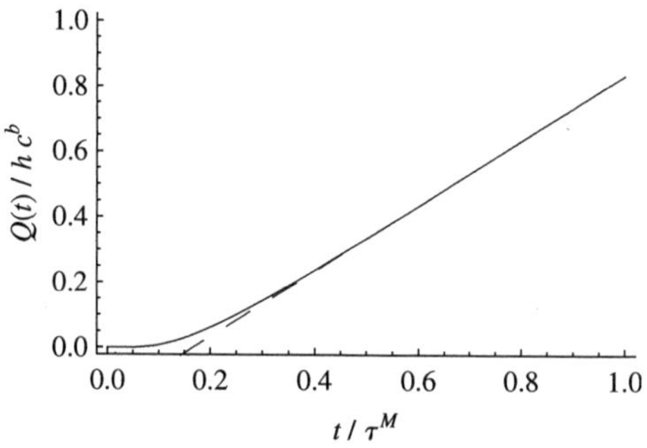

Fig. 4.5.
Cumulative flux density, in hc^b units, against time according to eqn (4.29); the first 100 terms have been computed in the series. The dashed line corresponds to the approximate behaviour at large times described by eqn (4.30).

The flux density represented in Fig. 4.4 is not measured directly. Instead, the concentration in compartment β can be measured as a function of time and used to evaluate, through the relation $Q = V^\beta c^\beta/A$, the cumulative flux

$$Q(t) \equiv \int_0^t j(h,t')\mathrm{d}t'$$

$$= \frac{Dc^{\mathrm{b}}}{h}\left[t - \frac{\tau^{\mathrm{M}}}{6} - \frac{2\tau^{\mathrm{M}}}{\pi^2}\sum_{n=1}^{\infty}\frac{(-1)^n}{n^2}\exp\left(-\frac{n^2\pi^2 Dt}{h^2}\right)\right], \tag{4.29}$$

where we have used that $\sum_{n=1}^{\infty}(-1)^n/n^2 = -\pi^2/12$. Equation (4.29) reduces at large times to

$$Q(t) \approx \frac{Dc^{\mathrm{b}}}{h}(t - \tau_{\mathrm{lag}}) = hc^{\mathrm{b}}\frac{t - \tau_{\mathrm{lag}}}{\tau^{\mathrm{M}}}, \tag{4.30}$$

which is the same as would be expected if the steady-state flux density Dc^{b}/h were established after a lag time $\tau_{\mathrm{lag}} \equiv \tau^{\mathrm{M}}/6 = h^2/6D$. Figure 4.5 shows a comparison of the graphical representations of eqns (4.29) and (4.30).

In conclusion, we have shown that the quasi-steady-state approximation is indeed very good when $\tau^{\mathrm{M}} \ll \tau^\alpha, \tau^\beta$ and that, at times $t \geq \tau^{\mathrm{M}}$, the concentration profile is approximately linear and the flux density is independent of position.

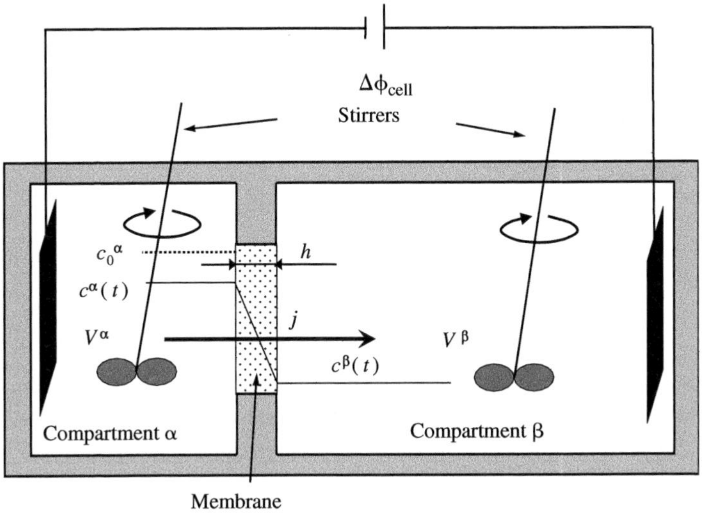

Fig. 4.6.
A cell divided in two compartments by a porous membrane. The compartments are well stirred so that concentration gradients only appear inside the membrane. The solute concentration in the compartments can vary with time due to the mass transport across the membrane. A potential difference is applied between two electrodes in compartments α and β. A fraction $\Delta\phi$ of this potential difference drops inside the membrane and influences the transport of the charged solute.

4.1.3 Iontophoretic enhancement

We consider now the transport of a charged solute in the experimental set-up shown in Fig. 4.6, where a potential difference $\Delta\phi_{\text{cell}}$ is applied between the electrodes. A fraction $\Delta\phi$ of this potential difference drops inside the membrane and influences the transport of the charged solute. We aim here to determine the migrational contribution to the steady-state solute flux density across the membrane.

The solute flux density j is given by the Nernst–Planck equation

$$j = -D\left(\frac{\mathrm{d}c}{\mathrm{d}x} + zcf\frac{\mathrm{d}\phi}{\mathrm{d}x}\right), \tag{4.31}$$

where $f \equiv F/RT$ and z is the solute charge number, and we consider in detail those cases in which migration enhances the solute flux in the positive x direction (i.e. from compartment α to β). This requires that $zf\,\mathrm{d}\phi/\mathrm{d}x < 0$. For the sake of simplicity, we use the Goldman constant-field assumption,[2] $\mathrm{d}\phi/\mathrm{d}x = \Delta\phi/h$. Since the steady-state flux density is independent of position, the Nernst–Planck equation then becomes a first-order, linear, ordinary differential equation that can be integrated over the membrane, extending from $x = 0$, where the solute concentration is $c(0) = c^{\alpha}$, to $x = h$, where it is $c(h) = c^{\beta}$. This leads to an exponential concentration profile inside the membrane

$$c(x) = c^{\beta} + (c^{\alpha} - c^{\beta})\frac{\mathrm{e}^{-zf\Delta\phi} - \mathrm{e}^{-zf\Delta\phi x/h}}{\mathrm{e}^{-zf\Delta\phi} - 1}, \tag{4.32}$$

[2] In principle, the local electric field is an unknown variable that needs to be determined from the local electroneutrality requirement. This field acts on every charged solute and hence is responsible for the coupling of the ionic fluxes. In the Goldman approach, the electric field is not determined and hence the coupling between the ionic flux equations is eliminated, and we can solve the transport equation of any charged solute without considering the other solutes.

and to the Goldman equation for the flux density

$$j = -\frac{D}{h}\frac{zf\,\Delta\phi}{e^{zf\,\Delta\phi}-1}(c^{\beta}e^{zf\,\Delta\phi} - c^{\alpha}) = \frac{D}{h}E(c^{\alpha} - c^{\beta}e^{zf\,\Delta\phi}), \tag{4.33}$$

where

$$E \equiv \frac{h}{\int_0^h e^{zf[\phi(x)-\phi(0)]}dx} = \frac{zf\,\Delta\phi}{e^{zf\,\Delta\phi}-1} \tag{4.34}$$

is the so-called iontophoretic enhancement factor. Figure 4.7 shows the effect of a dimensionless membrane potential on the concentration profile. Obviously, when $\Delta\phi = 0$ there is no migrational contribution and the concentration profile is linear, $c = c^{\alpha} + (c^{\beta} - c^{\alpha})x/h$.

To illustrate the importance of the iontophoretic enhancement factor (Fig. 4.8), it can be mentioned that in drug-delivery problems, compartment α might represent a drug patch and compartment β the body circulation, so that $V^{\beta} \gg V^{\alpha}$ and $c^{\beta} \approx 0$ [see eqn (4.3)]. That is, compartment β behaves as a perfect sink for the solute. The solute flux density is then $j = DEc^{\alpha}/h$, so that E tells us how much the applied electric potential difference (such that $-zf\,\Delta\phi > 0$) enhances the flux of the charged drug across the membrane (i.e. across the human skin). At high potential differences $-zf\,\Delta\phi \gg 1$, $E \approx -zf\,\Delta\phi$ and, since $1/f \approx 25$mV, a potential difference $\Delta\phi = -1$V makes $E \approx 40$ for a singly charged drug, $z = 1$. This explains why iontophoretic drug delivery has received quite a lot of attention during recent decades.

In closing this section, we analyse how the (time-independent) membrane potential $\Delta\phi$ influences the time evolution of the solute concentration in compartments α and β. These vary, respectively, from their initial values c_0^{α} and 0 to the final equilibrium values. Setting $j = 0$ in eqn (4.33), it is deduced that the equilibrium concentrations satisfy the Nernst equation

$$\Delta\phi = \phi^{\beta} - \phi^{\alpha} = \frac{1}{zf}\ln\frac{c_{\infty}^{\alpha}}{c_{\infty}^{\beta}}. \tag{4.35}$$

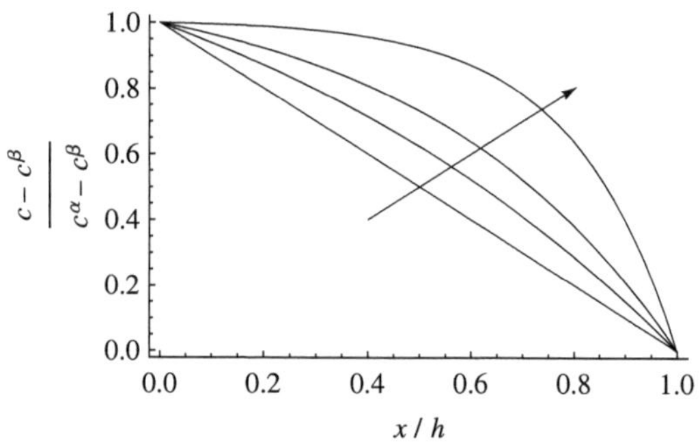

Fig. 4.7.
Dimensionless stationary concentration profiles calculated from eqn (4.32) for $-zf\,\Delta\phi = 0, 1, 2,$ and 5 (increasing in the arrow direction).

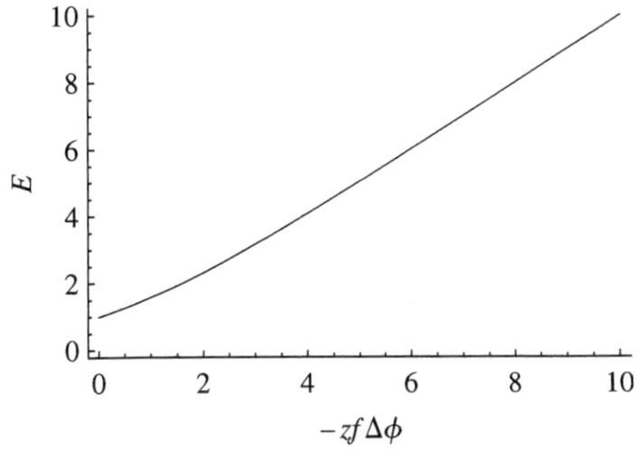

Fig. 4.8.
The iontophoretic enhancement factor $E \equiv zf\,\Delta\phi/(e^{zf\,\Delta\phi}-1)$ equals one in the absence of membrane potential ($\Delta\phi = 0$) and increases linearly with $-zf\,\Delta\phi > 0$ at large membrane potentials.

This equilibrium condition, together with the mass balance $V^{\alpha}(c_0^{\alpha} - c_{\infty}^{\alpha}) = V^{\beta}c_{\infty}^{\beta}$, determines the final concentrations

$$c_{\infty}^{\alpha} = c_0^{\alpha}\frac{V^{\alpha}}{V^{\alpha} + V^{\beta}e^{-zf\,\Delta\phi}} \tag{4.36}$$

$$c_{\infty}^{\beta} = c_0^{\alpha}\frac{V^{\alpha}}{V^{\alpha}e^{zf\,\Delta\phi} + V^{\beta}}. \tag{4.37}$$

From eqns (4.1), (4.3) and (4.33), the time variation of the solute concentration in compartment α is given by

$$\frac{dc^{\alpha}}{dt} + \left(\frac{1}{\tau^{\alpha}} + \frac{e^{zf\,\Delta\phi}}{\tau^{\beta}}\right)Ec^{\alpha} = \frac{c_0^{\alpha}Ee^{zf\,\Delta\phi}}{\tau^{\beta}}, \tag{4.38}$$

where τ^{α} and τ^{β} are the characteristic times defined in eqns (4.8) and (4.9). The solution of this equation with the initial condition $c^{\alpha}(0) = c_0^{\alpha}$ is

$$c^{\alpha}(t) = c_0^{\alpha}e^{-t/\tau(\Delta\phi)} + c_{\infty}^{\alpha}[\,1 - e^{-t/\tau(\Delta\phi)}], \tag{4.39}$$

$$c^{\beta}(t) = c_{\infty}^{\beta}[\,1 - e^{-t/\tau(\Delta\phi)}], \tag{4.40}$$

$$j = \frac{DEc_0^{\alpha}}{h}e^{-t/\tau(\Delta\phi)}, \tag{4.41}$$

where

$$\frac{1}{\tau(\Delta\phi)} \equiv E\left(\frac{1}{\tau^{\alpha}} + \frac{e^{zf\,\Delta\phi}}{\tau^{\beta}}\right). \tag{4.42}$$

The time evolution of the solute concentration in the compartments calculated from eqns (4.39) and (4.40) is presented in Fig. 4.9 for the case $\tau^{\alpha} = \tau^{\beta}$. It is observed that rather modest values of $-zf\,\Delta\phi > 0$ empty compartment α very fast and effectively. That is, the effect of a migrational contribution to the solute flux (from compartment α to β) is to decrease both the characteristic time of the process $\tau(\Delta\phi)$ and the final concentration c_{∞}^{α}, as can be deduced from eqns (4.36) and (4.42).

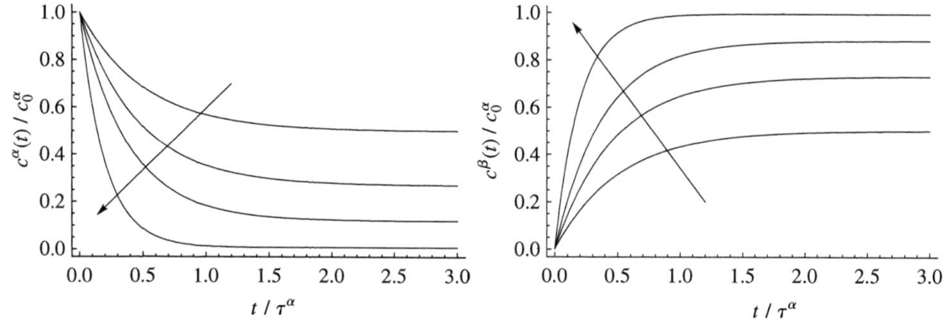

Fig. 4.9.
Time variation of the solute concentration in compartments α and β for $\tau^\alpha = \tau^\beta$ and $-zf\,\Delta\phi = 0, 1, 2,$ and 5 (increasing in the arrow direction).

4.1.4 Lag time in electrodiffusion

Figure 4.9 and eqn (4.42) show that an increase in the membrane potential reduces the characteristic time for the observation of the concentration changes in the compartments. The theoretical approach that has led to Fig. 4.9 is based on the use of the quasi-steady-state assumption and we can expect, from the arguments worked out in Section 4.1.2, that this assumption is valid only when $\tau(\Delta\phi) \gg \tau^{\mathrm{M}}$. In other words, a sufficiently large increase in the membrane potential might have the effect of breaking the validity of the quasi-steady-state assumption. It is therefore convenient to extend the analysis carried out in Section 4.1.2 to take into account the effect of migration.

The actual time and spatial variation of the solute concentration inside the membrane should be obtained from the combination of the Nernst–Planck equation and the continuity equation. Under the Goldman constant-field assumption, this equation is

$$\frac{\partial c}{\partial t} + \frac{\partial j}{\partial x} = \frac{\partial c}{\partial t} - D\left(\frac{\partial^2 c}{\partial x^2} + \frac{zf\,\Delta\phi}{h}\frac{\partial c}{\partial x}\right) = 0. \tag{4.43}$$

We solve this equation next for a situation in which the concentrations in compartments α and β are kept constant (by external circulation) to the values c^α and c^β, respectively. At the beginning of the experiment, the solute concentration inside the membrane is $c(x,0) = c^\beta$ and the boundary conditions for eqn (4.43) are then $c(0,t) = c^\alpha$ and $c(h,t) = c^\beta$. Using an approach similar to that worked out in Section 4.1.2, we write the solution of eqn (4.43) in the form

$$c(x,t) = c^\alpha + (c^\beta - c^\alpha)\frac{\mathrm{e}^{-zf\,\Delta\phi\, x/h} - 1}{\mathrm{e}^{-zf\,\Delta\phi} - 1} + X(x)T(t), \tag{4.44}$$

and split it into the following two linear, ordinary differential equations

$$\frac{1}{DT}\frac{\mathrm{d}T}{\mathrm{d}t} = -\lambda^2, \tag{4.45}$$

$$\frac{1}{X}\left(\frac{\mathrm{d}^2 X}{\mathrm{d}x^2} + \frac{zf\,\Delta\phi}{h}\frac{\mathrm{d}X}{\mathrm{d}x}\right) = -\lambda^2, \tag{4.46}$$

whose solutions are

$$T = e^{-\lambda^2 Dt} \tag{4.47}$$

$$X = e^{-zf\,\Delta\phi\, x/2h}(A \sin rx + B \cos rx), \tag{4.48}$$

where $r \equiv \sqrt{\lambda^2 - (zf\,\Delta\phi/2h)^2}$. The boundary condition $c(0,t) = c^\alpha$ requires that $B = 0$ and the boundary condition $c(h,t) = c^\beta$ requires that $\sin rh = 0$, which can be satisfied if r is of the form $r_n = n\pi/h$ with n integer. Therefore, the complete solution is

$$c(x,t) = c^\alpha + (c^\beta - c^\alpha)\frac{e^{-zf\,\Delta\phi\, x/h} - 1}{e^{-zf\,\Delta\phi} - 1} + e^{-zf\,\Delta\phi\, x/2h}$$
$$\times \sum_{n=1}^{\infty} A_n \sin\frac{n\pi x}{h} \exp\left\{-\frac{Dt}{h^2}[n^2\pi^2 + (zf\,\Delta\phi/2)^2]\right\}. \tag{4.49}$$

Multiplying both sides of eqn (4.49) by $e^{zf\,\Delta\phi\, x/2h}\sin(m\pi x/h)$, integrating over the membrane thickness and making use of the initial condition $c(x,0) = c^\beta$ the coefficients A_n are determined as

$$A_n = 2(c^\beta - c^\alpha)\frac{n\pi}{n^2\pi^2 + (zf\,\Delta\phi/2)^2}, \tag{4.50}$$

and the final solution is then

$$c(x,t) = c^\alpha + (c^\beta - c^\alpha)\frac{e^{-zf\,\Delta\phi\, x/h} - 1}{e^{-zf\,\Delta\phi} - 1} + 2(c^\beta - c^\alpha)e^{-zf\,\Delta\phi\, x/2h}$$
$$\times \sum_{n=1}^{\infty} \frac{n\pi}{n^2\pi^2 + (zf\,\Delta\phi/2)^2} \sin\frac{n\pi x}{h}$$
$$\times \exp\left\{-\frac{Dt}{h^2}[n^2\pi^2 + (zf\,\Delta\phi/2)^2]\right\}. \tag{4.51}$$

Figure 4.10 shows the graphical representation of eqn (4.51) and it is observed that migration modifies the final concentration profile but the time required to achieve it is still of the order of $\tau^M = h^2/D$, like in the absence of migration.

The flux density that enters the compartment β is

$$j(h,t) = -D\left[\left(\frac{\partial c}{\partial x}\right)_{x=h} + c^\beta \frac{zf\,\Delta\phi}{h}\right] = -\frac{D}{h}E(c^\beta e^{zf\,\Delta\phi} - c^\alpha)$$
$$-2\frac{D}{h}(c^\beta - c^\alpha)e^{-zf\,\Delta\phi/2}\sum_{n=1}^{\infty}\frac{(-1)^n n^2\pi^2}{n^2\pi^2 + (zf\,\Delta\phi/2)^2}$$
$$\times \exp\left\{-\frac{t}{\tau^M}[n^2\pi^2 + (zf\,\Delta\phi/2)^2]\right\} \tag{4.52}$$

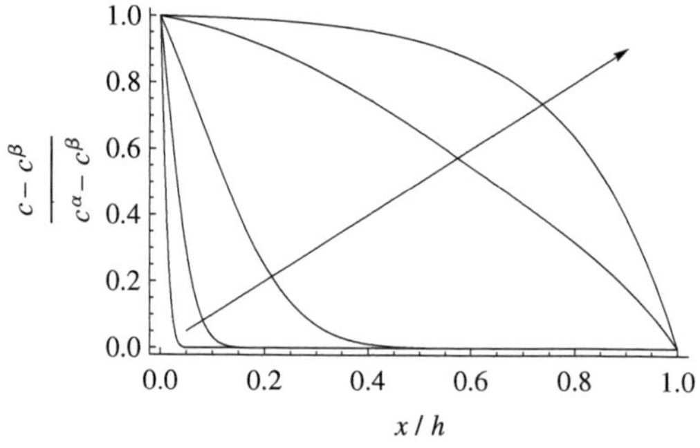

Fig. 4.10. Concentration profiles inside the membrane according to eqn (4.51) for $-zf\,\Delta\phi = 5$ and times $t/\tau^M = Dt/h^2 = 10^{-4}, 10^{-3}, 10^{-2}, 10^{-1}$, and 1 (increasing in the arrow direction); the first 100 terms have been computed in the series.

and the cumulative flux is

$$Q(t) \equiv \int_0^t j(h,t')\mathrm{d}t' = -hE(c^\beta \mathrm{e}^{zf\,\Delta\phi} - c^\alpha)\,\frac{t}{\tau^{\mathrm{M}}}$$

$$+ 2h(c^\beta - c^\alpha)\mathrm{e}^{-zf\,\Delta\phi/2}\sum_{n=1}^{\infty}\frac{(-1)^n n^2\pi^2}{[n^2\pi^2 + (zf\,\Delta\phi/2)^2]^2}$$

$$\times\left(\exp\left\{-\frac{t}{\tau^{\mathrm{M}}}[n^2\pi^2 + (zf\,\Delta\phi/2)^2]\right\} - 1\right). \tag{4.53}$$

At large times, the cumulative flux can be approximated by

$$Q(t) = hE(c^\alpha - c^\beta \mathrm{e}^{zf\,\Delta\phi})\,\frac{t - \tau_{\mathrm{lag}}}{\tau^{\mathrm{M}}}, \tag{4.54}$$

where

$$\tau_{\mathrm{lag}} \equiv \tau^{\mathrm{M}}\frac{c^\alpha - c^\beta}{c^\alpha - c^\beta \mathrm{e}^{zf\,\Delta\phi}}\,\frac{zf\,\Delta\phi\coth(zf\,\Delta\phi/2) - 2}{(zf\,\Delta\phi)^2}, \tag{4.55}$$

and we have used the result [1, 2]

$$\sum_{n=1}^{\infty}\frac{(-1)^n n^2\pi^2}{[n^2\pi^2 + (zf\,\Delta\phi/2)^2]^2} = -\frac{1}{4}\,\frac{zf\,\Delta\phi\coth(zf\,\Delta\phi/2) - 2}{zf\,\Delta\phi\sinh(zf\,\Delta\phi/2)}. \tag{4.56}$$

The lag time in eqn (4.55) takes its maximum value $\tau^{\mathrm{M}}/6$ when the membrane potential vanishes and decreases with increasing $-zf\,\Delta\phi > 0$ so that at large potential differences $\tau_{\mathrm{lag}} \approx \tau^{\mathrm{M}}/\,|zf\,\Delta\phi|$. Figure 4.11 shows the time evolution of the cumulative flux and its approximate behaviour at large times for the case $c^\beta = 0$.

In conclusion, we have analysed the effect of migration (in cases where it has the same direction as diffusion) and shown that it increases the flux density by an amount given by the iontophoretic enhancement factor. In this case of flux enhancement, migration also reduces both the lag time and the

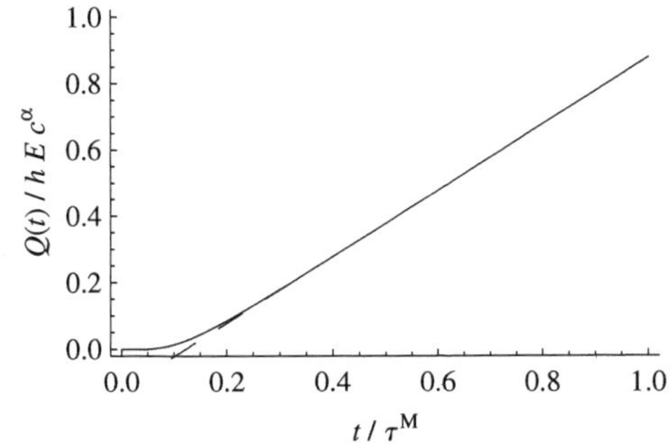

Fig. 4.11.
Cumulative electrodiffusive flux density, in hEc^{α} units, against time for $-zf\,\Delta\phi = 5$ and $c^{\beta} = 0$ according to eqn (4.53). The first 100 terms have been computed in the series. The dashed line corresponds to the approximate behaviour at large times described by eqn (4.54). The lag time is $\tau_{\text{lag}} \approx 0.123\tau^{\text{M}}$ for $-zf\,\Delta\phi = 5$ while in the absence of migration it is $\tau_{\text{lag}} \approx \tau^{\text{M}}/6$.

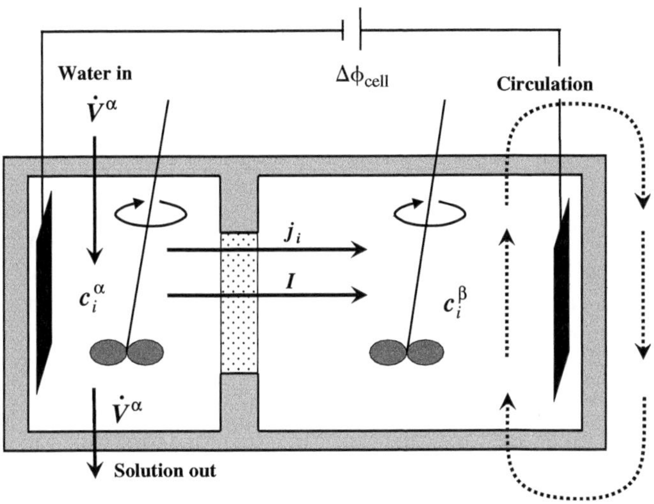

Fig. 4.12.
Schematic drawing of the cell used to study the transport of species i through a porous membrane with circulation of the solution in compartment β. A volume flow rate $\dot{V}^{\alpha}$ of pure water is pumped into compartment α and the same flow rate is taken as an outflow, so that no convection takes place across the membrane. The arrows for the flux density of species i and the electric current density indicate the positive direction, not the actual flow direction.

characteristic time for the changes in the compartment concentrations. The conditions of validity of the quasi-steady transport approximation are not significantly modified (at the potential differences considered here).

4.1.5 Circulation of compartments

In Section 4.1.3 we have studied the electrodiffusion of a charged solute across a neutral porous membrane under quasi-steady-state conditions. We consider now a similar process in which the solutions in compartments α and β are changed as shown schematically in Fig. 4.12. The compartments are ideally mixed and have uniform concentrations of the same binary electrolyte. A volume flow rate $\dot{V}^{\alpha}$ of pure water is pumped into compartment α and the same flow rate is taken as an outflow, so that no convection takes place across the membrane. The electrolyte concentration in this outflow is c_{12}^{α}. The volume of compartment β is circulated and, after a transient period, the system reaches a state with approximately constant concentrations c_{12}^{α} and c_{12}^{β}. This is so in spite of the fact that we are pumping water and taking solution out from compartment α, because the volume of compartment β is very large, $V^{\beta} \gg V^{\alpha}$, and its solute concentration is not affected in practice by the flow through the membrane towards compartment α. In other words, the system can be considered to be in a steady state.

The mass balance for an ionic species $i (i = 1, 2)$ in compartment α

$$V^\alpha \frac{dc_i^\alpha}{dt} = -j_i A - \dot{V}^\alpha c_i^\alpha \tag{4.57}$$

reduces at steady state to

$$j_i = -\frac{\dot{V}^\alpha c_i^\alpha}{A} < 0, \tag{4.58}$$

which allows us to evaluate the molar flux density j_i across the membrane. In the diffusion–conduction approach, this flux density is given by

$$j_i = -D_{12}\frac{dc_i}{dx} + \frac{t_i I}{z_i F} = -D_{12}\frac{\Delta c_i}{h} + \frac{t_i I}{z_i F}, \tag{4.59}$$

where $\Delta c_i \equiv c_i^\beta - c_i^\alpha$ and we have made use of the fact that the concentration profile is linear because j_i is a constant under steady-state conditions. Note that D_{12} and t_i take approximately the same value in the porous membrane as in the external solutions.

As mentioned in Section 4.1.1, the membrane constant A/h is an important membrane parameter that has to be determined experimentally. In the cell of Fig. 4.12, the membrane constant can be determined by analysing the outflow concentrations as

$$\frac{A}{h} = \frac{c_i^\alpha \dot{V}^\alpha + t_i IA/z_i F}{D_{12}(c_i^\beta - c_i^\alpha)}, \tag{4.60}$$

where eqns (4.58) and (4.59) have been used. It must be emphasized that the membrane constant cannot be evaluated from the thickness and porosity provided by the manufacturer because, among other effects, the membrane swells in solution. In addition, the effect of the assumptions introduced in the theoretical modelling or possible inaccuracies of the experimental set-up are also hidden in the value of the membrane constant.

4.1.6 Convective electrophoresis

Continuing with the same experimental set-up depicted in Fig. 4.12, let us now consider that the outflow rate from compartment α, $\dot{V}_{\text{out}}^\alpha$, is different from the inflow rate $\dot{V}_{\text{in}}^\alpha$ of pure water. The mass balance for the whole cell shows that a convective motion is now created across the membrane (Fig. 4.13) and that the solution velocity is given by

$$v = \frac{\dot{V}_{\text{in}}^\alpha - \dot{V}_{\text{out}}^\alpha}{A}. \tag{4.61}$$

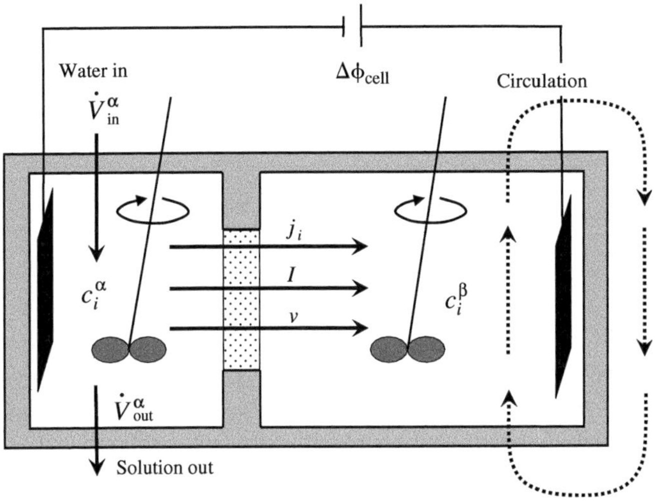

Fig. 4.13.
Schematic drawing of the cell used to study the transport of species i through a porous membrane in the presence of simultaneous diffusion, conduction and convection. A volume flow rate $\dot{V}^{\alpha}_{in}$ of pure water is pumped into compartment α and a different flow rate $\dot{V}^{\alpha}_{out}$ is taken as an outflow. The solution in compartment β is circulated. The arrows for the flux density of species i, the convective velocity and the electric current density indicate the positive direction, not the actual flow direction.

The importance of this convective motion in relation to the electrolyte diffusion across the membrane is characterized by the dimensionless Peclet number

$$\mathrm{Pe} \equiv \frac{vh}{D_{12}}, \tag{4.62}$$

so that the influence of convection is expected to be significant when $\mathrm{Pe} \gg 1$.

Adding the convection term to the diffusion–conduction equation, the flux density of a solute species i (in the membrane-fixed reference system) is now given by

$$j_i = c_i v - D_{12}\frac{\mathrm{d}c_i}{\mathrm{d}x} + \frac{t_i I}{z_i F}, \tag{4.63}$$

or

$$\frac{\mathrm{d}c_i}{\mathrm{d}x} = \frac{\mathrm{Pe}}{h}(c_i - C_i), \tag{4.64}$$

where

$$C_i \equiv \frac{1}{v}\left(j_i - \frac{t_i I}{z_i F}\right). \tag{4.65}$$

The solution of this equation under the boundary conditions $c_i(0) = c_i^{\alpha}$ and $c_i(h) = c_i^{\beta}$ is

$$c_i(x) = C_i + (c_i^{\alpha} - C_i)\,\mathrm{e}^{\mathrm{Pe}\,x/h} = c_i^{\beta} + (c_i^{\alpha} - c_i^{\beta})\frac{\mathrm{e}^{\mathrm{Pe}\,x/h} - \mathrm{e}^{\mathrm{Pe}}}{1 - \mathrm{e}^{\mathrm{Pe}}}. \tag{4.66}$$

It is interesting to observe that the concentration profile is only determined by the Peclet number Pe and the external concentrations, although the external concentration c_i^{α} is determined by the electric current density as explained

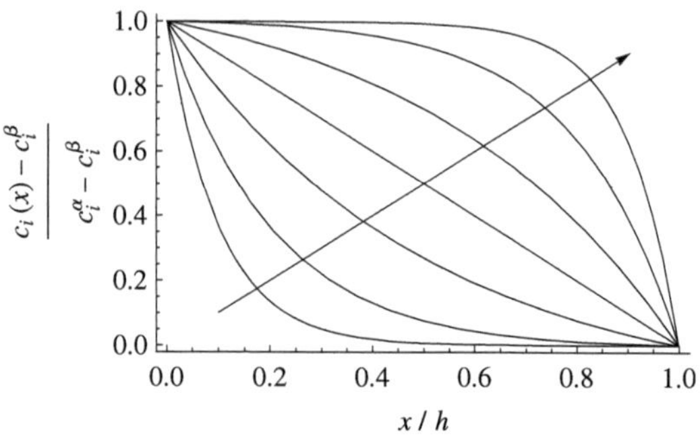

Fig. 4.14.
Concentration profiles in convective diffusion calculated from eqn (4.66). The curves correspond to Peclet numbers: Pe $= -10, -5, -2, 0, 2, 5,$ and 10 (increasing in the arrow direction).

below. Figure 4.14 shows the concentration profiles corresponding to different values of Pe. In the absence of convection, Pe = 0, the concentration profile is linear and the concentration gradient inside the membrane is $(c_i^\alpha - c_i^\beta)/h$. When Pe is positive and large, the concentration inside the membrane is equal to that in compartment α, except in the close vicinity of the boundary with compartment β where the concentration drops sharply to c_i^β and the concentration gradient is $-\text{Pe}(c_i^\alpha - c_i^\beta)/h$. When Pe is negative and large in magnitude, the solution in compartment β is pushed towards the membrane phase, the concentration inside the membrane is equal to c_i^β except in the close vicinity of the boundary with compartment α where the concentration gradient is $-\text{Pe}(c_i^\alpha - c_i^\beta)/h$.

The steady-state flux density is a constant that can be evaluated from the mass balance in compartment α as

$$j_i = -\frac{\dot{V}_{\text{out}}^\alpha c_i^\alpha}{A} < 0, \tag{4.67}$$

and related to the solute concentration in compartments α and β by

$$j_i = \frac{D_{12}}{h} \frac{\text{Pe}}{1 - \text{e}^{-\text{Pe}}} (c_i^\alpha - c_i^\beta \text{e}^{-\text{Pe}}) + \frac{t_i I}{z_i F}. \tag{4.68}$$

Curiously, the factor $\text{Pe}/(1 - \text{e}^{-\text{Pe}})$ in eqn (4.68) has a close mathematical similarity to the iontophoretic enhancement factor introduced in Section 4.1.3. This is due to the fact that both eqn (4.31) under the Goldman constant-field assumption and eqn (4.63) are similar first-order, linear, ordinary differential equations. The important fact, however, is that eqns (4.66) and (4.68) establish a relation between the solute concentration in the compartments, the convective flow velocity, and the electric current density. Therefore, for given experimental values of I, v and c_i^β, these equations allow us to evaluate the concentration c_i^α that we should expect to measure in compartment α when the steady state is reached.

An interesting particular case of the above situation is that with $\dot{V}_{\text{out}}^\alpha = 0$. The transport mechanisms inside the membrane are then coupled in such a way that the flux density of species i given by eqn (4.68) is zero. This has been studied here for the case of a binary solution, but it can also be achieved in multi-ionic solutions.

4.1.7 Liquid-junction potential

We know from Section 2.3.3 that the transport of a charged species is linked to the transport of other charged species due to the long-range electrostatic forces. This coupling implies that the transport of ionic species often gives rise to internal electric fields that must be evaluated from the solution of the transport equations. In this section we evaluate the electric potential drop $\Delta\phi_{\text{dif}} = \phi^{\beta} - \phi^{\alpha}$ that is established between the membrane boundaries due to the transport of ionic species under open-circuit conditions (i.e. in the absence of electric current passing through the membrane).[3] We do it first for the case of a binary electrolyte, and then for electrolyte mixtures.

The local electroneutrality assumption accounts for the electrostatic coupling between ionic species. In the simple case of a binary electrolyte this is

$$z_1 c_1 + z_2 c_2 = 0, \tag{4.69}$$

and the diffusion potential gradient

$$-f\frac{\mathrm{d}\phi_{\text{dif}}}{\mathrm{d}x} = \frac{t_1}{z_1}\frac{\mathrm{d}\ln c_1}{\mathrm{d}x} + \frac{t_2}{z_2}\frac{\mathrm{d}\ln c_2}{\mathrm{d}x} \tag{4.70}$$

can then be easily integrated over the membrane to give

$$f\,\Delta\phi_{\text{dif}} = -\left(\frac{t_1}{z_1} + \frac{t_2}{z_2}\right)\ln\frac{c_{12}^{\beta}}{c_{12}^{\alpha}} = \frac{D_2 - D_1}{z_1 D_1 - z_2 D_2}\ln\frac{c_{12}^{\beta}}{c_{12}^{\alpha}}. \tag{4.71}$$

This implies that $\Delta\phi_{\text{dif}} = \phi^{\beta} - \phi^{\alpha}$ has the same sign as $\Delta c_{12} = c_{12}^{\beta} - c_{12}^{\alpha}$ when the anion has a larger diffusion coefficient than the cation, and vice versa. That is, the diffusion potential gives a migrational contribution to the ionic flux densities that enhances the diffusive contribution of the ion with lower mobility and opposes the diffusive contribution of the ion with higher mobility.

In multi-ionic systems the situation is much more complicated because the ionic transport numbers vary with position and the expression for the diffusion potential gradient

$$-f\frac{\mathrm{d}\phi_{\text{dif}}}{\mathrm{d}x} = \sum_i \frac{t_i}{z_i}\frac{\mathrm{d}\ln c_i}{\mathrm{d}x} = \frac{\sum_i z_i D_i (\mathrm{d}c_i/\mathrm{d}x)}{\sum_j z_j^2 D_j c_j} \tag{4.72}$$

cannot be integrated analytically because the concentration profiles are not known in general. In the case of mixtures of symmetric z:z electrolytes, the transport equations can be solved and the diffusion potential obtained from the solution of two algebraic equations. The integration procedure is explained in detail in Section 4.3.7 and we outline here the most important steps (for the

[3] The evaluation of the membrane potential in the presence of an electric current can be done as a particular case of that considered in Section 4.3.7 when the fixed-charge concentration vanishes.

case of a neutral membrane). Using the boundary conditions $c_i(0) = c_i^\alpha$ and $c_i(h) = c_i^\beta$, the diffusion potential is

$$\Delta\phi_{\mathrm{dif}} = \frac{\Gamma}{f} \ln \frac{c_{\mathrm{T}}^\beta}{c_{\mathrm{T}}^\alpha}, \tag{4.73}$$

and the flux densities of cations and anions are

$$\frac{j_i}{D_i} = -(1 + z\Gamma)\frac{c_{\mathrm{T}}^\beta - c_{\mathrm{T}}^\alpha}{h} \frac{c_i^\beta \mathrm{e}^{zf\Delta\phi_{\mathrm{dif}}} - c_i^\alpha}{c_{\mathrm{T}}^\beta \mathrm{e}^{zf\Delta\phi_{\mathrm{dif}}} - c_{\mathrm{T}}^\alpha}, \quad \text{if } z_i = z, \tag{4.74}$$

$$\frac{j_i}{D_i} = -(1 - z\Gamma)\frac{c_{\mathrm{T}}^\beta - c_{\mathrm{T}}^\alpha}{h} \frac{c_i^\beta \mathrm{e}^{-zf\Delta\phi_{\mathrm{dif}}} - c_i^\alpha}{c_{\mathrm{T}}^\beta \mathrm{e}^{-zf\Delta\phi_{\mathrm{dif}}} - c_{\mathrm{T}}^\alpha}, \quad \text{if } z_i = -z, \tag{4.75}$$

where $c_{\mathrm{T}} \equiv \sum_i c_i$ is the total ionic concentration and Γ is an unknown constant. The open-circuit condition, $I = F\sum_i z_i j_i = 0$, requires that

$$\frac{\mathrm{e}^{zf\Delta\phi_{\mathrm{dif}}} \sum\limits_{+} D_i c_i^\beta - \sum\limits_{+} D_i c_i^\alpha}{c_{\mathrm{T}}^\beta \mathrm{e}^{zf\Delta\phi_{\mathrm{dif}}} - c_{\mathrm{T}}^\alpha} = \frac{1 - z\Gamma}{1 + z\Gamma} \frac{\mathrm{e}^{-zf\Delta\phi_{\mathrm{dif}}} \sum\limits_{-} D_i c_i^\beta - \sum\limits_{-} D_i c_i^\alpha}{c_{\mathrm{T}}^\beta \mathrm{e}^{-zf\Delta\phi_{\mathrm{dif}}} - c_{\mathrm{T}}^\alpha}, \tag{4.76}$$

and further elimination of the diffusion potential using eqn (4.73) leads to

$$\frac{(c_{\mathrm{T}}^\beta/c_{\mathrm{T}}^\alpha)^{z\Gamma} \sum\limits_{+} D_i c_i^\beta - \sum\limits_{+} D_i c_i^\alpha}{(c_{\mathrm{T}}^\beta/c_{\mathrm{T}}^\alpha)^{-z\Gamma} \sum\limits_{-} D_i c_i^\beta - \sum\limits_{-} D_i c_i^\alpha} = \frac{1 - z\Gamma}{1 + z\Gamma} \frac{(c_{\mathrm{T}}^\beta/c_{\mathrm{T}}^\alpha)^{1+z\Gamma} - 1}{(c_{\mathrm{T}}^\beta/c_{\mathrm{T}}^\alpha)^{1-z\Gamma} - 1}. \tag{4.77}$$

This transcendental equation must be solved numerically to obtain the value of Γ, and then the diffusion potential can be evaluated from eqn (4.73). The $+$ and $-$ signs under the sums in eqns (4.76) and (4.77) indicate that they are restricted to cations and anions, respectively.

Although the above procedure is not too complicated, there are two alternative methods that are more popular because they allow for a much simpler approximate evaluation of the diffusion potential. The first one is Henderson's method [3, 4], which assumes that the concentration profiles of all ionic species have the same functional form

$$c_i = c_i^\alpha + \Delta c_i \delta(x/h), \tag{4.78}$$

where $\Delta c_i \equiv c_i^\beta - c_i^\alpha$ and $\delta(x/h)$ is an undetermined function that satisfies $\delta(0) = 0$ and $\delta(1) = 1$. Integration of eqn (4.72) gives then

$$f\Delta\phi_{\mathrm{dif}} \approx -\frac{\sum_i z_i D_i \Delta c_i}{\sum_j z_j^2 D_j \Delta c_j} \ln \frac{\sum_k z_k^2 D_k c_k^\beta}{\sum_l z_l^2 D_l c_l^\alpha}. \tag{4.79}$$

In the case of mixtures of $z : z$ symmetric electrolytes, Henderson's equation reduces to

$$zf\,\Delta\phi_{\text{dif}} \approx -\frac{\sum_{+} D_i \Delta c_i - \sum_{-} D_i \Delta c_i}{\sum_j D_j \Delta c_j} \ln \frac{\sum_k D_k c_k^\beta}{\sum_l D_l c_l^\alpha}. \tag{4.80}$$

In the case of a binary electrolyte, eqn (4.79) is exact and reduces to eqn (4.71).

The second method is based on the Goldman constant-field assumption. This amounts to considering that the electric potential gradient in the Nernst–Planck flux equation is independent of position and can be written as $d\phi/dx = \Delta\phi_{\text{dif}}/h$ (in the case $I = 0$). The integration of the flux equation for species i then leads to the Goldman flux equation

$$\frac{j_i}{D_i} = -\frac{z_i f\,\Delta\phi_{\text{dif}}}{e^{z_i f\,\Delta\phi_{\text{dif}}} - 1} \frac{c_i^\beta e^{z_i f\,\Delta\phi_{\text{dif}}} - c_i^\alpha}{h}. \tag{4.81}$$

The use of eqn (4.81) and the open-circuit condition, $I = F\sum_i z_i j_i = 0$, leads in general to an algebraic equation in $e^{f\,\Delta\phi_{\text{dif}}}$ that allows for the determination of $\Delta\phi_{\text{dif}}$. In the case of mixtures of symmetric, $z : z$ electrolytes, this determination is very simple because the open-circuit condition requires

$$\frac{e^{zf\,\Delta\phi_{\text{dif}}} \sum_{+} D_i c_i^\beta - \sum_{+} D_i c_i^\alpha}{e^{zf\,\Delta\phi_{\text{dif}}} - 1} = -\frac{e^{-zf\,\Delta\phi_{\text{dif}}} \sum_{-} D_i c_i^\beta - \sum_{-} D_i c_i^\alpha}{e^{-zf\,\Delta\phi_{\text{dif}}} - 1} \tag{4.82}$$

and multiplication of both the numerator and the denominator of the fraction in the right-hand side of the above equation by $e^{zf\,\Delta\phi_{\text{dif}}}$ leads immediately to the Goldman equation for the diffusion potential

$$zf\,\Delta\phi_{\text{dif}} = \ln \frac{\sum_{+} D_i c_i^\alpha + \sum_{-} D_i c_i^\beta}{\sum_{+} D_i c_i^\beta + \sum_{-} D_i c_i^\alpha}. \tag{4.83}$$

The $+$ and $-$ signs under the sums indicate, once again, that they are restricted to cations and anions, respectively.

To compare these three alternative ways of evaluating the diffusion potential in neutral membranes, we consider a solution formed by mixing two 1:1 electrolytes with a common anion (which is denoted as species 3). Equation (4.77) then reduces to

$$\frac{(c_3^\beta/c_3^\alpha)^{1+\Gamma}(r_{13}^\beta + r_{23}^\beta) - (r_{13}^\alpha + r_{23}^\alpha)}{(c_3^\beta/c_3^\alpha)^{1+\Gamma} - 1} = \frac{1-\Gamma}{1+\Gamma}, \tag{4.84}$$

where $r_{13}^{\alpha} \equiv D_1 c_1^{\alpha}/D_3 c_3^{\alpha}, r_{13}^{\beta} \equiv D_1 c_1^{\beta}/D_3 c_3^{\beta}, r_{23}^{\alpha} \equiv D_2 c_2^{\alpha}/D_3 c_3^{\alpha}$, and $r_{23}^{\beta} \equiv D_2 c_2^{\beta}/D_3 c_3^{\beta}$. Henderson's equation reduces in this case to

$$f\,\Delta\phi_{\text{dif}} = -\frac{(r_{13}^{\beta} + r_{23}^{\beta} - 1)(c_3^{\beta}/c_3^{\alpha}) - (r_{13}^{\alpha} + r_{23}^{\alpha} - 1)}{(r_{13}^{\beta} + r_{23}^{\beta} + 1)(c_3^{\beta}/c_3^{\alpha}) - (r_{13}^{\alpha} + r_{23}^{\alpha} + 1)} \times \ln\left(\frac{c_3^{\beta}}{c_3^{\alpha}}\frac{r_{13}^{\beta} + r_{23}^{\beta} + 1}{r_{13}^{\alpha} + r_{23}^{\alpha} + 1}\right). \tag{4.85}$$

And Goldman's equation becomes

$$f\,\Delta\phi_{\text{dif}} = -\ln\frac{(r_{13}^{\beta} + r_{23}^{\beta})(c_3^{\beta}/c_3^{\alpha}) + 1}{r_{13}^{\alpha} + r_{23}^{\alpha} + (c_3^{\beta}/c_3^{\alpha})}. \tag{4.86}$$

Finally, it is noted that the exact diffusion potential can be formally written as[4]

$$f\,\Delta\phi_{\text{dif}} = -\frac{(r_{13}^{\beta} + r_{23}^{\beta} - 1)(c_3^{\beta}/c_3^{\alpha})^{1+\Gamma} - (r_{13}^{\alpha} + r_{23}^{\alpha} - 1)}{(r_{13}^{\beta} + r_{23}^{\beta} + 1)(c_3^{\beta}/c_3^{\alpha})^{1+\Gamma} - (r_{13}^{\alpha} + r_{23}^{\alpha} + 1)} \ln\frac{c_3^{\beta}}{c_3^{\alpha}}, \tag{4.87}$$

which resembles eqn (4.85) and, hence, somehow evidences that the Henderson approximation is more accurate than Goldman's one. However, eqn (4.87) is not really useful because it contains the unknown Γ, which must be evaluated from eqn (4.84), just like the much simpler eqn (4.73) from which it has been derived.

Figure 4.15 shows the graphical representation of eqns (4.85)–(4.87) for a case with $0.2\,D_1 = D_2 = D_3$ and different values of the ratios r_{13}^{α} and r_{13}^{β}; the values of r_{23}^{α} and r_{23}^{β} are then determined from the electroneutrality condition $c_1 + c_2 = c_3$. It is concluded from these plots that Henderson's approximation is a much better approximation than Goldman's one. The former is exact in the case of binary solution (i.e. when either r_{13}^{α} or r_{13}^{β} vanishes). It is also exact when the two electrolytes are in the same ratio in compartments α and β, i.e. when $r_{13}^{\alpha} = r_{13}^{\beta}$, and both eqn (4.85) and eqn (4.87) simplify then to

$$f\,\Delta\phi_{\text{dif}} = -\frac{r_{13}^{\beta} + r_{23}^{\beta} - 1}{r_{13}^{\beta} + r_{23}^{\beta} + 1} \ln\frac{c_3^{\beta}}{c_3^{\alpha}}. \tag{4.88}$$

On the contrary, Goldman's equation only provides a reasonably good approximation to the exact diffusion potential when $c_3^{\beta} \approx c_3^{\alpha}$. In fact, it can be proved (see Section 4.3.7) that the electric field is indeed constant when

[4] Since Γ is given by eqn (4.84) and this is a transcendental equation, it is not possible to solve completely for Γ. To derive eqn (4.87) we must obtain a formal expression for Γ and substitute it in eqn (4.73). We have considered that the left-hand side of eqn (4.84) is known (even though it contains Γ) and that Γ in the right-hand side is the unknown we must solve for.

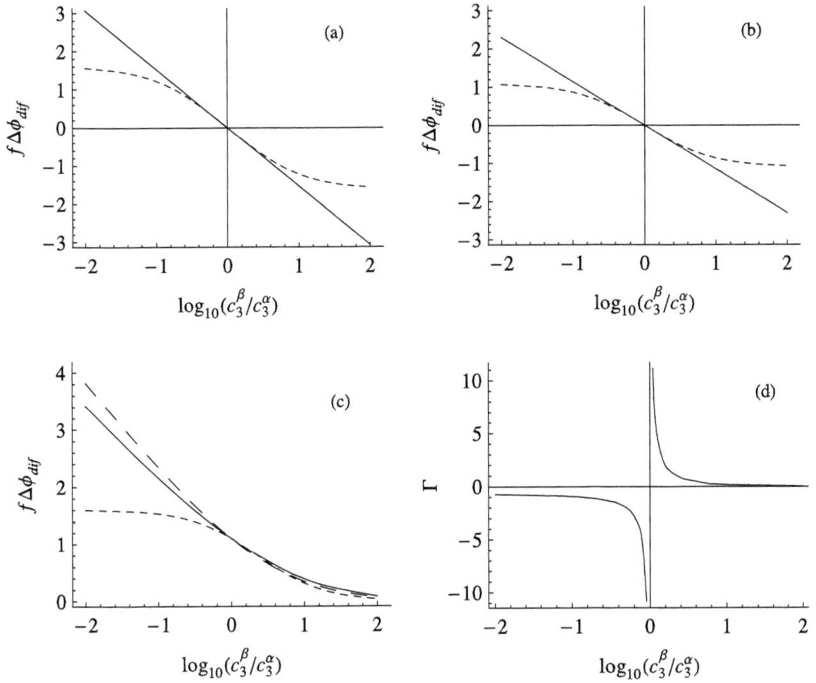

Fig. 4.15.
Diffusion potential (in *RT/F* units) vs. $\log_{10}(c_3^\beta/c_3^\alpha)$ in a ternary system formed by mixing two 1:1 electrolytes with a common anion (species 3). The diffusion coefficients satisfy the relation $0.2\,D_1 = D_2 = D_3$ and the concentrations in compartments α and β are: (a) $c_1^\alpha/c_3^\alpha = 1, c_1^\beta/c_3^\beta = 1$, (b) $c_1^\alpha/c_3^\alpha = 0.5, c_1^\beta/c_3^\beta = 0.5$, and (c) $c_1^\alpha/c_3^\alpha = 1, c_1^\beta/c_3^\beta = 0$. That is, plot (a) corresponds to a binary system, plot (b) to a mixture of electrolytes at equal concentrations, and plot (c) to bi-ionic conditions where the compartments contain different electrolytes. The solid lines represent exact results, the long dashed lines have been obtained from the Henderson approximation, and the short dashed lines from the Goldman approximation. Henderson's results are exact in cases (a) and (b), and hence the solid and long-dashed lines overlap. Regardless of the value of c_3^β/c_3^α, $\Gamma = (D_3 - D_1)/(D_1 + D_3) = -2/3$ and $\Gamma = (2D_3 - D_1 - D_2)/(2D_3 + D_1 + D_2) = -1/2$ in cases (a) and (b), respectively. In case (c), Γ varies as shown in plot (d).

$c_T^\beta = c_T^\alpha$,[5] and therefore the accuracy of Goldman's approximation decreases as the difference between c_T^β and c_T^α increases.

Plots (c) and (d) in Fig. 4.15 correspond to the case $c_{23}^\alpha = c_{13}^\beta = 0$ in which compartment α contains only the electrolyte 13 and compartment β the electrolyte 23. In contrast to plots (a) and (b), the diffusion potential is different from zero when $c_3^\alpha = c_3^\beta$. This is the so-called bi-ionic potential and amounts to

$$f\,\Delta\phi_{\text{dif}} = \ln\frac{D_1 + D_3}{D_2 + D_3}. \tag{4.89}$$

[5] It is also required that there are only two ion classes, i.e. the ions have either charge number z_1 or z_2.

When c_3^β/c_3^α is much larger than unity, the diffusion potential goes to zero because electrolyte 23 is then dominant and it creates no diffusion potential (since we have taken $D_2 = D_3$).[6] This explains the somewhat fortuitous accuracy of the Goldman approximation in this range. When c_3^β/c_3^α is much smaller than unity, the electrolyte 13 is then dominant and it creates a significant diffusion potential because we have taken $D_1 = 5D_3$. This potential is approximately given by

$$f\,\Delta\phi_{\text{dif}} \approx -\frac{D_1 - D_3}{D_1 + D_3} \ln \frac{c_3^\beta}{c_3^\alpha}, \quad c_3^\beta/c_3^\alpha \ll 1. \tag{4.90}$$

4.1.8 Uphill transport

The transport of a neutral solute across a neutral membrane takes place in the direction of decreasing concentration. Charged solutes must also respond to the electric field and, therefore, they can be transported in the direction of increasing concentration by application of an appropriate electric field. The transport of an ionic species against its concentration gradient can also be observed in the absence of an externally applied electric field. The reason is that an internal field can be generated due to the presence of concentration gradients and the differences in ionic mobilities [5, 6]. The transport against the concentration gradient under these conditions receives the name of uphill transport, and it deserves special attention because it nicely illustrates the coupling of ionic transport.

Uphill transport can be better understood by analysing the energetics of ion transport. The transport of one mol of species i from compartment α where its concentration is c_i^α to compartment β where its concentration is $c_i^\beta > c_i^\alpha$ requires a minimum work $\Delta\mu_i = RT\ln(c_i^\beta/c_i^\alpha) > 0$. This energy can be taken from an electric field: if the membrane potential $\Delta\phi = \phi^\beta - \phi^\alpha$ satisfies

$$RT\ln(c_i^\beta/c_i^\alpha) + z_i F\Delta\phi \le 0, \tag{4.91}$$

then uphill transport is possible. This condition can be written as $\Delta\tilde{\mu}_i = \tilde{\mu}_i^\beta - \tilde{\mu}_i^\alpha \le 0$ and highlights an important feature of uphill transport, namely, that an ionic species i can move down or up its concentration gradient, but it must move down the gradient of its electrochemical potential.[7]

Another interesting aspect is that, since the system is under open-circuit conditions ($I = 0$), the electrostatic energy $-z_i F\Delta\phi$ taken by species i during its uphill transport must be provided by other ionic species that move down their concentration gradient and generate a diffusion potential

[6] This constitutes the principle of the KCl salt bridge.

[7] In the Nernst–Planck formalism, the transport of an ionic species always takes place in the direction of decreasing electrochemical potential. The transport against the gradient of the electrochemical potential can take place across biological membranes, where additional sources of energy are available, and receives the name of active transport.

$\Delta\phi = \Delta\phi_{\text{dif}}$ that satisfies eqn (4.91). As we learned from Chapter 1, the second law of thermodynamics requires that the dissipation function must be positive, and therefore

$$-\sum_i j_i \Delta\tilde{\mu}_i = -\sum_i j_i \Delta\mu_i \geq 0, (I = 0). \tag{4.92}$$

For the species that moves against its concentration gradient, the contribution $-j_i \Delta\mu_i$ to the sum is negative, but this is compensated by positive contributions due to the ions that move down their concentration gradients.

Consider that the membrane separates ternary electrolyte solutions formed by mixing two 1:1 binary electrolytes AC and DC with a common anion C^-. The electrolytes are denoted by indices 13 and 23, respectively, and are assumed to be completely dissociated. The electrolyte concentrations in compartments α and β are, respectively, c_{13}^{α}, c_{23}^{α}, c_{13}^{β}, and $c_{23}^{\beta} > c_{23}^{\alpha}$. We aim at describing the uphill transport of the cation D^+, that is, those situations in which its flux density j_2 is positive (i.e. from α to β) in spite of the fact that $c_2^{\beta} > c_2^{\alpha}$ (see Fig. 4.16). This requires that $c_3^{\alpha} = c_{13}^{\alpha} + c_{23}^{\alpha} < c_{13}^{\beta} + c_{23}^{\beta} = c_3^{\beta}$ and therefore the ratio $r_3 \equiv c_3^{\alpha}/c_3^{\beta} < 1$ is a convenient parameter to be used. The case $c_3^{\beta} = c_3^{\alpha}$ is not considered here because no uphill transport can be then observed.

In order to obtain the ionic flux densities, the Nernst–Planck equations must be solved. The local electroneutrality assumption

$$c_1 + c_2 = c_3 \tag{4.93}$$

can be used to eliminate the electric field from the Nernst–Planck equations as follows

$$G_0 \equiv \frac{j_1}{D_1} + \frac{j_2}{D_2} + \frac{j_3}{D_3} = -2\frac{\mathrm{d}c_3}{\mathrm{d}x}. \tag{4.94}$$

This implies that the anion concentration profile is linear and that $G_0 = 2(c_3^{\alpha} - c_3^{\beta})/h$. This information can be used in turn to integrate the Nernst–Planck equation for anions and show that the electric potential profile is given by

$$f[\phi(x) - \phi^{\alpha}] = \Gamma \ln \frac{c_3(x)}{c_3^{\alpha}}, \tag{4.95}$$

where

$$\Gamma \equiv \frac{\frac{j_1}{D_1} + \frac{j_2}{D_2} - \frac{j_3}{D_3}}{\frac{j_1}{D_1} + \frac{j_2}{D_2} + \frac{j_3}{D_3}}. \tag{4.96}$$

The diffusion potential can then be written as $\Delta\phi = \Delta\phi_{\text{dif}} = -(\Gamma/f)\ln r_3$ and the parameter Γ is yet to be determined.

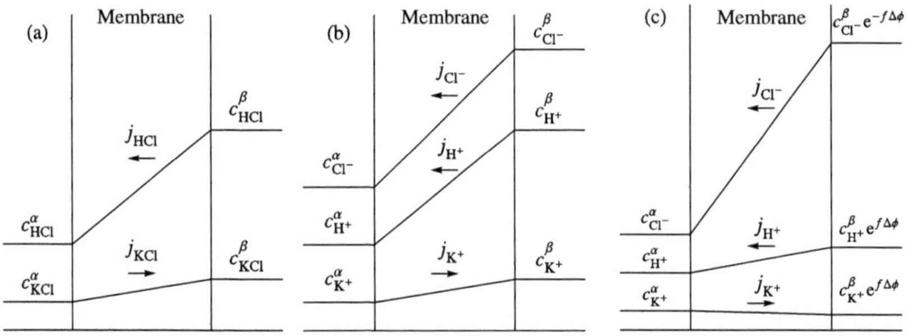

Fig. 4.16.
Schematic illustration of the uphill diffusion process. A neutral membrane separates two solutions of HCl and KCl. Both electrolytes have a larger concentration in compartment β than in compartment α. However, while HCl moves from β to α, it is possible to observe under some conditions that KCl moves from α to β, that is, against its concentration gradient (a). This is due to the fact that the electrolytes diffuse as ions and the electrodiffusion of the three ions involved is coupled through the condition $I = 0$. This creates a diffusion potential that maintains zero current. As shown in drawing (b), the potassium ions also move against their concentration gradient. However, all three ions move in the decreasing direction of their electrochemical potentials (c).

The anion flux density is

$$j_3 = \frac{D_3}{h}(1-\Gamma)(c_3^\alpha - c_3^\beta) = \frac{D_3 c_3^\beta}{h}(1-\Gamma)(r_3 - 1), \tag{4.97}$$

and those of the cations are

$$j_1 = \frac{D_1}{h} E\,(c_{13}^\alpha - c_{13}^\beta e^{f\Delta\phi}), \tag{4.98}$$

$$j_2 = \frac{D_2}{h} E\,(c_{23}^\alpha - c_{23}^\beta e^{f\Delta\phi}), \tag{4.99}$$

where[8]

$$E \equiv \frac{h}{\int_o^h e^{f[\phi(x)-\phi^\alpha]}\mathrm{d}x} = \frac{\mathrm{d}c_3}{\mathrm{d}x}\frac{h(c_3^\alpha)^\Gamma}{\int_{c_3^\alpha}^{c_3^\beta}(c_3)^\Gamma \mathrm{d}c_3} = (1+\Gamma)\frac{(r_3)^{1+\Gamma} - (r_3)^\Gamma}{(r_3)^{1+\Gamma} - 1}. \tag{4.100}$$

The condition $I/zF = j_1 + j_2 - j_3 = 0$ and eqns (4.97)–(4.99) lead to the following transcendental equation for Γ

$$e^{f\,\Delta\phi} = \left(\frac{c_3^\beta}{c_3^\alpha}\right)^\Gamma = \frac{\sum_i z_i D_i c_i^\alpha + \Gamma \sum_i D_i c_i^\alpha}{\sum_i z_i D_i c_i^\beta + \Gamma \sum_i D_i c_i^\beta}. \tag{4.101}$$

[8] In deriving eqn (4.100) it is implicitly assumed that $\Gamma \neq -1$. Note the similarity with the factor E defined in eqn (4.34). When $c_3^\alpha \approx c_3^\beta$, Goldman's constant field is a good approximation and eqn (4.34) yields approximate values of E very close to those obtained from the exact eqn (4.100).

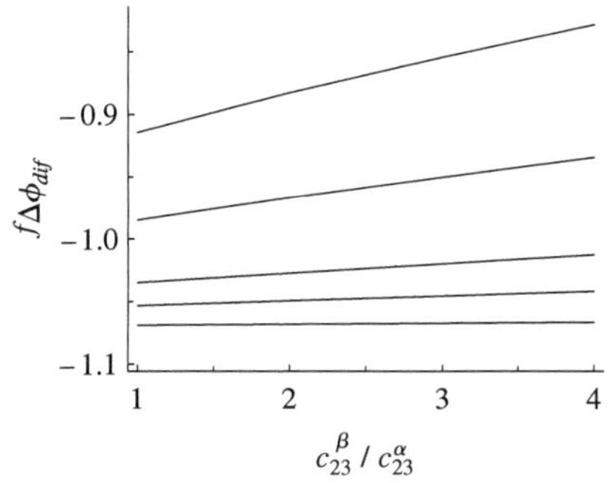

Fig. 4.17. Dimensionless diffusion potential established across a neutral membrane that separates solutions of a mixture of two 1:1 binary electrolytes 13 and 23 against $c_{23}^{\beta}/c_{23}^{\alpha}$ when $c_{13}^{\beta}/c_{13}^{\alpha} = 5$, and $c_{13}^{\alpha}/c_{23}^{\alpha} = 1, 2, 5, 10,$ and 50 (from top to bottom). The diffusion coefficients satisfy the relation $0.2D_1 = D_2 = D_3$.

[Note that eqn (4.101) is the same as eqn (4.84), and it can also be obtained as a particular case of eqn (4.77).] An initial value that can be used to find iteratively the numerical solution of this equation is $\Gamma_0 = (D_3 - D_1)/(D_3 + D_1)$, which corresponds to the absence of electrolyte 23.

Consider that the membrane separates two HCl-KCl solutions such that $c_{\mathrm{HCl}}^{\beta} > c_{\mathrm{HCl}}^{\alpha}$, $c_{\mathrm{KCl}}^{\beta} > c_{\mathrm{KCl}}^{\alpha}$, and $c_{\mathrm{HCl}}^{\alpha} \gg c_{\mathrm{KCl}}^{\alpha}$. The last condition allows us to conclude that the diffusion potential established in this ternary system should be practically the same as in the absence of KCl. This diffusion potential is then approximately given by

$$f\,\Delta\phi_{\mathrm{dif}} = f\,\Delta\phi \approx \Gamma_0 \ln \frac{c_{\mathrm{HCl}}^{\beta}}{c_{\mathrm{HCl}}^{\alpha}} = \frac{D_3 - D_1}{D_3 + D_1} \ln \frac{c_{13}^{\beta}}{c_{13}^{\alpha}}, \qquad (4.102)$$

and, since $D_{\mathrm{Cl}^-} = D_3 < D_1 = D_{\mathrm{H}^+}$, it takes negative values. That is, since the hydrogen ions have a larger mobility than the chloride ions, they make the electric potential in compartment α positive (with respect to compartment β) when they move towards compartment α due to the concentration difference $c_{\mathrm{HCl}}^{\beta} - c_{\mathrm{HCl}}^{\alpha} > 0$ (see Fig. 4.16). Of course, when taking into account both the concentration and the electric potential gradient, it turns out that hydrogen and chloride ions move at the same velocity because the diffusion potential enhances the diffusive flux of chloride ions and retards that of hydrogen ions, and therefore $j_1 \approx j_3$. The transport of potassium ions is affected by the diffusion potential in a similar way to that of hydrogen ions. That is, the migrational contribution to the flux density of potassium ions is positive (i.e. from compartment α to β) and this contribution can overrule a moderate concentration gradient in the opposite direction (i.e. creating a negative diffusive contribution to the flux density of potassium ions when $c_{\mathrm{KCl}}^{\beta} > c_{\mathrm{KCl}}^{\alpha}$). When this happens, we talk of uphill transport of potassium ions or, equivalently, uphill transport of potassium chloride due to the countertransport of hydrochloric acid. Figure 4.16 schematically depicts this situation.

The above comments have referred to a situation in which the concentration of potassium chloride was much smaller than that of hydrochloric acid, $c_{\mathrm{HCl}}^{\alpha} \gg c_{\mathrm{KCl}}^{\alpha}$, and hence the diffusion potential was mainly governed by the diffusion of the latter. This helps us in understanding the nature of uphill transport but it is not a necessary condition. Thus, although the uphill flux of KCl decreases

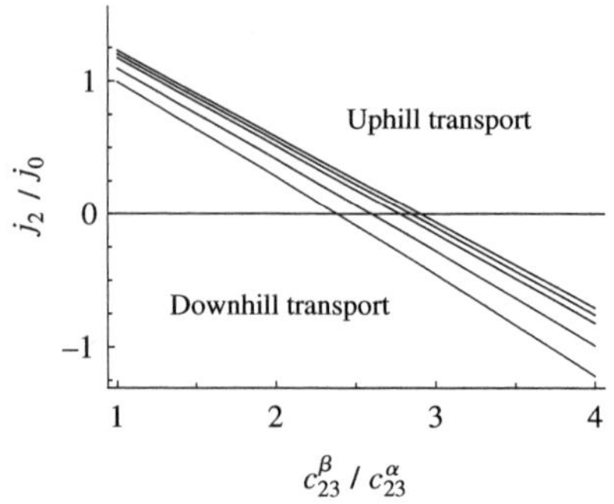

Fig. 4.18.
Flux density of ionic species 2 (in $j_0 \equiv D_3 c_{23}^{\alpha}/h$ units) against $c_{23}^{\beta}/c_{23}^{\alpha}$ when $c_{13}^{\beta}/c_{13}^{\alpha} = 5$, and $c_{13}^{\alpha}/c_{23}^{\alpha} = 50, 10, 5, 2,$ and 1 (from top to bottom). The diffusion coefficients satisfy the relation $0.2D_1 = D_2 = D_3$. The positive values correspond to flux direction from compartment α to β and hence to uphill transport.

in magnitude as $c_{\mathrm{HCl}}^{\alpha}/c_{\mathrm{KCl}}^{\alpha}$ decreases, Figs. 4.17 and 4.18 show that this uphill transport can also be observed when $c_{\mathrm{HCl}}^{\alpha} = c_{\mathrm{KCl}}^{\alpha}$ and ca. $c_{\mathrm{KCl}}^{\beta}/c_{\mathrm{KCl}}^{\alpha} < 2.4$ because the diffusion potential is still rather large (and negative).

4.2 Donnan equilibrium in charged membranes

4.2.1 Ion-exchange membranes

Ion-exchange membranes can be defined as ion-exchange resins that can be regenerated with electric current. They are used in, e.g., various modes of electrodialysis and polymer electrolyte fuel cells as well as to separate (unwanted) electrode reaction products from the rest of the process. Ion-exchange membranes are either cation or anion selective. This is implemented by insertion of fixed acidic or basic dissociating groups into the membrane matrix. These groups can be either weak (like carboxyl and amino groups) or strong (like sulphonate groups), thus leading to fixed-charge distributions that depend or not, respectively, on the composition of the solution filling the membrane phase. The membranes with negatively charged (or anionic) groups repel electrostatically the anions and attract the cations, so that their fixed charge is compensated. They are known as cation-selective, cation-exchange or anionic membranes, although the latter denomination is discouraged. Similarly, the membranes with positively charged (or cationic) groups are known as anion-selective, anion-exchange or cationic membranes. The ions in the solution filling the membrane are denoted either as counterions if they are of opposite sign to the fixed charge groups, or as co-ions if they have similar charge. In general, cation-exchange membranes are more selective and durable than anion-exchange membranes, but both types have lately been improved significantly and can resist elevated temperatures and strong acidic and alkaline solutions. A typical structure of a cation-exchange membrane is schematically depicted in Fig. 4.19.

4.2.2 Donnan equilibrium

The immobilized ion-exchange groups in the membrane, $-\mathrm{R}^z$, bring about an interesting equilibrium state between the electrolyte solutions internal and external to the membrane. Consider that these solutions are composed of the

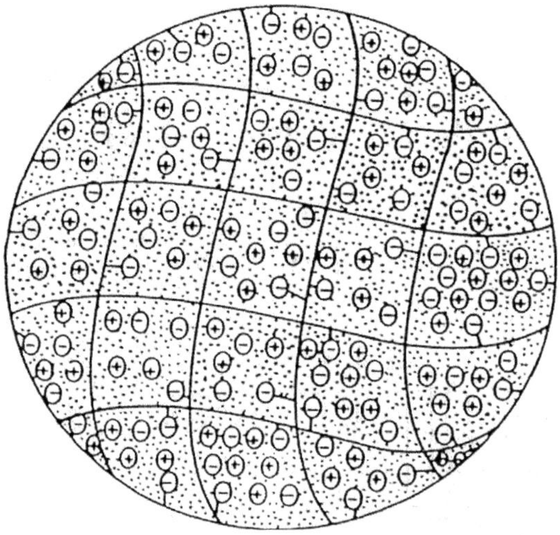

Fig. 4.19.
A schematic drawing of a cation-exchange membrane, with fixed negative groups (◯), counter-ions (⊕), co-ions (⊖) and water (. . .). Solid lines indicate reinforcing of the membrane matrix. (Reproduced with permission from Ref. [7].)

same strong binary electrolyte $A_{\nu_1}C_{\nu_2}$, dissociating completely into ν_1 ions A^{z_1} and ν_2 ions C^{z_2} such that $z_1\nu_1 + z_2\nu_2 = 0$. The molar concentration[9] of fixed groups is c_M and their charge number is z_M; in practice, all available ion-exchange membranes have univalently charged groups and $z_M = \pm 1$.

The Donnan equilibrium condition for the distribution of an ionic species i between the internal (phase superscript M) and external (superscript w) solutions establish that its electrochemical potential is the same in the two phases

$$\tilde{\mu}_i^{M} = \tilde{\mu}_i^{w}. \tag{4.103}$$

Similarly, the equilibrium condition for the electrolyte is

$$\mu_{12}^{M} = \nu_1\tilde{\mu}_1^{M} + \nu_2\tilde{\mu}_2^{M} = \nu_1\tilde{\mu}_1^{w} + \nu_2\tilde{\mu}_2^{w} = \mu_{12}^{w}. \tag{4.104}$$

Using the expression

$$\tilde{\mu}_i = \mu_i^{\circ} + RT\ln(\gamma_i c_i) + z_i F\phi \tag{4.105}$$

for the electrochemical potential of species i, the above distribution equilibrium conditions become[10]

$$\mu_i^{\circ,M} + RT\ln(\gamma_i^{M}c_i^{M}) + z_iF\phi^{M} = \mu_i^{\circ,w} + RT\ln(\gamma_i^{w}c_i^{w}) + z_iF\phi^{w}, \tag{4.106}$$

$$\mu_{12}^{\circ,M} + \nu_{12}RT\ln(\gamma_{12}^{M}c_{\pm,12}^{M}) = \mu_{12}^{\circ,w} + \nu_{12}RT\ln(\gamma_{12}^{w}c_{\pm,12}^{w}), \tag{4.107}$$

[9] Rather often, the product $z_M c_M$ is referred to as the fixed-charge concentration, which has the dimensions of a molar concentration and is positive for anion-exchange membranes and negative for cation-exchange membranes. Here, we use X for $z_M c_M / z_2 > 0$, where subscript 2 denotes the co-ion. Other literature sources may use X for $z_M c_M$ or $|z_M c_M|$.

[10] The effect of the pressure difference between the phases M and w is discussed later in this section.

where $c_{\pm,12} \equiv (c_1^{\nu_1} c_2^{\nu_2})^{1/\nu_{12}}$ and $\gamma_{12} \equiv (\gamma_1^{\nu_1} \gamma_2^{\nu_2})^{1/\nu_{12}}$ are the electrolyte mean molar concentration and mean activity coefficient, respectively. The partition coefficient is then defined as

$$K_i \equiv \frac{c_i^{\mathrm{M}}}{c_i^{\mathrm{w}}} = \frac{\gamma_i^{\mathrm{w}}}{\gamma_i^{\mathrm{M}}} \mathrm{e}^{-(\mu_i^{\mathrm{o,M}} - \mu_i^{\mathrm{o,w}})/RT} \mathrm{e}^{-z_i f \Delta\phi_{\mathrm{D}}}, \; i = 1, 2, \tag{4.108}$$

$$K_{12} \equiv \frac{c_{\pm,12}^{\mathrm{M}}}{c_{\pm,12}^{\mathrm{w}}} = \frac{\gamma_{12}^{\mathrm{w}}}{\gamma_{12}^{\mathrm{M}}} \mathrm{e}^{-(\mu_{12}^{\mathrm{o,M}} - \mu_{12}^{\mathrm{o,w}})/\nu_{12} RT}, \tag{4.109}$$

where $\Delta\phi_{\mathrm{D}} \equiv \phi^{\mathrm{M}} - \phi^{\mathrm{w}}$ is the Donnan potential. The first exponential factor in eqns (4.108) and (4.109), $K_{c,i} \equiv \mathrm{e}^{-(\mu_i^{\mathrm{o,M}} - \mu_i^{\mathrm{o,w}})/RT}$ $(i = 1, 2, 12)$, is known as the chemical partition coefficient and accounts for the difference in the solute environment (water concentration, dielectric constant, short-range interactions, etc.) in the two phases. The second exponential factor in eqn (4.108), $K_{\mathrm{e},i} \equiv \mathrm{e}^{-z_i f \Delta\phi_{\mathrm{D}}}$, is known as the electrostatic partition coefficient and accounts for the difference in electrostatic energy of the ions in the two phases due to the charge associated to the fixed groups. It is greater than unity for counterions and less than unity for co-ions. In symmetric binary electrolytes the relation $K_{\mathrm{e},1} K_{\mathrm{e},2} = 1$ is satisfied.

In membranes with high water content, it can be assumed that the activity coefficients and the standard chemical potentials are the same in both phases, and therefore the distribution equilibria are described by the simpler equations

$$c_i^{\mathrm{M}} = c_i^{\mathrm{w}} \mathrm{e}^{-z_i f \Delta\phi_{\mathrm{D}}}, \tag{4.110}$$

$$c_{\pm,12}^{\mathrm{M}} = c_{\pm,12}^{\mathrm{w}}. \tag{4.111}$$

In the case of multicomponent systems (under equilibrium conditions), similar equations are valid for all ionic species and neutral electrolytes that achieve the distribution equilibrium between the two phases.

The above description of the Donnan equilibrium has been based on the thermodynamically meaningful mean electrolyte concentration $c_{\pm,12}$. In the external solution, this concentration is related to the stoichiometric concentration

$$c_{12}^{\mathrm{w}} \equiv \frac{c_1^{\mathrm{w}}}{\nu_1} = \frac{c_2^{\mathrm{w}}}{\nu_2} \tag{4.112}$$

by the simple relation

$$c_{\pm,12}^{\mathrm{w}} = \nu_{\pm,12} c_{12}^{\mathrm{w}}, \tag{4.113}$$

where $\nu_{\pm,12} \equiv (\nu_1^{\nu_1} \nu_2^{\nu_2})^{1/\nu_{12}}$. In the electrolyte solution filling the membrane phase, however, eqn (4.112) is not satisfied, that is, $c_1^{\mathrm{M}}/\nu_1 \neq c_2^{\mathrm{M}}/\nu_2$, and therefore eqn (4.113) does not hold either, $c_{\pm,12}^{\mathrm{M}} \neq \nu_{\pm,12} c_{12}^{\mathrm{M}}$. In fact, the stoichiometric electrolyte concentration in the membrane phase c_{12}^{M} is yet to be defined.

Since the ion-exchange groups $-R^{z_M}$ participate in the ionic distribution equilibrium, the solution filling the membrane can be considered to be composed of two binary electrolytes with a common ion, the counterion A^{z_1}, which is denoted as species 1. The electrolyte formed by the counterions and the fixed-charge groups can be denoted as $A_{\nu_{1,M}}R_{\nu_M}$ and that formed with the co-ions is $A_{\nu_1}C_{\nu_2}$. The stoichiometric concentration of the latter is

$$c_{12}^{M} \equiv c_2^{M}/\nu_2. \tag{4.114}$$

This magnitude is the so-called Donnan electrolyte concentration and, contrarily to $c_{\pm,12}$ [see eqn (4.111)], it does not take the same value in the internal and external phases under equilibrium conditions, i.e. $c_{12}^{M} \neq c_{12}^{w}$.

The local electroneutrality condition inside the membrane

$$z_1 c_1^{M} + z_2 c_2^{M} + z_M c_M = 0 \tag{4.115}$$

and eqns (4.110) and (4.112) lead to the following equation for the Donnan potential

$$X \equiv z_M c_M / z_2 = \nu_2 c_{12}^{w} \left(e^{-z_1 f \Delta\phi_D} - e^{-z_2 f \Delta\phi_D}\right). \tag{4.116}$$

Since $X > 0$, because species 2 is the co-ion (and therefore $z_M z_2 > 0$), this equation implies that $z_2 f \Delta\phi_D \geq 0$. That is, the potential in the membrane phase is positive with respect to the external phase, $\Delta\phi_D \equiv \phi^M - \phi^w > 0$, if the fixed groups are positively charged and negative otherwise. Moreover, we conclude that the stoichiometric electrolyte concentration inside the membrane is smaller than in the external phase

$$c_{12}^{M} = c_2^{M}/\nu_2 = c_{12}^{w}\, e^{-z_2 f \Delta\phi_D} \leq c_{12}^{w}. \tag{4.117}$$

This phenomenon is known as Donnan exclusion (or co-ion exclusion) and it is responsible for the permselectivity of the ion-exchange membranes, that is, for their selectivity with respect to the transfer of charged species across them.

The value of the Donnan potential drop must be obtained from the solution of eqn (4.116). In general, this is an algebraic, linear equation of order $|z_1 - z_2|$, which has to be solved numerically. Figures 4.20 (a) and (b) show the graphical representation of eqn (4.116) for different electrolytes. There are some features that can be easily identified. In strongly charged membranes, $X \gg c_{12}^{w}$, the co-ions are excluded and their charge number does not affect the value of the Donnan potential; the magnitude represented in the abscissa axis apparently includes the co-ion charge number, but $z_2 c_{12}^{w}$ can also be rewritten as the counterion concentration in the external bulk solution. The lines represented in Fig. 4.20 (a) are then linear with a slope of $(60/|z_1|)$ mV/decade. In weakly charged membranes, $X \ll c_{12}^{w}$, the Donnan potential is so small that the exponentials in eqn (4.116) can be linearized. This approximately linear relation between the Donnan potential and the fixed-charge concentration is apparent in Fig. 4.20 (b).

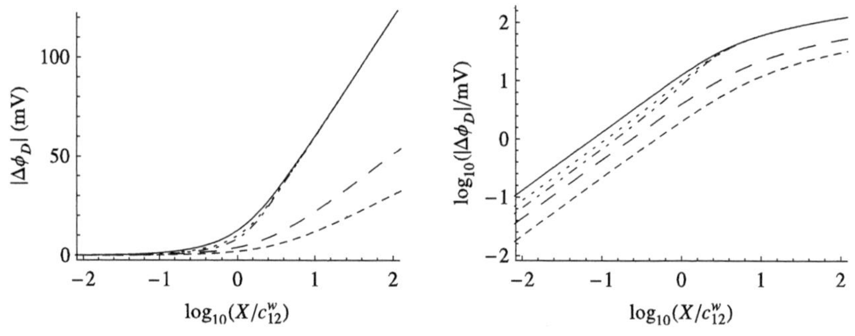

Fig. 4.20.
Donnan potential $|\Delta\phi_D|$ vs. X/c^{w}_{12} in semi-logarithmic (a) and double-logarithmic plot (b) calculated from eqn (4.116) for different binary electrolytes: $|z_1| : |z_2| = 1 : 1$ (solid line), 2:1 (long dashed line), 3:1 (short dashed line), 1:2 (dotted line), and 1:3 (dot-dashed line). The charge numbers have been identified by the ratio of the absolute values corresponding to the counterion and coion, regardless whether they are cations or anions. In positively charged membranes, the Donnan potential is positive, the counterion is an anion ($z_1 < 0$) and the coion is a cation ($z_2 > 0$). In negatively charged membranes, $\Delta\phi_D < 0, z_1 > 0$, and $z_2 < 0$. The value 26 mV has been used for $1/f = RT/F$.

In the case of symmetric electrolytes, eqns (4.115) and (4.116) simplify to

$$c_1^{\mathrm{M}} = c_2^{\mathrm{M}} + X, \tag{4.118}$$

$$z_2 f\,\Delta\phi_{\mathrm{D}} = \operatorname{arcsinh}\frac{X}{2c^{\mathrm{w}}_{12}} = \ln\left\{\frac{X}{2c^{\mathrm{w}}_{12}} + \left[\left(\frac{X}{2c^{\mathrm{w}}_{12}}\right)^2 + 1\right]^{1/2}\right\}. \tag{4.119}$$

From eqns (4.117) and (4.119), the stoichiometric electrolyte concentration in the membrane phase can be evaluated as

$$c^{\mathrm{M}}_{12} = -(X/2) + [(X/2)^2 + (c^{\mathrm{w}}_{12})^2]^{1/2}. \tag{4.120}$$

In weakly charged membranes, $X \ll c^{\mathrm{w}}_{12}$, eqn (4.120) yields the expected result $c^{\mathrm{M}}_{12} \approx c^{\mathrm{w}}_{12}$. In strongly charged membranes, $X \gg c^{\mathrm{w}}_{12}$, this can be approximated by $c^{\mathrm{M}}_{12} \approx (c^{\mathrm{w}}_{12})^2/X \ll c^{\mathrm{w}}_{12}$. The membranes that are so strongly charged that c^{M}_{12} can be neglected are known as ideally selective membranes. Real membranes, however, are never ideally selective. This inability to exclude completely the Donnan electrolyte (that is, the co-ions) is known as Donnan failure.[11]

Similarly, the ionic concentrations in the membrane phase can be obtained as $c_2^{\mathrm{M}} = c^{\mathrm{M}}_{12}$ and

$$c_1^{\mathrm{M}} = c_1^{\mathrm{w}}\,\mathrm{e}^{-z_1 f\,\Delta\phi_{\mathrm{D}}} = c_2^{\mathrm{M}} + X = (X/2) + [(X/2)^2 + (c^{\mathrm{w}}_{12})^2]^{1/2}, \tag{4.121}$$

[11] Under process conditions, the Donnan electrolyte concentration is a function of position inside the membrane. The ability to exclude co-ions is then also affected by the extent of the concentration polarization at the boundary layers flanking the membrane. In operating commercial membranes, the fraction of counterions is seldom higher than 95%.

and the total ionic concentration inside the membrane is

$$c_1^{\mathrm{M}} + c_2^{\mathrm{M}} = [X^2 + (2c_{12}^{\mathrm{w}})^2]^{1/2}. \tag{4.122}$$

Remember that eqns (4.118)–(4.122) are only valid for symmetric electrolytes.

In closing this section, we comment on the equilibrium condition for the solvent. In strongly charged membranes, $X \gg c_{12}^{\mathrm{w}}$, eqn (4.122) implies that the total ionic concentration inside the membrane is much larger than outside the membrane, $c_1^{\mathrm{M}} + c_2^{\mathrm{M}} \gg 2c_{12}^{\mathrm{w}}$. If not only the solutes but also the solvent is equilibrated between the internal and external phases, a significant pressure difference may exist between them. Indeed, the Gibbs–Duhem equation $c_0 \mathrm{d}\mu_0 + \sum_i c_i \mathrm{d}\mu_i = \mathrm{d}p$ and the Euler equation for the solution volume $c_0 \upsilon_0 + \sum_i c_i \upsilon_i = 1$ lead to $\mathrm{d}\mu_0 = \upsilon_0(\mathrm{d}p - \mathrm{d}\pi)$, where π is the osmotic pressure. The equilibrium condition for water, $\mu_0^{\mathrm{M}} = \mu_0^{\mathrm{w}}$, implies that the Donnan pressure difference must follow the osmotic pressure difference, $\Delta p_{\mathrm{D}} \equiv p^{\mathrm{M}} - p^{\mathrm{w}} = \pi^{\mathrm{M}} - \pi^{\mathrm{w}} \equiv \Delta\pi_{\mathrm{D}}$, where it has been assumed that the water partial molar volume is approximately the same in both phases. The osmotic pressure is given by the differential $\mathrm{d}\pi \equiv \sum_i c_i \mathrm{d}\mu_i^c / c_0 \upsilon_0 = (RT/c_0\upsilon_0)\mathrm{d}\sum_i c_i \approx RT\mathrm{d}\sum_i c_i$, and therefore $\Delta\pi_{\mathrm{D}} \approx RT(\sum_i c_i^{\mathrm{M}} - \sum_i c_i^{\mathrm{w}}) \approx RTX$ [see eqn (4.121)], which is of the order of 26 atm at 300 K for $X = 1$ M. If the $p\upsilon_i$ contribution to the electrochemical potential of the ions needs to be included in the equilibrium conditions, these become

$$\frac{c_i^{\mathrm{M}}}{c_i^{\mathrm{w}}} = \frac{\gamma_i^{\mathrm{w}}}{\gamma_i^{\mathrm{M}}} \mathrm{e}^{-(\mu_i^{\mathrm{o,M}} - \mu_i^{\mathrm{o,w}})/RT} \mathrm{e}^{-z_i f \Delta\phi_{\mathrm{D}}} \mathrm{e}^{-\Delta\pi_{\mathrm{D}}\upsilon_i/RT}, \tag{4.123}$$

$$\frac{c_{\pm,12}^{\mathrm{M}}}{c_{\pm,12}^{\mathrm{w}}} = \frac{\gamma_{12}^{\mathrm{w}}}{\gamma_{12}^{\mathrm{M}}} \mathrm{e}^{-(\mu_{12}^{\mathrm{o,M}} - \mu_{12}^{\mathrm{o,w}})/\nu_{12}RT} \mathrm{e}^{-\Delta\pi_{\mathrm{D}}\upsilon_{12}/\nu_{12}RT}, \tag{4.124}$$

where $\upsilon_{12} = \nu_1\upsilon_1 + \nu_2\upsilon_2$ is the electrolyte partial molar volume, and we have assumed that all partial molar volumes are approximately the same in both phases. In eqn (4.124), for instance, the pressure correction is approximately equal to $\mathrm{e}^{-X\upsilon_{12}}$, which is only significant when the volume fraction occupied by the fixed-charged groups and their counterions is of the order of unity.

4.2.3 Ion-exchange equilibrium

The ion-exchange membranes owe their name to their ability to equilibrate with the bathing solution and replace the counterions inside by those present in the solution. This ion-exchange process requires some time because the counterions have to diffuse in and out of the membrane. Later in this section we describe the time evolution of the counterion concentration in the bathing solution within the context of a particular example of practical interest, the drug-release kinetics from a reservoir. First, we describe the ion-exchange equilibrium that is attained at large times.

Consider an ion-exchange membrane with fixed-charged groups $-\mathrm{R}^z$. Initially the membrane is in A^{z_1} form, that is, the counterions are A^{z_1} ions.

The membrane is then immersed in a very large volume of a bathing solution containing counterions D^{z_2}. The achievement of the thermodynamic distribution equilibrium requires that some of the D^{z_2} ions must enter the membrane and some of the A^{z_1} ions must leave it. This ion-exchange process is described by either of the two following forms

$$\nu_1 A^{z_1}(M) + \nu_2 D^{z_2}(w) \rightleftarrows \nu_2 D^{z_2}(M) + \nu_1 A^{z_1}(w), \tag{4.125}$$

$$-RA_{\nu_1}(M) + \nu_2 D^{z_2}(w) \rightleftarrows -RD_{\nu_2}(M) + \nu_1 A^{z_1}(w), \tag{4.126}$$

where the stoichiometric relation $z_1\nu_1 = z_2\nu_2$ must be satisfied. Assuming that the activity coefficients satisfy the relation $(\gamma_1^{\mathrm{w}}/\gamma_1^{\mathrm{M}})^{\nu_1} = (\gamma_2^{\mathrm{w}}/\gamma_2^{\mathrm{M}})^{\nu_2}$, the thermodynamic equilibrium condition is represented by the mass action law

$$K_{12} = \mathrm{e}^{-\Delta_r G^\circ/RT} = \frac{(c_1^{\mathrm{w}})^{\nu_1}(c_2^{\mathrm{M}})^{\nu_2}}{(c_1^{\mathrm{M}})^{\nu_1}(c_2^{\mathrm{w}})^{\nu_2}} \tag{4.127}$$

where the standard Gibbs potential is $\Delta_r G^\circ = \nu_1\mu_1^{\circ,\mathrm{w}} + \nu_2\mu_2^{\circ,\mathrm{M}} - \nu_1\mu_1^{\circ,\mathrm{M}} - \nu_2\mu_2^{\circ,\mathrm{w}}$. The ion-exchange equilibrium constant can be expressed in terms of the ionic chemical partition coefficients $K_{c,i} \equiv \mathrm{e}^{-(\mu_i^{\circ,\mathrm{M}}-\mu_i^{\circ,\mathrm{w}})/RT}$ $(i = 1, 2)$ as

$$K_{12} = (K_{c,2})^{\nu_2}(K_{c,1})^{-\nu_1}. \tag{4.128}$$

When the volume of the bathing solution is much larger than that of the membrane, the final (equilibrium) concentration of A^{z_1} ions is practically zero in both phases, while that of D^{z_2} ions in the external solution is practically unaffected. The ion-exchange process can then be described as the exchange of the amount n_1^0 of moles of of A^{z_1} ions initially present in the membrane by an equivalent amount $(\nu_1/\nu_2)n_1^0$ of D^{z_2} ions.

When the volumes of the bathing solution and the membrane, V^{w} and V^{M}, respectively, are comparable, eqn (4.127) must be combined with the mass balances

$$n_1^0 \equiv c_1^{\mathrm{M},0}V^{\mathrm{M}} = c_1^{\mathrm{w}}V^{\mathrm{w}} + c_1^{\mathrm{M}}V^{\mathrm{M}}, \tag{4.129}$$

$$n_2^0 \equiv c_2^{\mathrm{w},0}V^{\mathrm{w}} = c_2^{\mathrm{w}}V^{\mathrm{w}} + c_2^{\mathrm{M}}V^{\mathrm{M}}, \tag{4.130}$$

in order to solve for the final equilibrium concentrations. Note, however, that we have four unknown concentrations and only three equations. The other equation is the electric charge balance.

In ideally selective membranes, when no co-ions can enter the membrane, the charge balance simplifies to

$$z_1 c_1^{\mathrm{M},0} = z_1 c_1^{\mathrm{M}} + z_2 c_2^{\mathrm{M}}. \tag{4.131}$$

If $\nu_1 = \nu_2 = 1$ and the membrane shows no preference for any of the counterions, $K_{12} = 1$, this system of equations reduces to

$$c_1^{\mathrm{M}} = x_1 c_1^{\mathrm{M},0}, \; c_2^{\mathrm{M}} = (1 - x_1) c_1^{\mathrm{M},0},$$
$$c_1^{\mathrm{w}} = x_1 c_2^{\mathrm{w},0}, \; c_2^{\mathrm{w}} = (1 - x_1) c_2^{\mathrm{w},0}, \tag{4.132}$$

where $x_1 \equiv n_1^0/(n_1^0 + n_2^0)$. For other values of K_{12}, the system can be reduced to a linear, second-order algebraic equation.

In non-ideal selective membranes, the amount of co-ions in the membrane phase is neither negligible nor constant, and their equilibrium distribution also needs to be taken into account. For the sake of simplicity, we take in this description all stoichiometric and activity coefficients equal to one. Consider a membrane that is initially equilibrated with a bathing solution of concentration c_{13}^0 of electrolyte AC. In this initial situation, the ionic concentrations inside the membrane are

$$c_3^{\mathrm{M},0} = c_1^{\mathrm{M},0} - X = -(X/2) + [(X/2)^2 + K_{13}(c_{13}^0)^2]^{1/2}, \tag{4.133}$$

where K_{13} is the partition coefficient of the electrolyte AC and $X \equiv z_{\mathrm{M}} c_{\mathrm{M}}/z_3$. The membrane is then immersed in another bathing solution of the same volume with a concentration c_{23}^0 of electrolyte DC. After waiting for the subsequent equilibration, the ionic concentrations inside the membrane are

$$c_1^{\mathrm{M}} = x_{13}\{(X/2) + [(X/2)^2 + K_{13}(c_{\pm 13}^{\mathrm{w}})^2 + K_{23}(c_{\pm 23}^{\mathrm{w}})^2]^{1/2}\}, \tag{4.134}$$
$$c_2^{\mathrm{M}} = x_{23}\{(X/2) + [(X/2)^2 + K_{13}(c_{\pm 13}^{\mathrm{w}})^2 + K_{23}(c_{\pm 23}^{\mathrm{w}})^2]^{1/2}\}, \tag{4.135}$$
$$c_3^{\mathrm{M}} = -(X/2) + [(X/2)^2 + K_{13}(c_{\pm 13}^{\mathrm{w}})^2 + K_{23}(c_{\pm 23}^{\mathrm{w}})^2]^{1/2}, \tag{4.136}$$

where $c_{\pm,13}^{\mathrm{w}} \equiv (c_1^{\mathrm{w}} c_3^{\mathrm{w}})^{1/2}$ and $c_{\pm,23}^{\mathrm{w}} \equiv (c_2^{\mathrm{w}} c_3^{\mathrm{w}})^{1/2}$ are the mean electrolyte concentrations, and

$$x_{13} \equiv \frac{K_{13}(c_{\pm 13}^{\mathrm{w}})^2}{K_{13}(c_{\pm 13}^{\mathrm{w}})^2 + K_{23}(c_{\pm 23}^{\mathrm{w}})^2} = 1 - x_{23}. \tag{4.137}$$

Equations (4.134)–(4.136) are obtained from the conditions of chemical equilibria

$$K_{13} = \frac{c_1^{\mathrm{M}} c_3^{\mathrm{M}}}{c_1^{\mathrm{w}} c_3^{\mathrm{w}}}, \quad K_{23} = \frac{c_2^{\mathrm{M}} c_3^{\mathrm{M}}}{c_2^{\mathrm{w}} c_3^{\mathrm{w}}}, \quad K_{12} = \frac{c_1^{\mathrm{w}} c_2^{\mathrm{M}}}{c_1^{\mathrm{M}} c_2^{\mathrm{w}}} = \frac{K_{23}}{K_{13}}, \tag{4.138}$$

and local electroneutrality

$$c_1^{\mathrm{w}} + c_2^{\mathrm{w}} = c_3^{\mathrm{w}}, \tag{4.139}$$
$$c_1^{\mathrm{M}} + c_2^{\mathrm{M}} - c_3^{\mathrm{M}} = X. \tag{4.140}$$

A difficulty associated with the use of eqns (4.134)–(4.136) is that the electrolyte concentrations in the external solutions are not known. In fact, the final equilibrium concentrations in both phases must be determined from the solution of eqns (4.134)–(4.136) and the mass balances

$$n_1^0 \equiv c_1^{M,0} V^M = c_1^w V^w + c_1^M V^M, \tag{4.141}$$

$$n_2^0 \equiv c_2^{w,0} V^w = c_2^w V^w + c_2^M V^M, \tag{4.142}$$

$$n_3^0 \equiv c_3^{w,0} V^w + c_3^{M,0} V^M = c_3^w V^w + c_3^M V^M, \tag{4.143}$$

where $c_2^{w,0} = c_3^{w,0} = c_{23}^0$.

4.2.4 Electrical double layer at the membrane/ external solution interface

The Donnan potential given by eqn (4.116) is the difference in electric potential between two homogeneous phases that have uniform potentials ϕ^M and ϕ^w and are locally electroneutral. Somewhere in between the two phases, the potential must show a non-linear spatial variation from ϕ^M to ϕ^w and, according to the Poisson equation of electrostatics, this requires a non-zero space-charge density. In order words, the Donnan potential drop must take place in a non-electroneutral (or electrified) interfacial region, which is known as the electrical double layer. Since the local electroneutrality assumption plays a key role in the solution of the ionic transport equations, it is interesting to determine the thickness of such electrified interfacial region where the local electroneutrality condition does not apply. This is calculated below from the equilibrium electrical potential distribution in the interfacial region between a charged membrane occupying the region $x < 0$ and a binary electrolyte solution in the region $x > 0$. Far from the interface, the electric potential in the membrane phase is ϕ^M, the electric field is zero, and the space-charge density is zero. This is known as the bulk of the membrane phase. Similarly, in the bulk of the external solution, the electric potential is ϕ^w, the electric field is zero, and the space-charge density is zero. The membrane is supposed to have a uniform concentration c_M of fixed groups, the electrolyte concentration in the bulk of the water phase is c_{12}^w, and both phases are assumed to have the same dielectric permittivity ε. The case $z_1 = -z_2$ and $K_{12} = 1$ is considered, and a convenient dimensionless electric potential variable is defined as follows

$$\varphi(x) \equiv z_2 f[\phi(x) - \phi^w], \tag{4.144}$$

where species 2 is the co-ion. This function is continuous and positively defined over the whole system. It tends to zero in the bulk external phase and to

$$\varphi_D \equiv z_2 f \Delta\phi_D = \operatorname{arcsinh}(X/2c_{12}^w) > 0 \tag{4.145}$$

in the bulk membrane phase.

The ionic distributions over the whole system are given by the Boltzmann equilibrium equations

$$c_1(x) = c_{12}^{\mathrm{w}}\, \mathrm{e}^{-z_1 f[\phi(x)-\phi^{\mathrm{w}}]} = c_{12}^{\mathrm{w}}\, \mathrm{e}^{\varphi}, \tag{4.146}$$

$$c_2(x) = c_{12}^{\mathrm{w}}\, \mathrm{e}^{-z_2 f[\phi(x)-\phi^{\mathrm{w}}]} = c_{12}^{\mathrm{w}}\, \mathrm{e}^{-\varphi}. \tag{4.147}$$

When eqns (4.145)–(4.147) are substituted in the Poisson equation in the membrane phase

$$\frac{\mathrm{d}^2\phi}{\mathrm{d}x^2} = \frac{z_2 F}{\varepsilon}(c_1 - c_2 - X) \tag{4.148}$$

this becomes the dimensionless Poisson–Boltzmann equation

$$\frac{\mathrm{d}^2\varphi}{\mathrm{d}\xi^2} = \sinh\varphi - \sinh\varphi_{\mathrm{D}} \tag{4.149}$$

where $\xi \equiv \kappa_{\mathrm{D}}^{\mathrm{w}} x$ is a dimensionless position variable and

$$\kappa_{\mathrm{D}}^{\mathrm{w}} \equiv \left(\frac{2z_2^2 F^2 c_{12}^{\mathrm{w}}}{\varepsilon RT}\right)^{1/2} \tag{4.150}$$

is the reciprocal Debye length. Similarly, in the aqueous phase the dimensionless Poisson–Boltzmann equation is

$$\frac{\mathrm{d}^2\varphi}{\mathrm{d}\xi^2} = \sinh\varphi. \tag{4.151}$$

Multiplying eqn (4.151) by $2\mathrm{d}\varphi/\mathrm{d}\xi$, it can be integrated between the bulk external phase and a position $x > 0$ to give

$$\left(\frac{\mathrm{d}\varphi}{\mathrm{d}\xi}\right)^2 = 2(\cosh\varphi - 1) = 4\sinh^2(\varphi/2). \tag{4.152}$$

Taking the square root (with negative sign) of eqn (4.152) and rewriting it as

$$\frac{\cosh(\varphi/4)}{\sinh(\varphi/4)}\, \frac{1}{\cosh^2(\varphi/4)}\, \frac{\mathrm{d}\varphi}{\mathrm{d}\xi} = -4, \tag{4.153}$$

the electric potential distribution in the external solution can be obtained by integration between the interface $x = 0$, where φ takes the value φ_{s}, and a position $x > 0$, as

$$\varphi(x) = 4\,\mathrm{arctanh}[\tanh(\varphi_{\mathrm{s}}/4)\, \mathrm{e}^{-\kappa_{\mathrm{D}}^{\mathrm{w}} x}],\ (x > 0). \tag{4.154}$$

Similarly, multiplying eqn (4.149) by $2\mathrm{d}\varphi/\mathrm{d}\xi$, it can be integrated between the bulk membrane phase and a position $x < 0$ to give

$$\left(\frac{\mathrm{d}\varphi}{\mathrm{d}\xi}\right)^2 = 4\sinh^2(\varphi/2) - 4\sinh^2(\varphi_\mathrm{D}/2) - 2\sinh\varphi_\mathrm{D}\,(\varphi - \varphi_\mathrm{D}). \qquad (4.155)$$

The condition of continuity of the electric displacement at the interface requires that eqns (4.152) and (4.155) must take the same value at $x = 0$. The interfacial value φ_s can then be determined as

$$\varphi_\mathrm{s} = \varphi_\mathrm{D} - \tanh(\varphi_\mathrm{D}/2). \qquad (4.156)$$

We can take advantage of the information provided by eqn (4.156) to obtain an approximate analytical integration of eqn (4.155) and hence an expression for the electric potential distribution in the membrane phase. Since the tanh function is bound to 1, the difference $|\varphi - \varphi_\mathrm{D}|$ is always smaller than 1 in the membrane phase and the Poisson–Boltzmann equation can be approximated there by its linear form

$$\frac{\mathrm{d}^2\varphi}{\mathrm{d}\xi^2} = \sinh\varphi - \sinh\varphi_\mathrm{D} \approx \cosh\varphi_\mathrm{D}\,(\varphi - \varphi_\mathrm{D})$$
$$= \left(\frac{\kappa_\mathrm{D}^\mathrm{M}}{\kappa_\mathrm{D}^\mathrm{w}}\right)^2 (\varphi - \varphi_\mathrm{D}), \qquad (4.157)$$

where

$$\kappa_\mathrm{D}^\mathrm{M} \equiv \kappa_\mathrm{D}^\mathrm{w}(\cosh\varphi_\mathrm{D})^{1/2} = \left(\frac{z_2^2F^2}{\varepsilon RT}\right)^{1/2} [X^2 + (2c_{12}^\mathrm{w})^2]^{1/4} \qquad (4.158)$$

is the reciprocal Debye length that characterizes the electrical potential distribution in the membrane phase. The solution to eqn (4.157) is

$$\varphi(x) = \varphi_\mathrm{D} - \tanh(\varphi_\mathrm{D}/2)\,\mathrm{e}^{\kappa_\mathrm{D}^\mathrm{M} x}, \quad x < 0. \qquad (4.159)$$

Equations (4.154) and (4.159) show that the electrical double layer extends over a region with a thickness of the order of $1/\kappa_\mathrm{D}^\mathrm{w}$ in the external phase and $1/\kappa_\mathrm{D}^\mathrm{M} < 1/\kappa_\mathrm{D}^\mathrm{w}$ in the membrane phase. These equations have been represented in Fig. 4.21 (a). It is clear there that the electric potential drop in the membrane phase takes place over a shorter distance than in the external solution. It is also clear in this figure that the magnitude of the potential drop in the membrane is smaller than in the external phase. Fig. 4.21 (b) shows the variation of the electrical charge density with position in the interfacial region. The space-charge density inside the membrane has the same sign as the fixed-charge groups, and that in the external phase has the opposite sign. Obviously, the magnitude of the total charge is the same at either side of the interface but the distribution takes

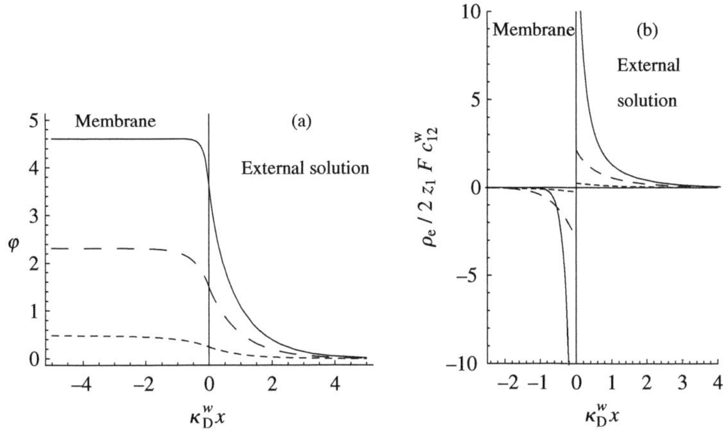

Fig. 4.21.
(a) Dimensionless electric potential $\varphi \equiv z_2 f(\phi - \phi^w)$ vs. $\kappa_D^w x$ calculated from eqns (4.52) and (4.57) and
(b) dimensionless space charge density $\rho_e/2z_1 F c_{12}^w$ vs. $\kappa_D^w x$ for different values of the ratio fixed charge to external concentration: $X/c_{12}^w = 100$ (solid line), 10 (long dashed line), and 1 (short dashed line). The charged membrane and the external solution occupy the regions $x < 0$ and $x > 0$, respectively. Note that the Debye length in the external solution has been used to scale the position axis.

place with a slightly different functional dependence. Finally, the solid curve corresponding to the strongly charge membrane, $X/c_{12}^w = 100$, shows that the charge accumulation close to the interface extends over a smaller thickness in the membrane than in the external solution.

Since Fig. 4.21 (a) has been presented in dimensionless units, it is convenient to estimate the values of the relevant magnitudes. The dimensionless electric potential can be converted to mV by using that $1/f \approx 26$ mV (which corresponds to 25° C). The dimensionless Donnan potential values in Fig. 4.21 (a) can then be compared to those in Fig. 4.20 (solid line) for the same values of X/c_{12}^w.

The position axis has been scaled with the Debye length in the external phase. Figure 4.22 shows the values of the Debye lengths $1/\kappa_D^w$ (solid line) and $1/\kappa_D^M$ (dashed lines) at 25 °C as a function of the external electrolyte concentration. The Debye length describes the screening of the interfacial electric fields and is larger in the external solution than inside the membrane because the total ionic concentration is larger inside the membrane.

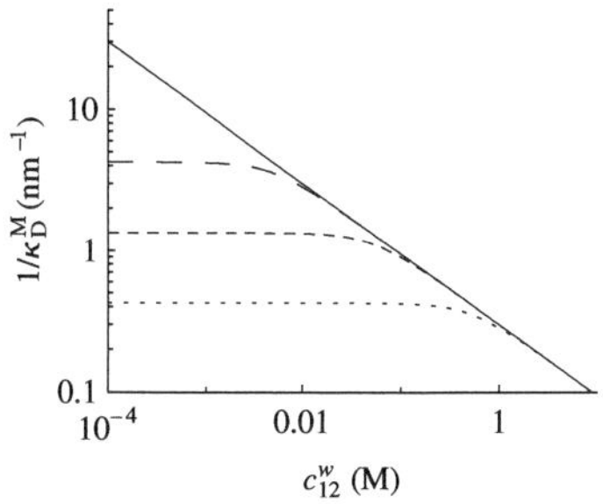

Fig. 4.22.
Debye length $1/\kappa_D^M$ as a function of the external electrolyte concentration c_{12}^w for different values of the membrane fixed charge concentration: $X = 0.01$ M (long dashed line), 0.1 M (short dashed line), and 1 M (dotted line). The solid line corresponds to $X = 0$ and represents the Debye length in the external solution, $1/\kappa_D^w$. As a typical value, the Debye length in the external phase is ca. 10 nm when $c_{12}^w = 1$ mM. When decreasing the external electrolyte concentration, the Debye length $1/\kappa_D^M$ saturates to a value determined by the fixed charge concentration. On the contrary, when the external concentration is very high, the Debye length is the same in the internal and external phases.

4.2.5 Influence of the membrane heterogeneities

So far, the membrane phase has been considered to be homogeneous and all variables are independent of position within the membrane under equilibrium conditions. Even though this is a reasonable assumption that allows for a simple description of the ion-exchange membrane equilibrium, it should be realized that real membranes have a certain degree of heterogeneity. This could result from an uneven distribution of fixed-charge groups throughout the membrane volume [8–10] or from the internal structure of the membrane. For instance, in charged porous membranes the charge groups are distributed on the pore walls and the solution inside the pore is not homogeneous [11]. These heterogeneities give rise to deviations from the behaviour predicted in Sections 4.2.1–4.2.4, and obviously this has practical importance when describing transport through the membrane. As a rule, the membrane heterogeneities give rise to a loss of permselectivity. This was first observed experimentally by Glueckauf and co-workers [12, 13] and discussed at length by Petropoulos [14]. We prove below that this is the case under conditions close to equilibrium. However, in the presence of a relatively large electric current, the heterogeneities may also give rise to permselectivity enhancement [8–10].

Consider, for instance, that the membrane is immersed in a solution with a concentration c_{12}^{w} of a symmetric binary electrolyte, and that the relations $\gamma_i^{\mathrm{w}} = \gamma_i^{\mathrm{M}}$ and $\mu_i^{\mathrm{o,M}} = \mu_i^{\mathrm{o,w}}$ hold for both ions. It was shown in Section 4.2.2 that the electrostatic partition coefficients of the ions satisfy the relation $K_{e,1}K_{e,2} = \mathrm{e}^{-zf\Delta\phi_{\mathrm{D}}}\mathrm{e}^{zf\Delta\phi_{\mathrm{D}}} = 1$. However, this was based on the idea that the electric potential inside the membrane takes the same value ϕ^{M} throughout the membrane. Imagine now that the electric potential inside the membrane is a function of position, $\phi^{\mathrm{M}}(\vec{r})$. The ionic concentrations inside the membrane are then given by

$$c_i^{\mathrm{M}}(\vec{r}) = c_{12}^{\mathrm{w}}\mathrm{e}^{-z_i f(\phi^{\mathrm{M}}-\phi^{\mathrm{w}})}, \tag{4.160}$$

and their average values inside the membrane can be written as

$$\left\langle c_i^{\mathrm{M}}\right\rangle = c_{12}^{\mathrm{w}}\mathrm{e}^{z_i f\phi^{\mathrm{w}}}\left\langle \mathrm{e}^{-z_i f\phi^{\mathrm{M}}}\right\rangle, \tag{4.161}$$

where the brackets $\langle\rangle$ denote spatial averaging over the membrane volume. In these equations ϕ^{w} is the electric potential in the bulk external solution.

Using the Cauchy–Schwartz–Buniakowski inequality

$$\frac{\left\langle c_1^{\mathrm{M}}\right\rangle\left\langle c_2^{\mathrm{M}}\right\rangle}{(c_{12}^{\mathrm{w}})^2} = \left\langle \mathrm{e}^{-zf\phi^{\mathrm{M}}}\right\rangle\left\langle \mathrm{e}^{zf\phi^{\mathrm{M}}}\right\rangle \geq 1, \tag{4.162}$$

where the equal sign corresponds to the homogenous case, and the global electroneutrality condition inside the membrane

$$\left\langle c_1^{\mathrm{M}}\right\rangle = \left\langle c_2^{\mathrm{M}}\right\rangle + \langle X\rangle, \tag{4.163}$$

we conclude that the ionic concentrations must be larger than in a homogeneous membrane with a uniform fixed-charge concentration $\langle X\rangle$. That is, due to the membrane heterogeneities, both ions increase their concentration in the same amount $\langle c_{12}\rangle_{\text{heter}} \geq 0$, so that

$$\left\langle c_1^{\mathrm{M}}\right\rangle = \left\langle c_1^{\mathrm{M}}\right\rangle_{\text{hom}} + \langle c_{12}\rangle_{\text{heter}}, \tag{4.164}$$

$$\left\langle c_2^{\mathrm{M}}\right\rangle = \left\langle c_2^{\mathrm{M}}\right\rangle_{\text{hom}} + \langle c_{12}\rangle_{\text{heter}}, \tag{4.165}$$

where the ionic concentration in a homogenous membrane is given by

$$\left\langle c_1^{\mathrm{M}}\right\rangle_{\text{hom}} = \left\langle c_2^{\mathrm{M}}\right\rangle_{\text{hom}} + \langle X\rangle = (\langle X\rangle/2) + [(\langle X\rangle/2)^2 + (c_{12}^{\mathrm{w}})^2]^{1/2}, \tag{4.166}$$

and subscript 2 denotes the co-ion. Equation (4.165) proves that the Donnan theory provides an upper bound for co-ion exclusion, but the actual exclusion is lower due to the membrane heterogeneities.

As mentioned above, the heterogeneities can have different origin (non-uniform distribution of fixed-charge groups along the flow direction, charged porous membrane structure, etc.). We close this section with an estimation of the increase in the co-ion uptake (or absorption), $\langle c_{12}\rangle_{\text{heter}}$, due to the non-uniform electric potential distribution in the radial direction of a charged porous membrane. We consider a membrane in equilibrium with a bathing solution of concentration c_{12}^{w} and assume that the membrane can be described as an array of parallel, cylindrical capillaries of radius a with a uniform surface-charge concentration σ on the pore walls. The relation between the surface-charge density σ and the fixed-charge concentration c_{M} used above is

$$z_{\text{M}}c_{\text{M}} = 2\sigma/aF. \tag{4.167}$$

Under these conditions, the electric potential inside the membrane is a function of a radial position co-ordinate r that measures the distance to the pore axis. This potential distribution must be obtained from the solution of the Poisson–Boltzmann equation in the membrane

$$\frac{1}{r}\frac{\text{d}}{\text{d}r}\left(r\frac{\text{d}\phi}{\text{d}r}\right) = \frac{z_2F}{\varepsilon}(c_1 - c_2) = \frac{2z_2Fc_{12}^{\text{w}}}{\varepsilon}\sinh\varphi, \tag{4.168}$$

where $\varphi(r) \equiv z_2f[\phi(r) - \phi^{\text{w}}]$ is the dimensionless electric potential. In terms of this potential, the Poisson–Boltzmann equation is

$$\frac{1}{\xi}\frac{\text{d}}{\text{d}\xi}\left(\xi\frac{\text{d}\varphi}{\text{d}\xi}\right) = \sinh\varphi, \tag{4.169}$$

where $\xi \equiv \kappa_{\text{D}}^{\text{w}}r$ is a dimensionless radial position variable and

$$\kappa_{\text{D}}^{\text{w}} \equiv \left(\frac{2z_2^2F^2c_{12}^{\text{w}}}{\varepsilon RT}\right)^{1/2} \tag{4.170}$$

is the reciprocal Debye length. The boundary conditions of eqn (4.169) are

$$\left(\frac{\text{d}\varphi}{\text{d}\xi}\right)_{\xi=0} = 0, \tag{4.171}$$

$$\left(\frac{\text{d}\varphi}{\text{d}\xi}\right)_{\xi=\kappa_{\text{D}}^{\text{w}}a} = \frac{z_2F\sigma}{\varepsilon RT\kappa_{\text{D}}^{\text{w}}}. \tag{4.172}$$

Note that the latter equation arises from Gauss' law at the pore walls and is equivalent to the global electroneutrality condition

$$\left\langle c_1^{\text{M}}\right\rangle = \left\langle c_2^{\text{M}}\right\rangle + X, \tag{4.173}$$

where the brackets $\langle\rangle$ denote spatial averaging over the pore cross-section, which in this case is equivalent to an average over the membrane volume.

For the sake of simplicity we consider that the surface-charge density is so low that $\varphi \ll 1$ and, hence, that the Poisson–Boltzmann equation can be linearized to

$$\frac{1}{\xi}\frac{\mathrm{d}}{\mathrm{d}\xi}\left(\xi\frac{\mathrm{d}\varphi}{\mathrm{d}\xi}\right) = \varphi. \tag{4.174}$$

The solution of this equation is

$$\begin{aligned}\varphi(\xi) &= \frac{z_2 F\sigma}{\varepsilon RT\kappa_{\mathrm{D}}^{\mathrm{w}}}\frac{I_0(\xi)}{I_1(\kappa_{\mathrm{D}}^{\mathrm{w}}a)} = \frac{X}{2c_{12}^{\mathrm{w}}}\frac{\kappa_{\mathrm{D}}^{\mathrm{w}}a}{2}\frac{I_0(\xi)}{I_1(\kappa_{\mathrm{D}}^{\mathrm{w}}a)}\\ &= \varphi(0)I_0(\xi),\end{aligned} \tag{4.175}$$

where $I_0(\xi)$ and $I_1(\xi)$ are the modified Bessel functions of orders 0 and 1, respectively. It is interesting to note that the average value of the potential is

$$\langle\varphi\rangle = \varphi(0)\,\langle I_0(\xi)\rangle = X/2c_{12}^{\mathrm{w}}, \tag{4.176}$$

where we have used the relation $\xi I_0 = \mathrm{d}(\xi I_1)/\mathrm{d}\xi$ to evaluate $\langle I_0(\xi)\rangle = 2I_1(\kappa_{\mathrm{D}}^{\mathrm{w}}a)/\kappa_{\mathrm{D}}^{\mathrm{w}}a$. Note that eqn (4.145) predicts that the dimensionless Donnan potential in the limit of weakly charged membranes is $\langle\varphi\rangle = X/2c_{12}^{\mathrm{w}}$.

The average ionic concentrations in the membrane can now be evaluated as

$$\begin{aligned}\left\langle c_1^{\mathrm{M}}\right\rangle &= c_{12}^{\mathrm{w}}\left\langle \mathrm{e}^{\varphi}\right\rangle \approx c_{12}^{\mathrm{w}}\left\langle 1+\varphi+(1/2)\varphi^2\right\rangle\\ &= c_{12}^{\mathrm{w}} + (X/2) + \langle c_{12}\rangle_{\mathrm{heter}},\end{aligned} \tag{4.177}$$

$$\begin{aligned}\left\langle c_2^{\mathrm{M}}\right\rangle &= c_{12}^{\mathrm{w}}\left\langle \mathrm{e}^{-\varphi}\right\rangle \approx c_{12}^{\mathrm{w}}\left\langle 1-\varphi+(1/2)\varphi^2\right\rangle\\ &= c_{12}^{\mathrm{w}} - (X/2) + \langle c_{12}\rangle_{\mathrm{heter}},\end{aligned} \tag{4.178}$$

where

$$\langle c_{12}\rangle_{\mathrm{heter}} = \frac{c_{12}^{\mathrm{w}}\varphi^2(0)}{2}\left\langle[I_0(\xi)]^2\right\rangle = \frac{X^2}{32c_{12}^{\mathrm{w}}}(\kappa_{\mathrm{D}}^{\mathrm{w}}a)^2\left\{\left[\frac{I_0(\kappa_{\mathrm{D}}^{\mathrm{w}}a)}{I_1(\kappa_{\mathrm{D}}^{\mathrm{w}}a)}\right]^2 - 1\right\} \tag{4.179}$$

is the additional amount of Donnan electrolyte that enters the membrane due to the non-uniform distribution of the electric potential along the radial direction. In the case of wide capillaries, $\kappa_{\mathrm{D}}^{\mathrm{w}}a \gg 1$, this can be estimated as

$$\langle c_{12}\rangle_{\mathrm{heter}} \approx \frac{X^2}{32c_{12}^{\mathrm{w}}}\kappa_{\mathrm{D}}^{\mathrm{w}}a \approx \frac{X}{8}\varphi(\kappa_{\mathrm{D}}^{\mathrm{w}}a) < \frac{X}{8}, \tag{4.180}$$

since we have assumed that $\varphi < 1$.

In narrow capillaries (of weakly charged membranes), $\kappa_{\mathrm{D}}^{\mathrm{w}} a$ is smaller than unity, so that $\varphi(0) \approx \varphi(\kappa_{\mathrm{D}}^{\mathrm{w}} a) \approx X/2c_{12}^{\mathrm{w}}$ and $\langle c_{12} \rangle_{\mathrm{heter}} \approx X^2/8c_{12}^{\mathrm{w}} < X/4$. In any case, the above estimation illustrates that the spatial variation of the electric potential inside the membrane leads to a poorer co-ion exclusion than predicted by the Donnan theory. Furthermore, it shows the actual co-ion exclusion can only be estimated if we know the potential distribution. Since the fixed-charge distribution inside a membrane cannot be known accurately, it should be apparent that the Donnan theory provides only a rough estimation of the electrolyte exclusion. In practical situations, the operating conditions (such as the electric current density passing through the membrane) also affect the co-ion exclusion, and the deviations from the equilibrium values can be significant [9, 15].

4.3 Steady-state transport across ion-exchange membranes

In this section we describe the electrodiffusion of electrolyte solutions across ion-exchange membrane systems in the absence of convection. By membrane system we refer to a three-layer system composed by the membrane itself and the two diffusion boundary layers flanking it. The effect of the diffusion boundary layers on the transport properties of the system is described in Sections 4.3.8 and 4.3.9. Homogeneous membranes with uniform concentration of fixed-charged groups are considered unless otherwise stated. As a rule, the transport equations are first formulated inside the membrane for the general case of asymmetric electrolytes and arbitrary geometry. The condition of electrochemical (or Donnan) equilibrium with the external solution is then applied at the membrane boundaries and the Nernst–Planck transport equations are finally solved (for planar geometry and, most often, for symmetric electrolytes) in order to determine the flux densities and the potential drop across the membrane, as well as the concentration and electric-potential distributions inside it.

4.3.1 Transport coefficients and their equilibrium values

The polymer nature, the presence of fixed-charge groups and internal structure influence the membrane transport properties and make them different from those in the external solution. We must therefore make explicit the phase in which the transport magnitudes are evaluated. In the next sections we use the following convention for the sake of clarity. The concentrations inside the membrane incorporate no superscript, and the concentrations in the external solution incorporate a superscript w, α or β. The ionic diffusion coefficients incorporate no superscript, and the symbols of other magnitudes in the internal and external phase include superscripts M and w, respectively.

Although the ionic diffusion coefficients may take, in practice, different values inside the membrane and in the external solution, we neglect here such a difference because we want to concentrate on the effect of the composition of the membrane phase on its transport properties. However, in membranes with low water content the diffusion coefficients are significantly smaller than in the external solutions. Moreover, electrostatic interactions also seem to be responsible for the observed reduction in the counterion diffusion coefficients in strongly charged membranes [16].

Consider the transport of a strong binary electrolyte $A_{\nu_1} C_{\nu_2}$ dissociated into ν_1 ions A^{z_1} and ν_2 ions C^{z_2} whose charge numbers z_1 and z_2 satisfy the stoichiometric relation $z_1 \nu_1 + z_2 \nu_2 = 0$. Fick's first law for the (Donnan) electrolyte diffusion

$$\vec{j}_{12} \equiv \frac{t_2^{\mathrm{M}}}{\nu_1} \vec{j}_1 + \frac{t_1^{\mathrm{M}}}{\nu_2} \vec{j}_2 = -D_{12}^{\mathrm{M}} \vec{\nabla} c_{12}, \qquad (4.181)$$

and the generalized Ohm's law for the electric conduction

$$\vec{I} \equiv F(z_1 \vec{j}_1 + z_2 \vec{j}_2) = -\kappa^{\mathrm{M}} (\vec{\nabla} \phi - \vec{\nabla} \phi_{\mathrm{dif}}) = -\kappa^{\mathrm{M}} \vec{\nabla} \phi_{\mathrm{ohm}} \qquad (4.182)$$

are valid inside the membrane, but the values of the electrolyte diffusion coefficient $D_{12}^{\mathrm{M}} \equiv t_2^{\mathrm{M}} D_1 + t_1^{\mathrm{M}} D_2$ and the electrical conductivity κ^{M} are different from those in the external solution (see Table 4.1).

The flux density of an ionic species inside a charged membrane can be written as

$$\vec{j}_i = -\frac{t_i^{\mathrm{M}} \kappa^{\mathrm{M}}}{z_i^2 F^2} \vec{\nabla} \tilde{\mu}_i. \qquad (4.183)$$

Taking eqn (4.183) to the expression for the electric current density $\vec{I} = F \sum_i z_i \vec{j}_i$ and comparing it to the generalized Ohm's law, it is obtained that the diffusion potential gradient inside the membrane is given

Table 4.1. Transport coefficients inside the membrane and in the external solution.

Transport coefficient	External solution	Membrane phase
Electrolyte diffusion coefficient (asymmetric electrolyte)	$D_{12}^{\mathrm{w}} = \dfrac{\nu_{12} D_1 D_2}{\nu_2 D_1 + \nu_1 D_2}$	$D_{12}^{\mathrm{M}} \equiv \dfrac{D_1 D_2 (z_1^2 c_1 + z_2^2 c_2)}{z_1^2 D_1 c_1 + z_2^2 D_2 c_2}$
Electrical conductivity (asymmetric electrolyte)	$\kappa^{\mathrm{w}} = \dfrac{z_1^2 \nu_1^2 F^2}{RT} \left(\dfrac{D_1}{\nu_1} + \dfrac{D_2}{\nu_2} \right) c_{12}^{\mathrm{w}}$	$\kappa^{\mathrm{M}} \equiv \dfrac{F^2}{RT} \sum_i z_i^2 D_i c_i$
Transport numbers (asymmetric electrolyte)	$t_i^{\mathrm{w}} = \dfrac{D_i / \nu_i}{D_1 / \nu_1 + D_2 / \nu_2}$	$t_i^{\mathrm{M}} \equiv \dfrac{z_i^2 D_i c_i}{\sum_j z_j^2 D_j c_j}$
Electrolyte diffusion coefficient (symmetric electrolyte)	$D_{12}^{\mathrm{w}} = \dfrac{2 D_1 D_2}{D_1 + D_2}$	$D_{12}^{\mathrm{M}} \equiv \dfrac{D_1 D_2 (c_1 + c_2)}{D_1 c_1 + D_2 c_2}$
Electrical conductivity (symmetric electrolyte)	$\kappa^{\mathrm{w}} = \dfrac{z_1^2 F^2}{RT} (D_1 + D_2) c_{12}^{\mathrm{w}}$	$\kappa^{\mathrm{M}} \equiv \dfrac{z_1^2 F^2}{RT} (D_1 c_1 + D_2 c_2)$
Transport numbers (symmetric electrolyte)	$t_i^{\mathrm{w}} = \dfrac{D_i}{D_1 + D_2}$	$t_i^{\mathrm{M}} \equiv \dfrac{D_i c_i}{D_1 c_1 + D_2 c_2}$

Table 4.2. Transport coefficients of a symmetric electrolyte inside an ion-exchange membrane under equilibrium conditions.

Weakly charged, $X \ll c_{12}^{\mathrm{w}}$	Moderately charged, $X > c_{12}^{\mathrm{w}}$	Strongly charged, $X \gg c_{12}^{\mathrm{w}}$
$D_{12}^{\mathrm{M}} \approx D_{12}^{\mathrm{w}}$	$D_{12}^{\mathrm{M}} \approx D_2\left[1+\left(1-\frac{D_2}{D_1}\right)\left(\frac{c_{12}^{\mathrm{w}}}{X}\right)^2\right]$	$D_{12}^{\mathrm{M}} \approx D_2$
$\kappa^{\mathrm{M}} \approx \kappa^{\mathrm{w}}$	$\kappa^{\mathrm{M}} \approx \frac{z_1^2 F^2 D_1 X}{RT}\left[1+\left(1+\frac{D_2}{D_1}\right)\left(\frac{c_{12}^{\mathrm{w}}}{X}\right)^2\right]$	$\kappa^{\mathrm{M}} \approx \frac{z_1^2 F^2 D_1 X}{RT}$
$t_i^{\mathrm{M}} \approx t_i^{\mathrm{w}}$	$t_1^{\mathrm{M}} = 1 - t_2^{\mathrm{M}} \approx 1 - \frac{D_2}{D_1}\left(\frac{c_{12}^{\mathrm{w}}}{X}\right)^2$	$t_1^{\mathrm{M}} = 1 - t_2^{\mathrm{M}} \approx 1$

by the expression

$$-f\,\vec{\nabla}\phi_{\mathrm{dif}} \equiv \sum_i \frac{t_i^{\mathrm{M}}}{z_i}\vec{\nabla}\ln c_i, \qquad (4.184)$$

where the transport numbers t_i^{M} are functions of the local composition and hence of position. Thus, for instance, when describing the transport of a binary electrolyte across a strongly charged membrane, it is satisfied that $c_1 \gg c_2$ and $t_1^{\mathrm{M}} \gg t_2^{\mathrm{M}}$, but $t_1^{\mathrm{M}}\vec{\nabla}\ln c_1 \approx t_2^{\mathrm{M}}\vec{\nabla}\ln c_2$.

As described in Section 4.2.2 under equilibrium conditions, the composition of the solution filling the membrane is determined by the influence of the bound fixed-charge groups on the electrostatic partitioning of ions. Thus, for instance, when the membrane is in equilibrium with a bathing solution of a symmetric binary electrolyte[12] of concentration c_{12}^{w}, the ionic molar concentrations in the membrane phase are

$$c_1 = c_2 + X = (X/2) + [(X/2)^2 + (c_{12}^{\mathrm{w}})^2]^{1/2}, \qquad (4.185)$$

where subscripts 1 and 2 denote the counterion and the co-ion, respectively. The use of eqn (4.185) allows us to find the approximate expressions of the transport coefficients shown in Table 4.2.

Figure 4.23 shows the dependence of these transport coefficients on the fixed-charge concentration for different values of the diffusion coefficient ratio D_2/D_1. As a rule, the transport numbers of the co-ions inside the membrane are smaller than in the external solutions while the transport numbers of the counterions show the opposite behaviour. In the limiting case of very low external electrolyte concentration (compared with that of the membrane fixed-charged groups) the co-ion exclusion is practically total, the electrical conductivity is determined by the counterions only, and the membrane is said to be ideally permselective. The real membrane systems, however, never

[12] The restriction of the transport equations to the case of a symmetric electrolyte can often be identified by the use of the quantity $X \equiv z_{\mathrm{M}}c_{\mathrm{M}}/z_2$. On the contrary, when asymmetric electrolytes are considered, the fixed-charge concentration c_{M} is used.

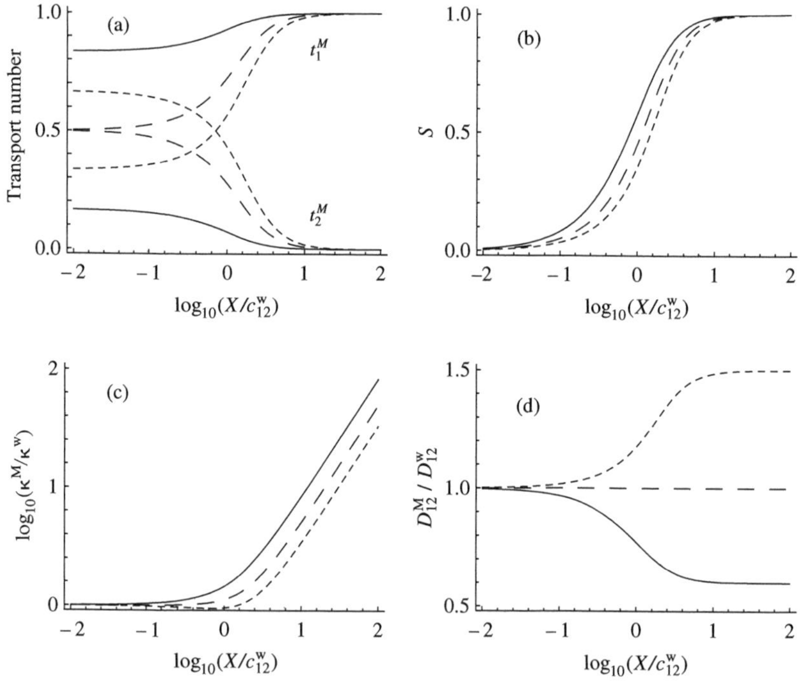

Fig. 4.23.
(a) Ionic transport numbers (1: counterion, 2: co-ion), (b) permselectivity, (c) electrical conductivity (referred to the external conductivity), and (d) electrolyte diffusion coefficient (referred to the external value) of an ion-exchange membrane with a fixed charge concentration X in equilibrium with a symmetric, binary electrolyte solution of concentration c_{12}^w. The ionic diffusion coefficient ratio has been given the values $D_2/D_1 = 0.2$ (solid lines), 1 (long dashed), and 2 (short dashed), which could be considered as characteristic of the electrolytes HCl, KCl, and TEACl (tetraethylammonium chloride), respectively. These transport coefficients take their external solution values when the membrane fixed charge concentration vanishes.

show ideal permselectivity due to incomplete co-ion exclusion (and the influence of the diffusion boundary layers). The term membrane permselectivity refers to its larger permeability to counterions than to co-ions, and it is often quantified by the ratio

$$S \equiv \frac{t_1^{\mathrm{M}} - t_1^{\mathrm{w}}}{1 - t_1^{\mathrm{w}}}. \tag{4.186}$$

This magnitude has also been represented in Fig. 4.23.

Note, finally, that the main difficulty in the accurate solution of the transport equations arises from the fact that the solution composition inside the membrane depends on position under non-equilibrium conditions. The transport numbers, the electrical conductivity and the electrolyte diffusion coefficient are determined by the composition and therefore they also vary with position.[13] Their local values are not known *a priori* and must be obtained from the solution of the transport equations.

4.3.2 The diffusion–conduction flux equation inside charged membranes

Consider the mass transport across a membrane that separates two uniform solutions of a symmetric, binary electrolyte with concentrations c_{12}^{α} and c_{12}^{β}

[13] In non-equilibrium systems, the membrane permselectivity does not vary with position, but its definition differs from that in eqn (4.186)[17].

under closed-circuit conditions, $I \neq 0$. In the diffusion-conduction approach, the flux density of species i ($i = 1, 2$) is written in terms of the electric current density $\vec{I}$ and the electrolyte flux density $\vec{j}_{12}$ in eqns (4.1) and (4.2) as

$$\vec{j}_i = \vec{j}_{12} + \frac{t_i^{\mathrm{M}}}{z_i}\frac{\vec{I}}{F} = -D_{12}^{\mathrm{M}}\vec{\nabla}c_i + \frac{t_i^{\mathrm{M}}}{z_i}\frac{\vec{I}}{F} \tag{4.187}$$

where we have used the fact that the concentration gradient of the Donnan electrolyte inside the membrane is $\vec{\nabla}c_{12} = \vec{\nabla}c_1 = \vec{\nabla}c_2$. Equation (4.187) is the diffusion–conduction flux equation for a symmetric, binary electrolyte inside a charged membrane. It is important to observe that the electrolyte diffusion coefficient D_{12}^{M} and the ionic transport numbers t_i^{M} are functions of the local ionic concentrations (see Table 4.1) and, therefore, they are position dependent under transport conditions. This makes the exact analytical integration of eqn (4.187) across the membrane difficult.

Approximate solutions can be obtained, however, when the external electrolyte concentrations are very similar to each other, i.e. when $\left|c_{12}^{\beta} - c_{12}^{\alpha}\right| \ll \overline{c_{12}^{\mathrm{w}}}$, where $\overline{c_{12}^{\mathrm{w}}} \equiv (c_{12}^{\beta} + c_{12}^{\alpha})/2$ is the average external concentration. The transport coefficients can then be approximated by the equilibrium (or membrane average) values

$$D_{12}^{\mathrm{M}} \approx \overline{D_{12}^{\mathrm{M}}} = \frac{D_1 D_2(\overline{c_1} + \overline{c_2})}{D_1\overline{c_1} + D_2\overline{c_2}} = \frac{D_{12}^{\mathrm{w}}[1 + (2\overline{c_{12}^{\mathrm{w}}}/X)^2]^{1/2}}{t_1^{\mathrm{w}} - t_2^{\mathrm{w}} + [1 + (2\overline{c_{12}^{\mathrm{w}}}/X)^2]^{1/2}}, \tag{4.188}$$

$$t_1^{\mathrm{M}} \approx \overline{t_1^{\mathrm{M}}} = \frac{D_1\overline{c_1}}{D_1\overline{c_1} + D_2\overline{c_2}} = \frac{t_1^{\mathrm{w}}\{1 + [1 + (2\overline{c_{12}^{\mathrm{w}}}/X)^2]^{1/2}\}}{t_1^{\mathrm{w}} - t_2^{\mathrm{w}} + [1 + (2\overline{c_{12}^{\mathrm{w}}}/X)^2]^{1/2}} = 1 - \overline{t_2^{\mathrm{M}}}, \tag{4.189}$$

where the average ionic concentrations are given by

$$\overline{c_1} = \overline{c_2} + X = (X/2) + [(X/2)^2 + (\overline{c_{12}^{\mathrm{w}}})^2]^{1/2}. \tag{4.190}$$

The ionic flux equation can then be integrated as

$$j_i \approx \overline{j_{12}} + \frac{\overline{t_i^{\mathrm{M}}}}{z_i}\frac{I}{F}, \tag{4.191}$$

where the electrolyte flux density is

$$\overline{j_{12}} \equiv -\overline{D_{12}^{\mathrm{M}}}\Delta c_{12}/h, \tag{4.192}$$

and the concentration difference inside the membrane, $\Delta c_{12} \equiv c_{12}(h) - c_{12}(0)$, is approximated by

$$\Delta c_{12} \approx \frac{\Delta c_{12}^{\mathrm{w}}}{[1 + (2\overline{c_{12}^{\mathrm{w}}}/X)^2]^{1/2}}, \tag{4.193}$$

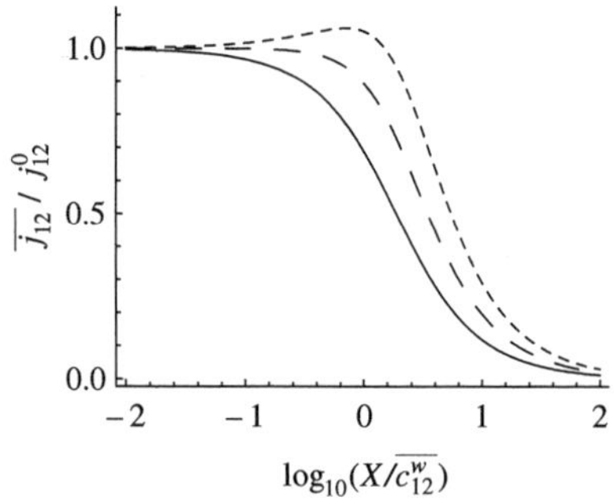

Fig. 4.24.
The effect of increasing the fixed concentration on the electrolyte transport as described by eqn (4.192). The electrolyte flux density has been scaled to the value corresponding to a neutral membrane, $j_{12}^0 = D_{12}^w(c_{12}^\alpha - c_{12}^\beta)/h$. The external concentration ratio has been fixed to $r_{12} \equiv c_{12}^\alpha/c_{12}^\beta = 1.2$ and the diffusion coefficient ratio D_2/D_1 has been given the values: 0.2 (solid lines), 1 (long dashed), and 2 (short dashed).

where $\Delta c_{12}^{\mathrm{w}} \equiv c_{12}^\beta - c_{12}^\alpha$ is the external concentration difference. Note that in neutral membranes ($c_{\mathrm{M}} = 0$ and superscript 0), $D_{12}^{\mathrm{M}} = D_{12}^{\mathrm{w}}$, $\Delta c_{12} = \Delta c_{12}^{\mathrm{w}}$, and the electrolyte flux density is

$$j_{12}^0 \approx -D_{12}^{\mathrm{w}} \Delta c_{12}^{\mathrm{w}}/h. \tag{4.194}$$

As shown in Section 4.3.8, eqn (4.191) can be useful in understanding the different transport mechanisms inside and outside the ion-exchange membrane.

Figure 4.24 shows the graphical representation of eqns (4.192) and (4.193), and evidences that, in general, the electrolyte flux density decreases with increasing fixed-charge concentration. This trend comes from the fact that the internal concentration difference is smaller than the external one, $|\Delta c_{12}| \leq |\Delta c_{12}^{\mathrm{w}}|$, as can be deduced from eqn (4.193). However, since D_{12}^{M} can be larger than D_{12}^{w} (see Fig. 4.23 (d)) it turns out that the electrolyte flux density can be larger than through a neutral membrane in moderately charged membranes when $t_1^{\mathrm{w}} < t_2^{\mathrm{w}}$ (or $D_1 < D_2$). It must be remembered that eqn (4.192) is only valid when $\left|c_{12}^\beta - c_{12}^\alpha\right| \ll \overline{c_{12}^{\mathrm{w}}}$; the general expression for j_{12} is obtained in the next section.

If we were able to drive an electric current density $\vec{I}$ through an ion-exchange membrane that separates two external solutions of identical composition while avoiding the development of concentration gradients inside the membrane, the flux density of ionic species i would then be given by[14]

$$\vec{j}_i = -z_i D_i c_i f\, \vec{\nabla}\phi = \frac{t_i^{\mathrm{M}}}{z_i} \frac{\vec{I}}{F}, \tag{4.195}$$

and the electric current density would satisfy Ohm's law $\vec{I} = -\kappa^{\mathrm{M}} \vec{\nabla}\phi$. Both t_i^{M} and κ^{M} would be independent of position under these conditions and Ohm's law could be straightforwardly integrated to $\Delta\phi = -I\,R^{\mathrm{M}}$ where $R^{\mathrm{M}} \equiv h/\kappa^{\mathrm{M}}$ is the membrane electrical resistance. Note the minus sign due to the sign convention for $\Delta\phi \equiv \phi(h) - \phi(0)$ and I, and that the SI units of R^{M} and κ^{M}

[14] In the absence of concentration gradients inside the membrane the transport number t_i^{M} is the fraction of electric current transported by species i, but in general it is defined as the contribution of this species to the electrical conductivity of the solution and depends on position.

are $\Omega\,m^2$ and $\Omega^{-1}\,m^{-1}$, respectively. It must be mentioned, however, that this case of pure electric conduction is very seldom applicable because concentration gradients develop even when the external solutions have the same composition (see Section 4.3.8).

A final interesting case is that of strongly charged membranes or low electrolyte concentrations, $c_{12}^{\alpha}, c_{12}^{\beta} \ll X$, where the ionic concentrations c_1 and c_2 are very different from each other and the co-ion transport number is negligible, $t_2^{\mathrm{M}} \ll t_1^{\mathrm{M}}$. The electrolyte diffusion is then determined by the co-ion because $D_{12}^{\mathrm{M}} \approx D_2$ and the ionic flux densities are approximated described by

$$\vec{j}_1 \approx -D_2\vec{\nabla}c_1 + \frac{\vec{I}}{z_1 F}, \tag{4.196}$$

$$\vec{j}_2 \approx -D_2\vec{\nabla}c_2. \tag{4.197}$$

These transport equations can be integrated immediately and evidence the fact that in these membranes co-ion transport takes place by diffusion, while counterion transport takes place by diffusion and migration. This conclusion is worked out in more detail at the end of the next section.

4.3.3 Diffusion of a binary electrolyte

Consider the diffusion process that occurs when a membrane separates two uniform solutions of a binary electrolyte $A_{\nu_1}C_{\nu_2}$ with concentrations c_{12}^{α} and c_{12}^{β} under open-circuit conditions, $I = 0$. The ionic flux densities are then related to the electrolyte flux density by the simple relation $\vec{j}_i = \nu_i\vec{j}_{12}$ $(i = 1, 2)$, and we aim to calculate the electrolyte flux density as a function of the external solution concentrations c_{12}^{α} and c_{12}^{β} and the membrane fixed-charge concentration c_{M}. As mentioned in the previous section, this cannot be achieved from Fick's equation $\vec{j}_{12} = -D_{12}^{\mathrm{M}}\vec{\nabla}c_{12}$ because of the dependence of the electrolyte diffusion coefficient D_{12}^{M} on the ionic concentrations, and hence, on position. Alternatively, the Nernst–Planck formalism is used. In the following paragraphs the transport equations are presented for the general case of an asymmetric electrolyte. The boundary conditions at the membrane/external solution interfaces require the solution of the equations describing the Donnan equilibria at the membrane/external solution interfaces, which only have a simple analytical form in the case of symmetric electrolytes, as described in Section 4.2.2. The final equations in this section are then restricted to symmetric electrolytes.

a) Electrolyte flux density

The steady-state Nernst–Planck equations for the ionic flux densities

$$-\vec{j}_1 = D_1(\vec{\nabla}c_1 + z_1 c_1 f\,\vec{\nabla}\phi) \tag{4.198}$$

$$-\vec{j}_2 = D_2(\vec{\nabla}c_2 + z_2 c_2 f\,\vec{\nabla}\phi) \tag{4.199}$$

constitute a system of two differential equations with three unknown variables c_1, c_2, and ϕ. The local electroneutrality assumption

$$z_1 c_1 + z_2 c_2 + z_M c_M = 0 \tag{4.200}$$

completes the equation system that allows for the evaluation of the electrolyte concentration and electric potential inside the membrane, as well as the membrane potential and the electrolyte flux density. In homogeneously charged membranes, the co-ion and counterion concentrations can differ by several orders of magnitude, but their gradients must be similar to each other because the local electroneutrality assumption implies that

$$z_1 \vec{\nabla} c_1 + z_2 \vec{\nabla} c_2 = \vec{0}. \tag{4.201}$$

This fact simplifies the mathematical problem and has interesting consequences.

Dividing eqn (4.198) by D_1 and eqn (4.199) by D_2, and adding them we get

$$\vec{j}_{12} = -D_{12}^{\mathrm{w}} \left(\vec{\nabla} c_{12} - \frac{z_M c_M}{\nu_{12}} f \, \vec{\nabla} \phi_{\mathrm{dif}} \right), \tag{4.202}$$

where the relation $\vec{j}_i = \nu_i \vec{j}_{12}$ $(i = 1, 2)$ has been used and $\nu_{12} \vec{\nabla} c_{12} = \vec{\nabla} c_1 + \vec{\nabla} c_2$ from eqn (4.201). Note that $\vec{\nabla} \phi = \vec{\nabla} \phi_{\mathrm{dif}}$ because there is no ohmic drop when $I = 0$. Equation (4.202) can be easily integrated because it involves constant coefficients. Integration over a planar membrane extending from position $x = 0$ to $x = h$, where h is the membrane thickness, yields

$$j_{12} = -\frac{D_{12}^{\mathrm{w}}}{h} \left(\Delta c_{12} - \frac{z_M c_M}{\nu_{12}} f \, \Delta \phi_{\mathrm{dif}} \right), \tag{4.203}$$

where $\Delta c_{12} \equiv c_{12}(h) - c_{12}(0)$ is the concentration drop inside the membrane. The electrolyte concentrations at the membrane boundaries are given by the Donnan equilibrium conditions, eqn (4.111). The diffusion potential drop, $\Delta \phi_{\mathrm{dif}} \equiv \phi(h) - \phi(0)$, is given by

$$\Delta \phi_{\mathrm{dif}} = \frac{\Gamma}{f} \ln \frac{c_1(h) + c_2(h) + z_M c_M [\Gamma + (z_1 + z_2)/(z_1 z_2)]}{c_1(0) + c_2(0) + z_M c_M [\Gamma + (z_1 + z_2)/(z_1 z_2)]}, \tag{4.204}$$

where

$$\Gamma \equiv -\left(\frac{t_1^{\mathrm{w}}}{z_1} + \frac{t_2^{\mathrm{w}}}{z_2} \right) = \frac{D_2 - D_1}{z_1 D_1 - z_2 D_2}, \tag{4.205}$$

and

$$t_1^{\mathrm{w}} \equiv \frac{z_1 D_1}{z_1 D_1 - z_2 D_2} = \frac{\nu_2 D_1}{\nu_2 D_1 + \nu_1 D_2} = 1 - t_2^{\mathrm{w}} \tag{4.206}$$

is the counterion transport number in the external solutions. Equation (4.204) has been obtained by transforming eqn (4.184) to

$$f\,\vec{\nabla}\phi_{\text{dif}} = \Gamma\vec{\nabla}\ln\left(c_1 + \frac{z_M c_M t_2^{\text{w}}}{z_1}\right) = \Gamma\vec{\nabla}\ln\left(c_2 + \frac{z_M c_M t_1^{\text{w}}}{z_2}\right)$$
$$= \Gamma\vec{\nabla}\ln\left[c_1 + c_2 + z_M c_M\left(\frac{t_1^{\text{w}}}{z_2} + \frac{t_2^{\text{w}}}{z_1}\right)\right] \quad (4.207)$$

and further integration.

It should be noted that the local electroneutrality equation, $z_1c_1(x)+z_2c_2(x)+z_Mc_M = 0$, implies that the spatial distribution of the counterion and co-ion concentrations must have the same functional dependence, $c_i(x) = c_i(0) + \Delta c_i\delta(x)$ $(i = 1, 2)$. Therefore, the Henderson equation for the diffusion potential

$$\Delta\phi_{\text{dif}} = \frac{\Gamma}{f}\ln\frac{z_1^2D_1c_1(h) + z_2^2D_2c_2(h)}{z_1^2D_1c_1(0) + z_2^2D_2c_2(0)} \quad (4.208)$$

is exact in this situation and, indeed, eqn (4.204) can be transformed to eqn (4.208).

In the case of a symmetric electrolyte,[15] the total electrolyte concentration at the membrane boundaries is

$$c_T(0) = c_1(0) + c_2(0) = [X^2 + (2c_{12}^{\alpha})^2]^{1/2}, \quad (4.209)$$
$$c_T(h) = c_1(h) + c_2(h) = [X^2 + (2c_{12}^{\beta})^2]^{1/2}, \quad (4.210)$$

where $X \equiv z_Mc_M/z_2$ and the electrolyte concentration drop inside the membrane is

$$\Delta c_{12} = \Delta c_2 = [(X/2)^2 + (c_{12}^{\beta})^2]^{1/2} - [(X/2)^2 + (c_{12}^{\alpha})^2]^{1/2}. \quad (4.211)$$

The electrolyte flux density can now be evaluated from eqn (4.203) as a function of the membrane fixed-charge and external solution concentrations. Since the reversal of the external concentration difference simply changes the flow direction, only the case $c_{12}^{\alpha} > c_{12}^{\beta}$ $(j_{12} > 0)$ is considered. Introducing the average value $\overline{c_{12}^{\text{w}}} \equiv (c_{12}^{\beta} + c_{12}^{\alpha})/2$ and the ratio $r_{12} \equiv c_{12}^{\alpha}/c_{12}^{\beta}$ of the external concentrations, eqns (4.203), (4.204), and (4.210) lead to

$$\frac{j_{12}}{j_{12,\max}^0} = A - B + C(t_1^{\text{w}} - t_2^{\text{w}})\ln\frac{B + C(t_1^{\text{w}} - t_2^{\text{w}})}{A + C(t_1^{\text{w}} - t_2^{\text{w}})}, \quad (4.212)$$

[15] Recall that the equations restricted to symmetric electrolytes can be identified by the use of symbol X. In contrast, the fixed-charge concentration c_M appears in the equations valid for general electrolytes.

where

$$A \equiv \frac{c_{\mathrm{T}}(0)}{4\overline{c_{12}^{\mathrm{w}}}} = \left[C^2 + \left(\frac{r_{12}}{1+r_{12}}\right)^2\right]^{1/2}, \tag{4.213}$$

$$B \equiv \frac{c_{\mathrm{T}}(h)}{4\overline{c_{12}^{\mathrm{w}}}} = \left[C^2 + \left(\frac{1}{1+r_{12}}\right)^2\right]^{1/2}, \tag{4.214}$$

$$C \equiv X/4\overline{c_{12}^{\mathrm{w}}}, \tag{4.215}$$

and $j_{12,\max}^{0} \equiv 2D_{12}^{\mathrm{w}}\overline{c_{12}^{\mathrm{w}}}/h$ is the maximum electrolyte flux density across a neutral membrane for a given value of $\overline{c_{12}^{\mathrm{w}}}$ (which corresponds to $c_{12}^{\beta} = 0$ and $c_{12}^{\alpha} = 2\overline{c_{12}^{\mathrm{w}}}$).

Equation (4.212) should be compared to the approximation expression in eqn (4.192), and to eqn (4.194) corresponding to neutral membranes. It was shown in Fig. 4.24 that the electrolyte flux density decreases with increasing fixed-charge concentration; a trend that is broken in the case of moderately charged membranes and $D_1 < D_2$. (Remember that $t_1^{\mathrm{w}} - t_2^{\mathrm{w}}$ is equal to $(D_1 - D_2)/(D_1 + D_2)$ for symmetric electrolytes.) Similarly, the exact eqn (4.212) also describes this trend, as shown in Fig. 4.25.

Contrarily to the expression $j_{12}^{0} = -D_{12}^{\mathrm{w}}\Delta c_{12}^{\mathrm{w}}/h$ for the electrolyte flux density through a neutral membrane, eqn (4.212) does not show clearly enough how the electrolyte flux density varies with the external concentration difference, $\Delta c_{12}^{\mathrm{w}}$. It can be observed in Fig. 4.26 that the flux density increases almost linearly with $-\Delta c_{12}^{\mathrm{w}} = c_{12}^{\alpha} - c_{12}^{\beta}$, and the effect of increasing the fixed-charge concentration is to reduce the coefficient of this approximately linear variation.

In Figs. 4.25 and 4.26 the ionic diffusion coefficients have been assumed to be equal to each other, $D_1 = D_2$, and hence the diffusion potential inside the membrane vanishes. The effect of varying the ratio of the diffusion coefficients was already shown in Fig. 4.24 and there arise no significant new features from the use of the exact eqn (4.212), except for the fact that the maximum that appears when $D_2 > D_1$ is slightly less pronounced (in relative terms) when increasing r_{12} (see Fig. 4.27).

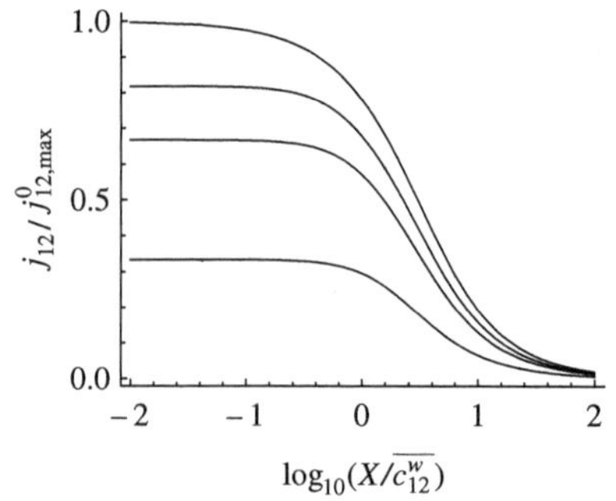

Fig. 4.25.
The electrolyte flux density decreases monotonously with the fixed charge concentration when $D_1 = D_2$. The external concentration ratio takes the values (from top to bottom) $r_{12} = 1000, 10, 5,$ and 2.

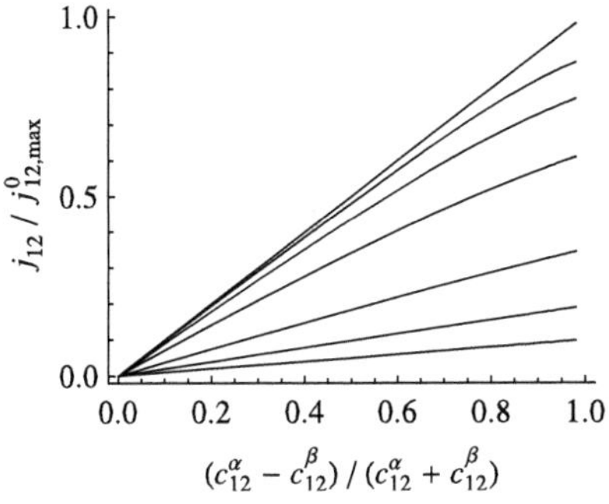

Fig. 4.26.
The electrolyte flux density increases monotonously with the external concentration difference. The diffusion coefficients have been assumed to be equal to each other, $D_1 = D_2$, and the ratio of the fixed charge concentration to the average external electrolyte concentration, $X/\overline{c^w_{12}}$, takes the values (from top to bottom) 0, 0.5, 1, 2, 5, 10, and 20.

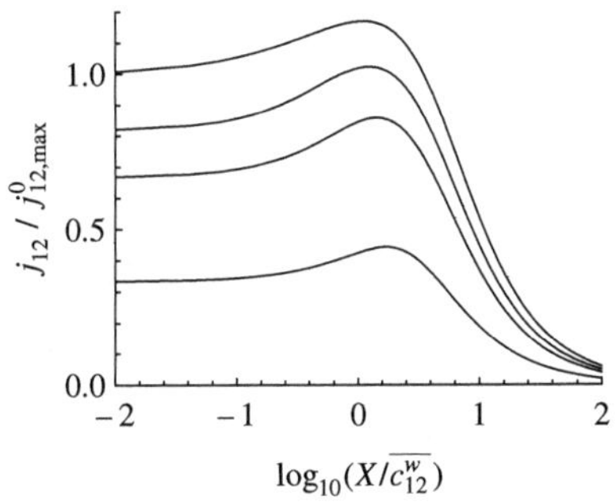

Fig. 4.27.
Variation of the electrolyte flux density with the fixed charge concentration for $D_2/D_1 = 5$. The external concentration ratio takes the values (from top to bottom) $r_{12} = 1000, 10, 5$, and 2. Note that $j_{12}/j^0_{12,\max} = (r_{12} - 1)/(r_{12} + 1)$ in the limit of weakly charged membranes.

b) Membrane potential

In the Teorell–Meyer–Sievers model [18], the membrane potential $\Delta\phi_M \equiv \phi^\beta - \phi^\alpha$, i.e. the potential drop across the external boundaries of the membrane, is evaluated as

$$\Delta\phi_M = \Delta\phi^\alpha_D - \Delta\phi^\beta_D + \Delta\phi_{\text{dif}}, \tag{4.216}$$

where $\Delta\phi^\alpha_D$ and $\Delta\phi^\beta_D$ are the Donnan potential drops, and $\Delta\phi_{\text{dif}}$ is the diffusion potential. For symmetric electrolytes, these are given by [see eqn (4.119)]

$$z_2 f\, \Delta\phi^\alpha_D = \ln\{(X/2c^\alpha_{12}) + [(X/2c^\alpha_{12})^2 + 1]^{1/2}\} = \ln\frac{(A+C)(1+r_{12})}{r_{12}}, \tag{4.217}$$

$$z_2 f\, \Delta\phi^\beta_D = \ln\{(X/2c^\beta_{12}) + [(X/2c^\beta_{12})^2 + 1]^{1/2}\} = \ln[(B+C)(1+r_{12})], \tag{4.218}$$

$$z_2 f\, \Delta\phi_{\text{dif}} = (t^w_1 - t^w_2)\ln\frac{B + C(t^w_1 - t^w_2)}{A + C(t^w_1 - t^w_2)}, \tag{4.219}$$

and the final expression for the membrane potential is

$$z_2 f\, \Delta\phi_M = -\ln r_{12} + \ln\frac{A+C}{B+C} + (t^w_1 - t^w_2)\ln\frac{B + C(t^w_1 - t^w_2)}{A + C(t^w_1 - t^w_2)}, \tag{4.220}$$

where A, B, and C are defined in eqns (4.213)–(4.215). This can also be written as

$$z_2 f\,\Delta\phi_M \equiv (\overline{t_1^M} - \overline{t_2^M}) \ln \frac{c_{12}^\beta}{c_{12}^\alpha}, \tag{4.221}$$

which constitutes the (implicit) definition of the potentiometric transport numbers $\overline{t_1^M}$ and $\overline{t_2^M} = 1 - \overline{t_1^M}$.

In weakly charged membranes, the membrane potential is approximately given by the diffusion potential, $\Delta\phi_M \approx \Delta\phi_{dif}^0$, and this can be easily obtained from eqn (4.204) as

$$f\,\Delta\phi_{dif}^0 = \Gamma \ln \frac{c_{12}^\beta}{c_{12}^\alpha} = -\Gamma \ln r_{12}. \tag{4.222}$$

The superscript 0 on the diffusion potential indicates the restriction $c_M \approx 0$.

In strongly charged membranes, $C \gg 1$, the membrane potential reduces to

$$z_2 f\,\Delta\phi_M \approx -\ln r_{12} = \ln \frac{c_{12}^\beta}{c_{12}^\alpha}, \tag{4.223}$$

which corresponds to the (Nernst) equilibrium potential for the counterion if the electrolyte is symmetric. That is, since the permeability of the membrane to the counterion is much larger than to the co-ion, and their fluxes are related through the open-circuit condition ($I = 0$), the deviation from equilibrium is much smaller for the counterion than for the co-ion.

Figure 4.28 shows the variation of the membrane potential with the external electrolyte concentration in compartment α (and constant concentration in compartment β) for different values of D_2/D_1. All the curves cross at $\Delta\phi_M = 0$ when $c_{12}^\alpha/X = 0.5 = c_{12}^\beta/X$, which corresponds to the equilibrium condition $j_{12} = 0$. In the region $c_{12}^\alpha < c_{12}^\beta$(right side of the plot), the flux density j_{12} is negative and very small in value. Correspondingly, the membrane potential tends in this region to the Nernstian slope of 60 mV/decade (particularly, when

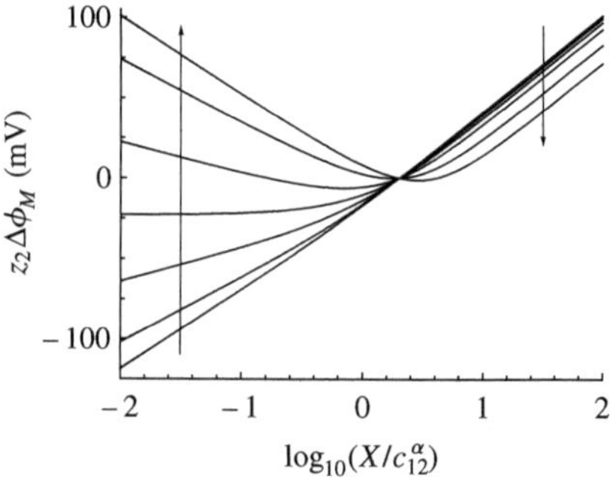

Fig. 4.28.
Membrane potential $\Delta\phi_M \equiv \phi^\beta - \phi^\alpha$ against electrolyte concentration in compartment α for constant concentration in compartment β, $c_{12}^\beta = 0.5X$, and different values of the ratio of diffusion coefficients: $D_2/D_1 = 0.1, 0.2, 0.5, 1, 2, 5,$ and 10 (increasing in the direction of the arrows). The membrane potential vanishes when the equilibrium condition $c_{12}^\alpha = c_{12}^\beta$ is satisfied.

$c_{12}^{\alpha} \ll X$). In the region $c_{12}^{\alpha} > c_{12}^{\beta}$ (left side of the plot), the flux density j_{12} is positive and increases almost linearly with c_{12}^{α}. The variation of the membrane potential with c_{12}^{α} is also logarithmic in the range $c_{12}^{\alpha} \gg X$: the slope is 60 mV/decade when $D_2/D_1 \ll 1$, –60 mV/decade when $D_2/D_1 \gg 1$, and differs from 60 mV/decade in a factor $t_1^{\mathrm{w}} - t_2^{\mathrm{w}} = (D_1 - D_2)/(D_1 + D_2)$ in other cases. Figure 4.28 can be used for both cation- and anion-exchange membranes. When used for cation-exchange membranes, the co-ion is an anion ($z_2 < 0$) and hence the quantity represented in the ordinate axis, $z_2 \Delta\phi_{\mathrm{M}}$, has opposite sign to the membrane potential. When used for anion-exchange membranes, there is no sign reversal, but the value of D_2/D_1 must be reversed, so that it continues to be the ratio of co-ion to counterion diffusion coefficient.

The experimental study of the membrane potential is often carried out by keeping constant the electrolyte concentration in, e.g., compartment β and varying that in compartment α. Obviously, the ratio D_2/D_1 then takes the value corresponding to the electrolyte under study (although it must be remembered that in practice the ionic diffusion coefficients inside the membrane are neither strictly constant nor exactly equal to those in the external solutions). The plots obtained in this way for different values of the concentration in compartment β are shown in Fig. 4.29 for $D_2/D_1 = 0.5$ and 2. These values of the ratio D_2/D_1 have been chosen to compare the membrane potential for the same electrolyte and external concentration in one anion- and one cation-exchange membrane that only differ in the sign of the fixed-charge groups. The lack of symmetry between these two families of curves is useful to identify the sign of the fixed

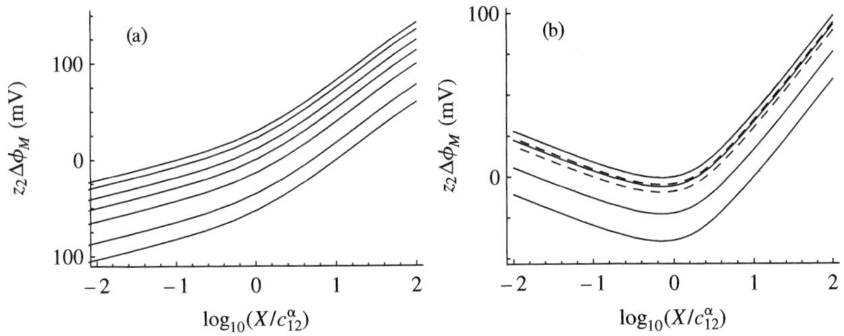

Fig. 4.29.
Variation of the membrane potential $\Delta\phi_{\mathrm{M}} \equiv \phi^{\beta} - \phi^{\alpha}$ with the external electrolyte concentration in compartment α while keeping constant that in compartment β. (a) The curves correspond to the values (from top to bottom) $c_{12}^{\beta}/X = 10, 5, 2, 1, 0.5, 0.2$, and 0.1, and the ratio of diffusion coefficients is $D_2/D_1 = 0.5$. In this case $z_2\Delta\phi_{\mathrm{M}}$ increases monotonously with X/c_{12}^{α} at constant c_{12}^{β}/X, and hence it also increases monotonously with c_{12}^{β}/X at constant X/c_{12}^{α}. (b) The solid lines correspond to (from top to bottom) $c_{12}^{\beta}/X = 1, 0.5, 0.2$, and 0.1, and $D_2/D_1 = 2$. The dashed lines correspond (from top to bottom) to $c_{12}^{\beta}/X = 2, 5$, and 10, and $D_2/D_1 = 2$. The presence of a minimum in the membrane potential curves in this case leads to a non-monotonous variation of $z_2\Delta\phi_{\mathrm{M}}$ with c_{12}^{β}/X at constant X/c_{12}^{α}. Note that $\Delta\phi_{\mathrm{M}}$ vanishes when the equilibrium condition $c_{12}^{\alpha} = c_{12}^{\beta}$ is satisfied.

charge when it is uncertain (note also the sign reversal of the membrane potential because the ordinate axis includes the co-ion transport number). Moreover, the values of the membrane potential along one of these curves can be used to estimate the fixed-charge concentration.

In order to derive the expressions for the electrolyte flux density and the diffusion potential, eqns (4.203) and (4.204), we have integrated the corresponding differential equations, eqns (4.202) and (4.207), over the membrane thickness. By performing these integrations from the boundary $x = 0$ to an arbitrary position x, the concentration $c_{12}(x) = c_2(x)$ and electric-potential distributions inside the membrane can be calculated. In the case of symmetric electrolytes these are given by the following equations

$$\frac{x}{h} = \frac{c_{12}(x) - c_{12}(0) - (X/2)z_2 f[\phi(x) - \phi(0)]}{c_{12}(h) - c_{12}(0) - (X/2)z_2 f[\phi(h) - \phi(0)]}, \tag{4.224}$$

$$z_2 f[\phi(x) - \phi(0)] = (t_1^{\mathrm{w}} - t_2^{\mathrm{w}}) \ln \frac{c_{12}(x) + t_1^{\mathrm{w}} X}{c_{12}(0) + t_1^{\mathrm{w}} X}, \tag{4.225}$$

$$c_{12}(0) = -(X/2) + [(X/2)^2 + (c_{12}^{\alpha})^2]^{1/2}, \tag{4.226}$$

$$c_{12}(h) = -(X/2) + [(X/2)^2 + (c_{12}^{\beta})^2]^{1/2}, \tag{4.227}$$

and have been represented in Figs. 4.30 and 4.31 for the case $c_{12}^{\alpha} = 2X = 4c_{12}^{\beta}$ and three values of the diffusion coefficient ratio D_2/D_1. We can observe there that the concentration and electric-potential profiles are almost linear under these conditions. Note that these figures are also valid for both anion- and cation-exchange membranes.

c) Co-ions move by diffusion in strongly charged membranes

We now discuss the relative importance of the diffusional and migrational contributions to the ionic flux densities in strongly charged membranes. The diffusional contributions of co-ions and counterions are similar in magnitude because of the local electroneutrality. The migrational contributions, on the contrary, are very different because they are proportional to the ionic concentrations and the counterion and co-ion concentrations are very different. It is then concluded that, when $\vec{j}_1$ and $\vec{j}_2$ are of similar magnitude, the migrational

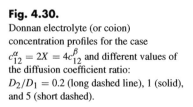

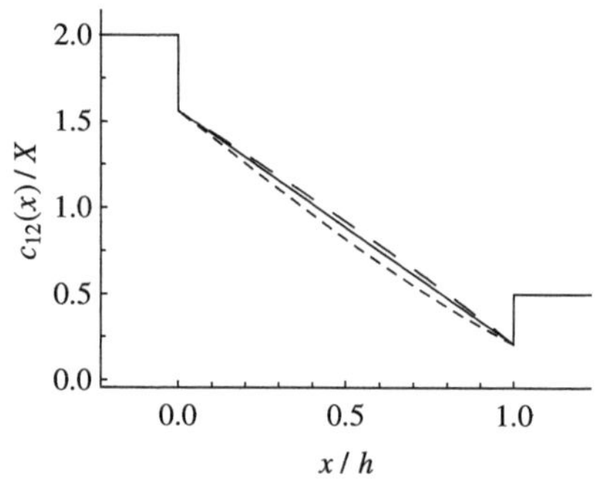

Fig. 4.30.
Donnan electrolyte (or coion) concentration profiles for the case $c_{12}^{\alpha} = 2X = 4c_{12}^{\beta}$ and different values of the diffusion coefficient ratio: $D_2/D_1 = 0.2$ (long dashed line), 1 (solid), and 5 (short dashed).

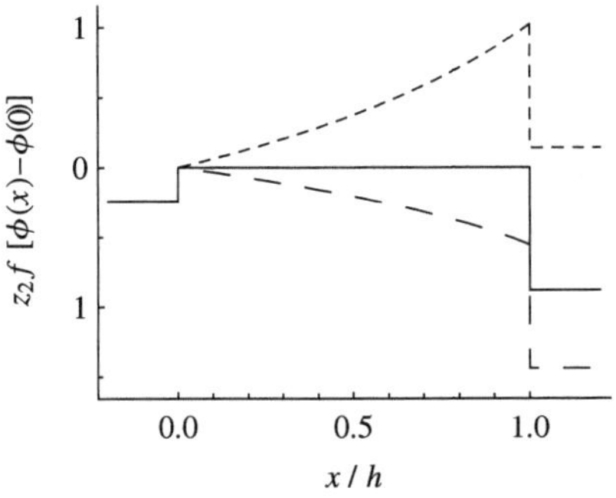

Fig. 4.31.
Electric potential profile for the case $c_{12}^{\alpha} = 2X = 4c_{12}^{\beta}$ and different values of the diffusion coefficient ratio: $D_2/D_1 = 0.2$ (long dashed line), 1 (solid), and 5 (short dashed).

contribution to the flux of co-ions must be negligible and $\vec{j}_2 \approx -D_2\vec{\nabla}c_2$. Thus, the main transport mechanism for co-ions under these conditions is diffusion, while counterion transport takes place both by diffusion and migration. Furthermore, the diffusion potential is negligible in strongly charged membranes. Indeed, since the co-ion concentration is then much smaller than the fixed-charge concentration, the argument of the logarithm in eqn (4.207) varies very slightly with position inside the membrane and the diffusion potential gradient is approximately given by

$$f\,\vec{\nabla}\phi_{\mathrm{dif}} \approx \frac{D_1 - D_2}{z_M c_M D_1}\vec{\nabla}c_1, \tag{4.228}$$

which is very small in strongly charged membranes. Note, however, that the migrational contribution to the counterion flux is not negligible, except for the trivial case $D_1 = D_2$ in which there is no potential gradient. This contribution is approximately given by $(D_1 - D_2)\vec{\nabla}c_1$ and eqn (4.198) becomes

$$\vec{j}_1 \approx -D_1(\vec{\nabla}c_1 - z_M c_M f\,\vec{\nabla}\phi_{\mathrm{dif}}) \approx -D_2\vec{\nabla}c_1, \tag{4.229}$$

which fully agrees with $\vec{j}_2 \approx -D_2\vec{\nabla}c_2$ because $\vec{j}_1 = (\nu_1/\nu_2)\vec{j}_2 = -(z_2/z_1)\vec{j}_2$.

Although these conclusions were also obtained in Section 4.3.2, we are now in position to work them out in more detail and deduce the conditions that make the migrational contribution to the coion flux negligible. Taking the (implicit) derivative of eqn (4.224) and making use of eqn (4.203) the concentration gradient inside the membrane can be written as

$$\frac{\mathrm{d}c_{12}}{\mathrm{d}x} = -\frac{j_{12}}{D_{12}^{\mathrm{w}}}\frac{c_{12}(x) + t_1^{\mathrm{w}}X}{c_{12}(x) + (X/2)}. \tag{4.230}$$

The relative contribution of diffusion to the co-ion flux density is then

$$\frac{j_{2,dif}(x)}{j_2} = -\frac{D_2}{j_2}\frac{\mathrm{d}c_2}{\mathrm{d}x} = \frac{(1/t_1^{\mathrm{w}})c_{12}(x) + X}{2c_{12}(x) + X}, \tag{4.231}$$

which has been represented at $x = 0$ in Fig. 4.32. This means that when the membrane fixed-charge concentration is larger than the external electrolyte concentration (in both compartments) by ca. a factor 10, eqn (4.203) can be approximated by the much simpler expression $j_{12} \approx -D_2\Delta c_{12}/h$.

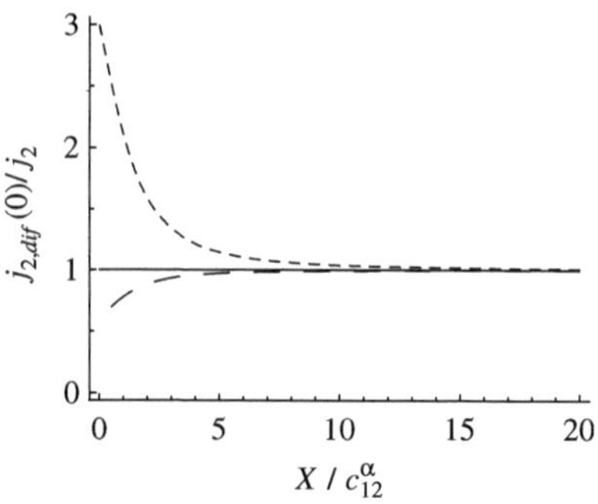

Fig. 4.32.
Variation of the diffusive contribution to the coion flux density, evaluated from eqn (4.231) at $x = 0$, with the fixed charge concentration for different values of the diffusion coefficient ratio: $D_2/D_1 = 0.2$ (long dashed line), 1 (solid line), and 5 (short dashed line).

d) Membrane permeability

We close this section with some additional comments on the electrolyte flux density. The membrane permeability to the electrolyte is defined as

$$P_{12} \equiv -\frac{j_{12}}{\Delta c_{12}^{\mathrm{w}}}. \tag{4.232}$$

The SI unit of this quantity is m/s, as can be easily checked from the expression corresponding to a neutral membrane

$$P_{12}^{0} \equiv -\frac{j_{12}^{0}}{\Delta c_{12}^{\mathrm{w}}} = \frac{D_{12}^{\mathrm{w}}}{h}. \tag{4.233}$$

We obtained in eqn (4.203) that j_{12} is the sum of $-D_{12}^{\mathrm{w}}\Delta c_{12}/h$ and another term proportional to $\Delta\phi_{\mathrm{dif}}$. Since the concentration drop of the Donnan electrolyte inside the membrane, $\Delta c_{12} \equiv c_{12}(h) - c_{12}(0)$, is generally smaller (in magnitude) than the difference of the external concentrations, $\Delta c_{12}^{\mathrm{w}} \equiv c_{12}^{\beta} - c_{12}^{\alpha}$, it is often the case that the permeability of a charged membrane to an electrolyte is significantly smaller than that of a neutral membrane, i.e. $P_{12} < P_{12}^{0}$. There are, however, exceptions to this rule.

The term proportional to $\Delta\phi_{\mathrm{dif}}$ in eqn (4.203) may have equal or opposite sign to that proportional to Δc_{12} depending on the values of the ionic transport numbers in the external solutions. In particular, we have observed in Figs. 4.24 and 4.27 that the electrolyte flux density may exhibit some maxima as a function of $X/\overline{c_{12}^{\mathrm{w}}}$ when the co-ion diffusion coefficient is larger than that of the counterion. It is then possible to find 'abnormal' situations in which a charged membrane is more permeable to the electrolyte than a neutral membrane of equal thickness in spite of the fact that the fixed charges exclude the (Donnan) electrolyte.

4.3.4 Membrane permselectivity

As an extension of the transport problem considered in Section 4.3.3, we consider here the conduction of an electric current density by a binary electrolyte

$$\vec{I} = F(z_1\vec{j}_1 + z_2\vec{j}_2) \tag{4.234}$$

across a membrane that separates two uniform solutions of a binary electrolyte with concentrations c_{12}^{α} and c_{12}^{β}. The main difference from the situation described in Section 4.3.3 is that the two ionic flux densities are no longer similar to each other. That is, the relation $\vec{j}_i = \nu_i \vec{j}_{12} (i = 1, 2)$ describing electrolyte diffusion does not hold and must be replaced by the diffusion–conduction equation, $\vec{j}_i = \nu_i \vec{j}_{12} + t_i^{\mathrm{M}} \vec{I}/(z_i F)$. In relation to the conduction term it must be observed that the transport numbers of the counterion and the co-ion inside charged membranes can be very different from each other and, therefore, the flux densities $\vec{j}_1$ and $\vec{j}_2$ can also be very different. As a limiting situation, we could think of an ideally permselective membrane in which $\vec{j}_1 \approx \vec{I}/z_1 F$ and $\vec{j}_2 \approx \vec{0}$. In general, however, both ions contribute to the transport of electric current and it becomes convenient to introduce the integral transport numbers $T_i (i = 1, 2)$ as

$$\vec{j}_i \equiv \frac{T_i}{z_i} \frac{\vec{I}}{F}. \tag{4.235}$$

Under steady-state conditions, the ionic flux densities and the electric current density are independent of position (or, at least, show no divergence in the case of non-planar geometry), and therefore the integral transport numbers are also independent of position. This is one of the main differences between T_i and the local migrational transport number t_i^{M} that appears in eqn (4.189). The other difference is that t_i^{M} is only related to electric conduction, while T_i is also affected by the electrolyte diffusion. In other words, the integral transport numbers should be rather understood as dimensionless values for the ionic flux densities. They are obtained by scaling the actual flux densities by the value that corresponds to the case in which only the species under consideration is responsible for electric current transport. Obviously, $T_i = t_i^{\mathrm{M}}$ in the absence of concentration gradients inside the membrane, but this situation occurs very seldom in practice.

Our aim in this section is to solve the Nernst–Planck equations and local electroneutrality approximation, eqns (4.198)–(4.200), and to evaluate the ionic flux densities, the membrane potential $\Delta\phi_{\mathrm{M}}$, and related magnitudes like the membrane permselectivity

$$S \equiv \frac{T_1 - t_1^{\mathrm{w}}}{1 - t_1^{\mathrm{w}}} \tag{4.236}$$

and the membrane electrical resistance

$$R^{\mathrm{M}} \equiv -\Delta\phi_{\mathrm{ohm}}/I. \tag{4.237}$$

A rather general solution procedure of the transport equations for asymmetric electrolytes is based on the use of a new set of concentration and flux-density variables. These are defined as

$$S_0 \equiv c_1 + c_2 \equiv c_T, \tag{4.238}$$

$$S_1 \equiv z_1 c_1 + z_2 c_2 = -z_M c_M, \tag{4.239}$$

$$S_2 \equiv z_1^2 c_1 + z_2^2 c_2 = (z_1 + z_2) S_1 - z_1 z_2 c_T, \tag{4.240}$$

$$\vec{G}_0 \equiv \frac{\vec{j}_1}{D_1} + \frac{\vec{j}_2}{D_2}, \tag{4.241}$$

$$\vec{G}_1 \equiv \frac{z_1 \vec{j}_1}{D_1} + \frac{z_2 \vec{j}_2}{D_2}. \tag{4.242}$$

Moreover, assuming that the vectors $\vec{G}_0$ and $\vec{G}_1$ have the same direction, a convenient dimensionless constant Γ can be defined implicitly through the relation

$$\vec{G}_1 \equiv -z_1 z_2 \Gamma \vec{G}_0. \tag{4.243}$$

The Nernst–Planck equations are then written in terms of these variables as

$$-\vec{G}_0 = \vec{\nabla} S_0 + S_1 f\, \vec{\nabla}\phi = \vec{\nabla} c_T - z_M c_M f\, \vec{\nabla}\phi, \tag{4.244}$$

$$-\vec{G}_1 = \vec{\nabla} S_1 + S_2 f\, \vec{\nabla}\phi = S_2 f\, \vec{\nabla}\phi, \tag{4.245}$$

where we have used that $\vec{\nabla} S_1 = -z_M \vec{\nabla} c_M = \vec{0}$. Since S_2 and c_T are related through eqn (4.240), eqns (4.244) and (4.245) can be considered as an equation system in the variables c_T and ϕ. This system can be easily integrated. Equation (4.244) has constant coefficients, and eqn (4.243) can be used to obtain the following expression for the electric potential gradient

$$f\, \vec{\nabla}\phi = \Gamma \vec{\nabla} \ln\{c_T + z_M c_M[\Gamma + (z_1 + z_2)/(z_1 z_2)]\}. \tag{4.246}$$

Finally, combining eqns (4.234), (4.241) and (4.242), the current density can be written as

$$\vec{I} = \left(\Gamma + \frac{t_1^{\mathrm{w}}}{z_1} + \frac{t_2^{\mathrm{w}}}{z_2}\right) \frac{z_1 z_2}{z_2 - z_1} (z_1 D_1 - z_2 D_2) F \vec{G}_0, \tag{4.247}$$

and this equation allows us to determine Γ; note that $\vec{G}_0$ is also known as a function of Γ after integration of eqns (4.244) and (4.246). The ionic flux densities are then evaluated as

$$\vec{j}_1 = \frac{z_2 D_1}{z_2 - z_1}(1 + z_1 \Gamma)\vec{G}_0 = \frac{\nu_1 D_{12}^{\mathrm{w}}}{\nu_{12}} \vec{G}_0 + \frac{t_1^{\mathrm{w}} \vec{I}}{z_1 F}, \tag{4.248}$$

$$\vec{j}_2 = \frac{z_1 D_2}{z_1 - z_2}(1 + z_2 \Gamma)\vec{G}_0 = \frac{\nu_2 D_{12}^{\mathrm{w}}}{\nu_{12}} \vec{G}_0 + \frac{t_2^{\mathrm{w}} \vec{I}}{z_2 F}. \tag{4.249}$$

In the case of a symmetric electrolyte and planar geometry, this solution procedure is applied as follows. Integration of eqns (4.244) and (4.246) over the membrane leads to

$$G_0 h = -\Delta c_{\mathrm{T}} + z_2 X f \,\Delta\phi, \tag{4.250}$$

$$f\,\Delta\phi = \Gamma \ln \frac{c_{\mathrm{T}}(h) + z_2 \Gamma X}{c_{\mathrm{T}}(0) + z_2 \Gamma X}, \tag{4.251}$$

where $c_{\mathrm{T}}(0)$ and $c_{\mathrm{T}}(h)$ are given by eqns (4.209) and (4.210), and $X \equiv z_{\mathrm{M}} c_{\mathrm{M}}/z_2$. The parameter Γ must be evaluated from the solution of the transcendental equation

$$I = (z_2\Gamma - t_1^{\mathrm{w}} + t_2^{\mathrm{w}}) \frac{z_1 F (D_1 + D_2)}{2h} \times \left[\Delta c_{\mathrm{T}} - z_2 \Gamma X \ln \frac{c_{\mathrm{T}}(h) + z_2 \Gamma X}{c_{\mathrm{T}}(0) + z_2 \Gamma X}\right], \tag{4.252}$$

which is obtained after substitution of eqn (4.250) into eqn (4.247). Note that Γ is related to the membrane permselectivity

$$S \equiv \frac{T_1 - t_1^{\mathrm{w}}}{1 - t_1^{\mathrm{w}}} = \frac{\nu_1 D_{12}^{\mathrm{w}}}{\nu_{12}(1 - t_1^{\mathrm{w}})} \frac{z_1 F G_0}{I} = \frac{z_1 F D_1 G_0}{I} = \frac{2 t_1^{\mathrm{w}}}{t_1^{\mathrm{w}} - t_2^{\mathrm{w}} - z_2 \Gamma}, \tag{4.253}$$

where eqns (4.247) and (4.248) have been used. Thus, when the membrane is strongly charged and exhibits a high permselectivity ($S \approx 1$), $z_2\Gamma$ takes the value -1.

The total ionic concentration $c_{\mathrm{T}}(x)$ and the electric-potential distributions inside the membrane can be calculated by integrating eqns (4.244) and (4.246) from $x = 0$ to an arbitrary position x. The expressions thus obtained are

$$\frac{x}{h} = \frac{c_{\mathrm{T}}(0) - c_{\mathrm{T}}(x) + z_2 X f[\phi(x) - \phi(0)]}{c_{\mathrm{T}}(0) - c_{\mathrm{T}}(h) + z_2 X f[\phi(h) - \phi(0)]}, \tag{4.254}$$

and

$$f[\phi(x) - \phi(0)] = \Gamma \ln \frac{c_{\mathrm{T}}(x) + z_2 \Gamma X}{c_{\mathrm{T}}(0) + z_2 \Gamma X}. \tag{4.255}$$

The ionic concentrations are then obtained as $c_1 = (c_{\mathrm{T}} + X)/2$ and $c_2 = (c_{\mathrm{T}} - X)/2$. The close similarity between eqns (4.254) and (4.255), on the one hand, and eqns (4.224) and (4.225), on the other hand, can be expected from the fact that all equations in Section 4.3.3 could have been obtained as particular cases of those in the present section after setting $I = 0$.

The potential drop inside the membrane, i.e. the electric potential difference between the internal boundaries of the membrane, can be written as the sum of ohmic and diffusional contributions, $\Delta\phi = \Delta\phi_{\text{dif}} + \Delta\phi_{\text{ohm}}$. The diffusional contribution is given by eqns (4.204) or (4.208) and is rewritten here as

$$z_2 f\, \Delta\phi_{\text{dif}} = (t_1^{\text{w}} - t_2^{\text{w}}) \ln \frac{c_{\text{T}}(h) + (t_1^{\text{w}} - t_2^{\text{w}})X}{c_{\text{T}}(0) + (t_1^{\text{w}} - t_2^{\text{w}})X}. \tag{4.256}$$

The ohmic contribution can be evaluated from eqns (4.251) and (4.256) as

$$z_2 f\, \Delta\phi_{\text{ohm}} = z_2\Gamma \ln \frac{c_{\text{T}}(h) + z_2\Gamma X}{c_{\text{T}}(0) + z_2\Gamma X} - (t_1^{\text{w}} - t_2^{\text{w}}) \ln \frac{c_{\text{T}}(h) + (t_1^{\text{w}} - t_2^{\text{w}})X}{c_{\text{T}}(0) + (t_1^{\text{w}} - t_2^{\text{w}})X}. \tag{4.257}$$

Remember that $t_1^{\text{w}} - t_2^{\text{w}}$ is the value of $z_2\Gamma$ when $I = 0$.

The membrane potential, i.e. the electric potential difference between the external boundaries of the membrane, is

$$\Delta\phi_{\text{M}} = \Delta\phi_{\text{D}}^{\alpha} - \Delta\phi_{\text{D}}^{\beta} + \Delta\phi \tag{4.258}$$

where $\Delta\phi_{\text{D}}^{\alpha}$ and $\Delta\phi_{\text{D}}^{\beta}$ are the Donnan potential drops given by eqns (4.217) and (4.218).

The graphical representation of the above transport magnitudes can be conveniently done after the introduction of the average value $\overline{c_{12}^{\text{w}}} \equiv (c_{12}^{\beta} + c_{12}^{\alpha})/2$ and the ratio $r_{12} \equiv c_{12}^{\alpha}/c_{12}^{\beta}$ of the external concentrations, as in Section 4.3.3. Equations (4.250)–(4.252) then take the form

$$\frac{G_0}{G_{0,\max}^0} = A - B + z_2\Gamma C \ln \frac{B + z_2\Gamma C}{A + z_2\Gamma C}, \tag{4.259}$$

$$f\, \Delta\phi = \Gamma \ln \frac{B + z_2\Gamma C}{A + z_2\Gamma C}, \tag{4.260}$$

$$z_2\Gamma = t_1^{\text{w}} - t_2^{\text{w}} - 2t_1^{\text{w}} t_2^{\text{w}} \frac{I}{I_0} \frac{G_{0,\max}^0}{G_0}, \tag{4.261}$$

where $G_{0,\max}^0 \equiv 4\overline{c_{12}^{\text{w}}}/h$ is the maximum value of G_0 in the case of neutral membranes, $I_0 \equiv 2z_1 F D_{12}^{\text{w}} \overline{c_{12}^{\text{w}}}/h$, and A, B, C are defined in eqns (4.213)–(4.215).

Figures 4.33 (a)–(f) show graphical representations of $G_0/G_{0,\max}^0$ vs. $\log_{10}(X/\overline{c_{12}^{\text{w}}}) = \log_{10}(4C)$ under different conditions. In very weakly charged membranes, $C \ll 1$, and

$$G_0 \approx -\frac{2\Delta c_{12}^{\text{w}}}{h} = \frac{2(c_{12}^{\alpha} - c_{12}^{\beta})}{h} = G_{0,\max}^0 \frac{r_{12} - 1}{r_{12} + 1}. \tag{4.262}$$

In very strongly charged membranes, co-ion exclusion is almost complete, so that $c_1 \approx X, \Delta c_{\text{T}} \approx 0$, and $j_2 \approx 0$. As a consequence, $z_2\Gamma \approx -1$, and

$$\frac{G_0}{G_{0,\max}^0} \approx t_2^{\text{w}} \frac{I}{I_0}. \tag{4.263}$$

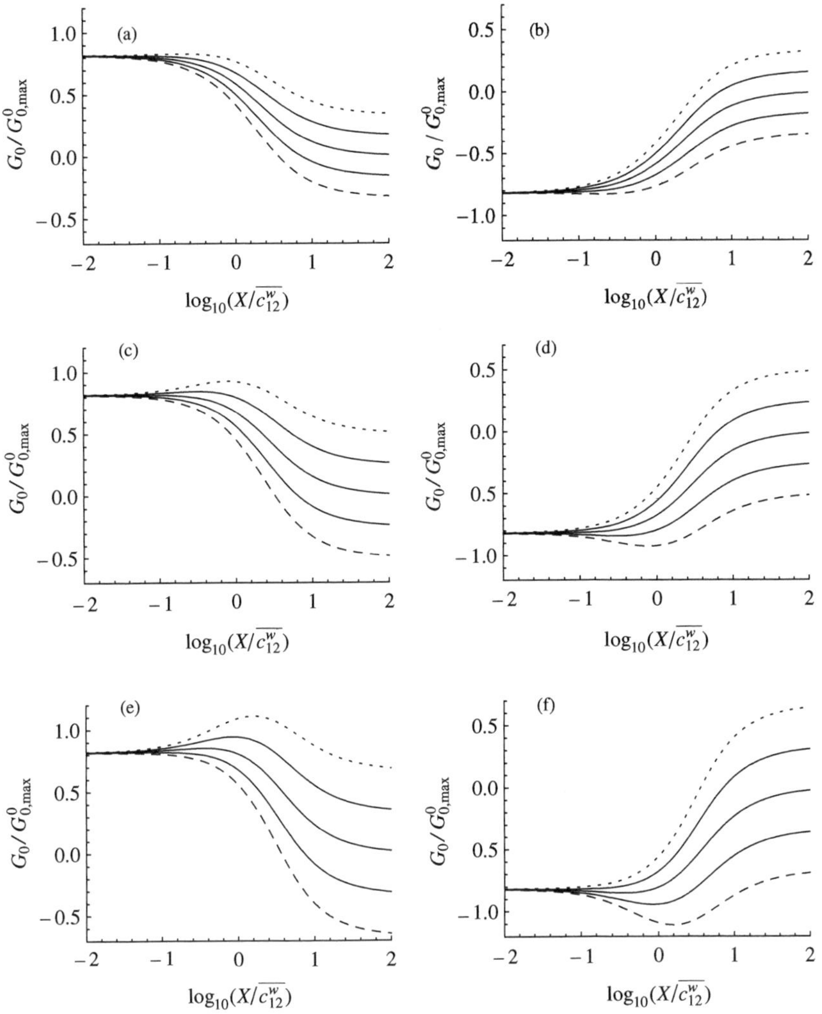

Fig. 4.33.
Variation of the magnitude $G_0/G^0_{0,\max}$ with the fixed charge concentration for different values (from top to bottom) of $I/I_0 = 1.0$ (dotted), 0.5, 0, -0.5, and -1.0 (dashed) and $D_2/D_1 = 0.5$ (a, b), 1.0 (c, d), and 2.0 (e, f). The external concentration ratio takes the values $r_{12} = 10$ (a, c, e) and 0.1 (b, d, f). Note that changing r_{12} to its reciprocal simply reverses the sign of $G_0/G^0_{0,\max}$.

In the absence of current density, $I = 0$, the results obtained in Section 4.3.3 are reproduced. In particular, $G_0/G^0_{0,\max}$ is then equal to $j_{12}/j^0_{12,\max}$ in Figs. 4.25 and 4.27.

Figures 4.34 (a)–(f) show the values of parameter $z_2\Gamma$ corresponding to Figs. 4.33 (a)–(f). In the absence of current density, this parameter takes the constant value $z_2\Gamma = t^w_1 - t^w_2$, as we deduced in Section 4.3.3. In the presence of current, the variation of $z_2\Gamma$ with $X/\overline{c^w_{12}}$ can be rather complicated and shows some singularities when G_0 vanishes. Since $G_0/G^0_{0,\max}$ takes the value given in eqn (4.262) in the limit of weakly charged membranes, the parameter $z_2\Gamma$ differs then from $t^w_1 - t^w_2$ in a value proportional to the current, $z_2\Gamma - t^w_1 + t^w_2 \propto I$. In the opposite limit of very strongly charged membranes, $z_2\Gamma$ tends to -1 in the presence of current.

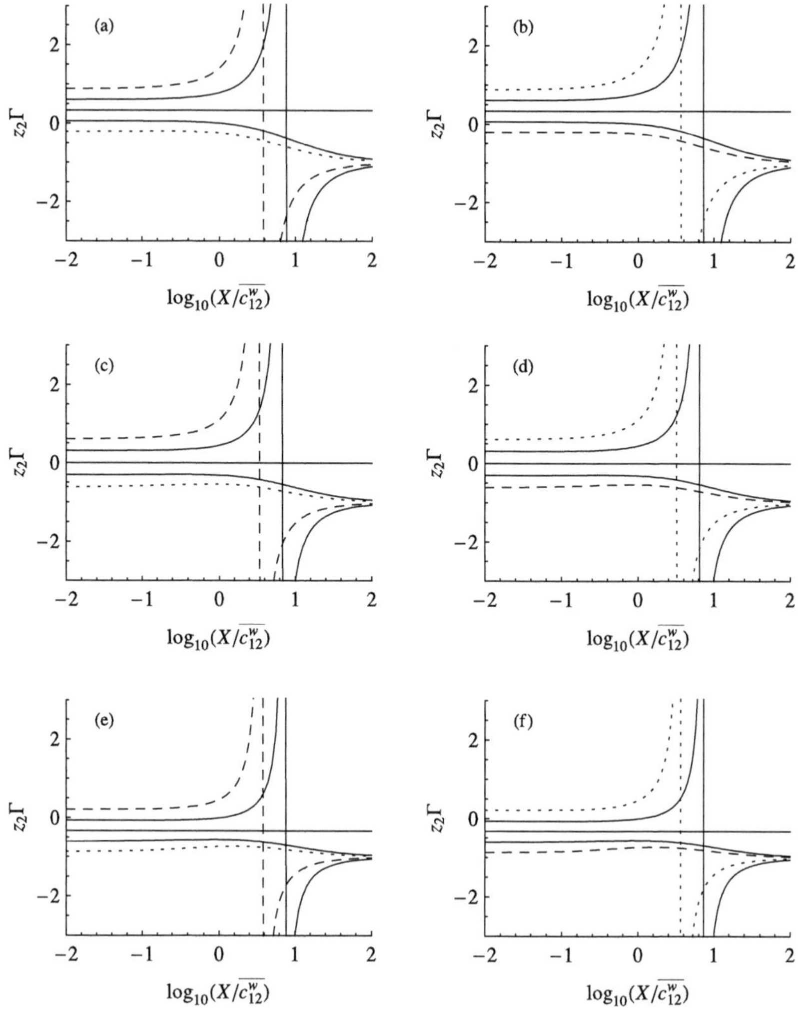

Fig. 4.34.
Variation of parameter Γ with the fixed charge concentration for different values of $I/I_0 = 1.0$ (dotted), 0.5, 0, -0.5, and -1.0 (dashed) and $D_2/D_1 = 0.5$ (a, b), 1.0 (c, d), and 2.0 (e, f). The external concentration ratio takes the values $r_{12} = 10$ (a, c, e) and 0.1 (b, d, f).

Figures 4.35 (a)–(f) show the graphical representations of the counterion and co-ion flux densities corresponding to Figs. 4.33 (a), (d), and (e). The counterion and co-ion flux densities are linked by the relation $(j_1 - j_2)/j^0_{12,\max} = I/I_0$, and they are equal to the electrolyte flux density in the absence of current, $j_1 = j_2 = j_{12}$. Note also that Figs. 4.35 (a), (d), and (e) and Figs. 4.33 (a), (d), and (e) contain similar information because the magnitudes represented there satisfy the relation $j_1/j^0_{12,\max} = G_0/G^0_{0,\max} + t_1^{\mathrm{w}} I/I_0$. Since the co-ions are excluded from very strongly charged membranes, $j_2 \approx 0$ and $j_1/j^0_{12,\max} = I/I_0$ in this limit.

Figures 4.36 (a)–(f) show some graphical representations of the membrane potential for electric current densities in the range, $|I/I_0| \leq 1$; values outside this range could have been considered as well because I_0 is merely a convenient unit. In very weakly charged membranes, the potential drops are $\Delta\phi_{\mathrm{D}}^{\alpha} - \Delta\phi_{\mathrm{D}}^{\beta} \approx 0$

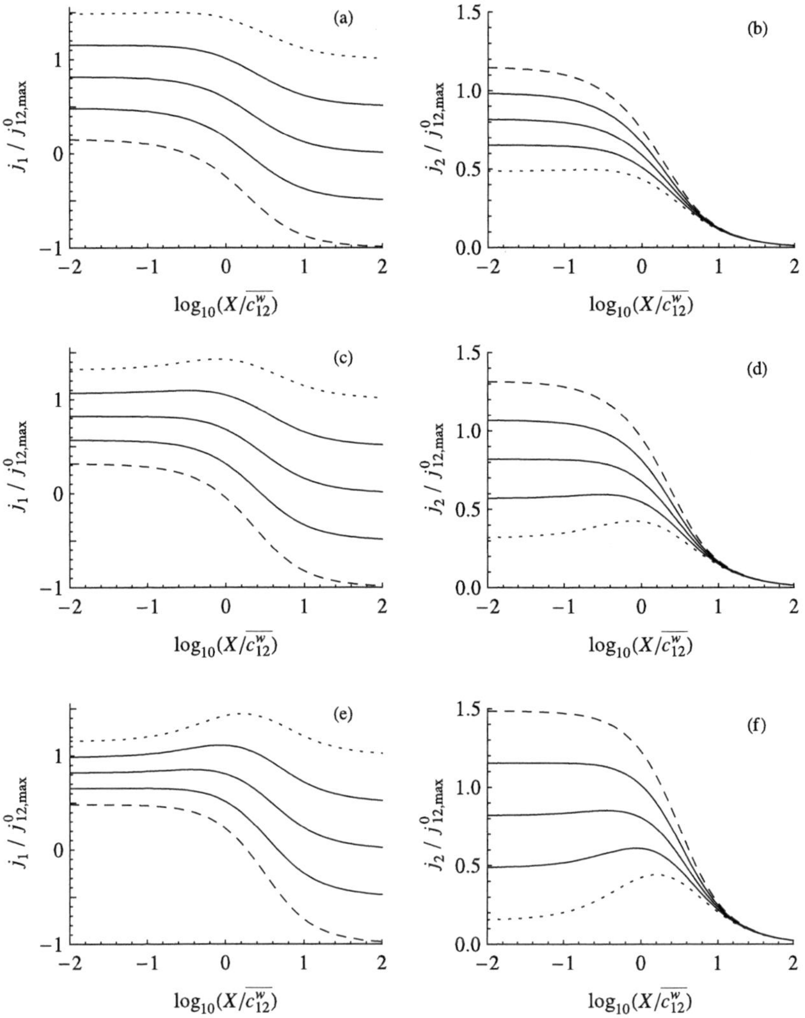

Fig. 4.35.
Variation of counterion (a, c, e) and co-ion flux densities (b, d, f), in $j^0_{12,\max}$ units, with the fixed charge concentration for an external concentration ratio $r_{12} = 10$, different values of the diffusion coefficient ratio $D_2/D_1 = 0.5$ (a, b), 1.0 (c, d), and 2.0 (e, f), and different values of $I/I_0 = 1.0$ (dotted), 0.5, 0, −0.5, and −1.0 (dashed).

and $\Delta\phi \approx (\Gamma/f)\ln(c^\beta_{12}/c^\alpha_{12}) = -(\Gamma/f)\ln r_{12}$. The latter can be written as the sum of a diffusional contribution

$$\Delta\phi_{\text{dif}} = -\frac{1}{f}\left(\frac{t^{\text{w}}_1}{z_1} + \frac{t^{\text{w}}_2}{z_2}\right)\ln\frac{c^\beta_{12}}{c^\alpha_{12}}, \tag{4.264}$$

which coincides with that in eqn (4.258), and an ohmic contribution $\Delta\phi_{\text{ohm}} \equiv -IR^{\text{M}}$, where

$$R^{\text{M}} \equiv \int_0^h \frac{\text{d}x}{\kappa^{\text{M}}} = \frac{RT}{z_1^2F^2(D_1+D_2)}\int_0^h \frac{\text{d}x}{c^{\text{M}}_{12}} = \frac{hRT}{z_1^2F^2(D_1+D_2)\Delta c^{\text{w}}_{12}}\ln\frac{c^\beta_{12}}{c^\alpha_{12}} \tag{4.265}$$

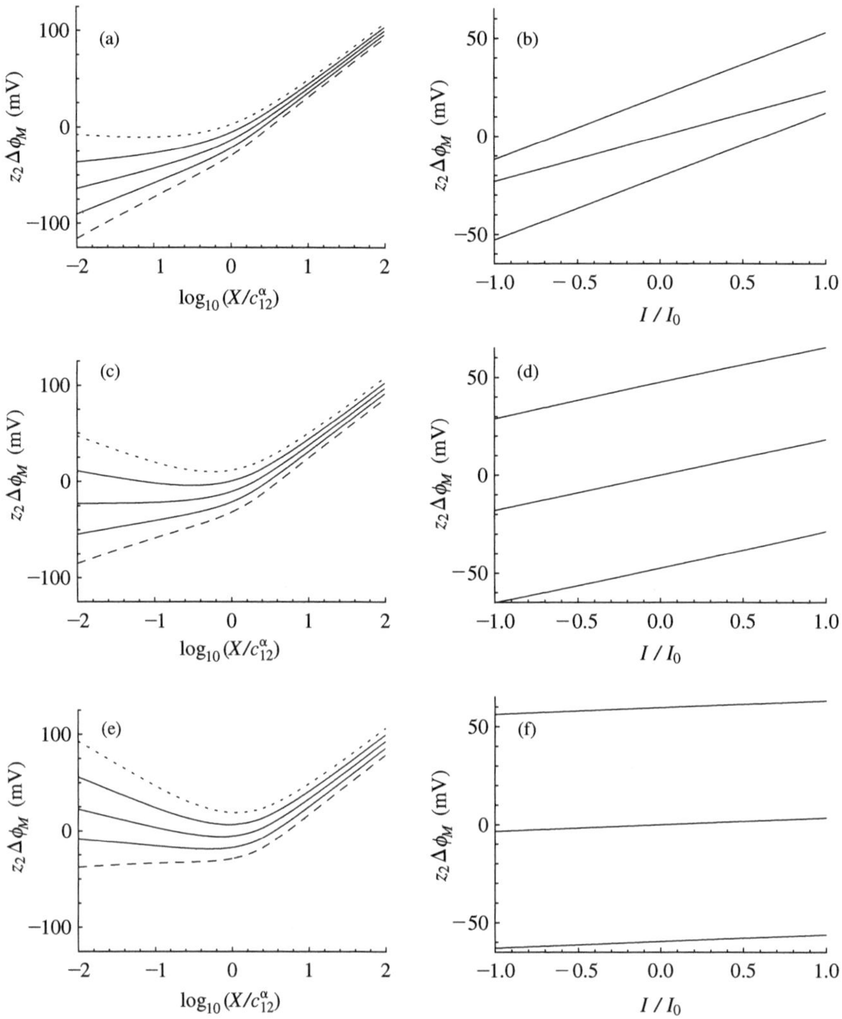

Fig. 4.36.
(a, c, e) Variation of the membrane potential $\Delta\phi_M$ with the electrolyte concentration in compartment α while keeping constant $c_{12}^{\beta} = 0.5X$. The electric current densities are $I/I_0 = 1.0$ (dotted), 0.5, 0, -0.5, and -1.0 (dashed), and the diffusion coefficient ratios are $D_2/D_1 = 0.5$ (a), 1 (c), and 2 (e). Note that $\Delta\phi_M = -I R^M$ when the equilibrium condition $c_{12}^{\alpha} = c_{12}^{\beta}$ is satisfied.
(b, d, f) Variation of $\Delta\phi_M$ with the electric current density I/I_0 for different values of the external concentration ratio $r_{12} = 0.1, 1$, and 10 (from top to bottom), and fixed diffusion coefficient ratio, $D_2/D_1 = 0.5$. Plot (b) corresponds to a weakly charged membrane with $X/\overline{c_{12}^{w}} = 0.01$, plot (d) to a moderately charged one with $X/\overline{c_{12}^{w}} = 1$, and plot (f) to a strongly charged one with $X/\overline{c_{12}^{w}} = 10$.

is the membrane electrical resistance. When evaluating the integral, we have used the fact that the distribution of the electrolyte concentration is linear inside the membrane.

In Fig. 4.36 (b), we have represented $z_2\Delta\phi_M \approx -(z_2\Gamma/f)\ln r_{12}$ against the current density at constant r_{12} for a weakly charged membrane. It is observed that $z_2\Delta\phi_M$ varies almost linearly with I. Since the Donnan potential drops are negligible then, and the diffusional contribution to the potential drop is constant under these conditions, the variation must be due to the ohmic potential drop, $z_2 f\,\Delta\phi_{ohm} \approx -(z_2\Gamma - t_1^w + t_2^w)\ln r_{12}$. Note that, according to eqns (4.261) and (4.262), $z_2\Gamma - t_1^w + t_2^w \propto I$ and this latter expression of $z_2\Delta\phi_{ohm}$ can be transformed to $\Delta\phi_{ohm} \equiv -I R^M$.

In very strongly charged membranes, co-ion exclusion is almost complete, the Donnan potential drops are $z_2 f(\Delta\phi_D^\alpha - \Delta\phi_D^\beta) \approx \ln(c_{12}^\beta/c_{12}^\alpha) = -\ln r_{12}$, and the potential drop in the membrane is ohmic, i.e. $\Delta\phi = -IR^M$ where $R^M \equiv h/\kappa^M$ is the membrane electrical resistance and $\kappa^M \approx z_1^2 F^2 D_1 X/RT$ is its conductivity.

In Fig. 4.36 (f), we have represented $z_2\Delta\phi_M \approx -(z_2\Gamma/f)\ln r_{12}$ against the current density at constant r_{12} for a strongly charged membrane. It is observed that $z_2\,\Delta\phi_M$ varies almost linearly with I. Since the diffusive contribution to the potential drop is negligible then, it is observed that $z_2 f\,\Delta\phi_M \approx -\ln r_{12}$ when $I = 0$. In the presence of electric current, the membrane potential varies linearly and the slope is smaller (i.e. the membrane resistance is lower) than in Figs. 4.36 (b) and (d), where the fixed-charge concentration is smaller. This reflects the fact that the membrane electrical conductivity increases with X. Moreover, using the values $X/\overline{c_{12}^w} = 10$ and $D_2/D_1 = 0.5$, corresponding to Fig. 4.36 (f), it can be obtained that $\Delta\phi = -IR^M \approx 3.5\,\text{mV}\,(I/I_0)$, in agreement with the slopes observed in this plot.

4.3.5 Counterion interdiffusion through an ideally permselective membrane

Consider a ternary electrolyte solution formed by two binary electrolytes with a common ion. The common ion is considered to be the co-ion and is denoted by index $i = 3$. The counterions are denoted by indices 1 and 2. The electrolytes $A_{\nu_1}C_{\nu_{3,1}}$ and $D_{\nu_2}C_{\nu_{3,2}}$ are denoted by indices 13 and 23, respectively, and are assumed to be completely dissociated according to

$$\mathrm{A}_{\nu_1}\mathrm{C}_{\nu_{3,1}} \rightleftarrows \nu_1\mathrm{A}^{z_1} + \nu_{3,1}\mathrm{C}^{z_3}, \tag{4.266}$$

$$\mathrm{D}_{\nu_2}\mathrm{C}_{\nu_{3,2}} \rightleftarrows \nu_2\mathrm{D}^{z_2} + \nu_{3,2}\mathrm{C}^{z_3}, \tag{4.267}$$

where the stoichiometric relations $z_1\nu_1 + z_3\nu_{3,1} = 0$ and $z_2\nu_2 + z_3\nu_{3,2} = 0$ are satisfied.

In the general case, the ion-exchange membrane separates two uniform solutions with electrolyte concentrations c_{13}^α, c_{23}^α and c_{13}^β, c_{23}^β under closed-circuit conditions, $I \neq 0$. For the sake of simplicity, the membrane is considered to be ideally selective and the co-ion ($i = 3$) is completely excluded from the membrane phase. In a binary electrolyte case, $c_{23}^\alpha = c_{23}^\beta = 0$, the only transport process that can occur across an ideally selective membrane is the conduction of electric current by the counterions, $j_1 = I/z_1F$ and $j_3 = 0$. Electrolyte diffusion cannot then take place because the electrolyte $\mathrm{A}_{\nu_1}\mathrm{C}_{\nu_{3,1}}$ is not a component in the membrane phase and $j_{13} = 0$. Similarly, in the ternary electrolyte case under consideration, the electrolytes $A_{\nu_1}C_{\nu_{3,1}}$ and $D_{\nu_2}C_{\nu_{3,2}}$ are not components of the membrane phase if this is ideally selective and there is no diffusion of neutral electrolytes, $j_{13} = 0$ and $j_{23} = 0$. Still, however, the interdiffusion of ions 1 and 2 can occur in addition to the conduction of electric current by the counterions.

Writing the electric potential gradient as the sum of ohmic and diffusion contributions

$$\vec{\nabla}\phi = \vec{\nabla}\phi_{\text{dif}} + \vec{\nabla}\phi_{\text{ohm}} = -\frac{1}{f}\left(\frac{t_1^{\text{M}}}{z_1}\vec{\nabla}\ln c_1 + \frac{t_2^{\text{M}}}{z_2}\vec{\nabla}\ln c_2\right) - \frac{\vec{I}}{\kappa}, \tag{4.268}$$

and substituting it in the Nernst–Planck equations for the ionic flux densities, these can be written in the diffusion–conduction form

$$\vec{j}_1 = -D_{12}^{\text{M}}\vec{\nabla}c_1 + \frac{t_1^{\text{M}}}{z_1}\frac{\vec{I}}{F}, \tag{4.269}$$

$$\vec{j}_2 = -D_{12}^{\text{M}}\vec{\nabla}c_2 + \frac{t_2^{\text{M}}}{z_2}\frac{\vec{I}}{F}, \tag{4.270}$$

where $D_{12}^{\text{M}} \equiv t_2^{\text{M}}D_1 + t_1^{\text{M}}D_2$, and

$$t_1^{\text{M}} = \frac{z_1^2 D_1 c_1}{z_1^2 D_1 c_1 + z_2^2 D_2 c_2}, \tag{4.271}$$

and $t_2^{\text{M}} = 1 - t_1^{\text{M}}$ are the local transport numbers of the counterions in the membrane. It should be stressed that the diffusion coefficient D_{12}^{M} does not describe the diffusion of any 'electrolyte' formed by the two counterions, but their interdiffusion across the membrane. The derivation of eqns (4.269) and (4.270) has made use of the relation $z_1\vec{\nabla}c_1 = -z_2\vec{\nabla}c_2$ that arises from the local electroneutrality assumption

$$z_1c_1 + z_2c_2 + z_{\text{M}}c_{\text{M}} = 0, \tag{4.272}$$

and the uniformity of the fixed-charge distribution. This relation implies that there is only one 'diffusional' driving force for the transport across the membrane (more exactly, a relation between the driving forces for the two ions) and therefore that a single diffusion coefficient, D_{12}^{M}, characterizes the interdiffusion process.

As commented in Section 4.3.2 for the case of a binary electrolyte, the diffusion–conduction equations, eqns (4.269) and (4.270), are not very useful for calculating the ionic flux densities as a function of the external solution concentrations and the electric current density because the transport coefficients D_{12}^{M} and $t_1^{\text{M}} = 1 - t_2^{\text{M}}$ are functions of the local concentrations and, hence, of position. Interestingly, the procedure worked out in Section 4.3.4 for the solution of the Nernst–Planck equations in a binary electrolyte can be directly applied to this case because there are two ionic species inside the ideally selective membrane and no assumption on the values of the charge numbers was used in Section 4.3.4.

If both counterions have the same charge number $z_1 = z_2 = z$, the parameter $\Gamma \equiv -G_1/(z_1 z_2 G_0) = -1/z$ and the concentrations gradients of the counterions are opposite and equal in magnitude, so that $\mathrm{d}c_{\mathrm{T}}/\mathrm{d}x = 0$. The electric field inside the membrane is then constant, and the Nernst–Planck equations can be integrated to give the Goldman equation

$$j_i = -\frac{D_i}{h}\frac{zf\,\Delta\phi}{\mathrm{e}^{zf\,\Delta\phi}-1}[c_i(h)\mathrm{e}^{zf\,\Delta\phi} - c_i(0)]. \tag{4.273}$$

The ionic concentrations at the membrane boundaries are given by the Donnan equilibrium conditions and the local electroneutrality condition as

$$\frac{c_1(h)}{c_1^{\beta}} = \frac{c_2(h)}{c_2^{\beta}} = -\frac{z_{\mathrm{M}}c_{\mathrm{M}}}{z(c_1^{\beta}+c_2^{\beta})} = -\frac{z_{\mathrm{M}}c_{\mathrm{M}}}{zc_{\mathrm{T}}^{\beta}}, \tag{4.274}$$

$$\frac{c_1(0)}{c_1^{\alpha}} = \frac{c_2(0)}{c_2^{\alpha}} = -\frac{z_{\mathrm{M}}c_{\mathrm{M}}}{z(c_1^{\alpha}+c_2^{\alpha})} = -\frac{z_{\mathrm{M}}c_{\mathrm{M}}}{zc_{\mathrm{T}}^{\alpha}}. \tag{4.275}$$

The electric potential drop can be determined by substituting eqn (4.273) in the equation for the electric current density, $I = zF(j_1 + j_2)$, as the solution to the following transcendental equation

$$\Delta\phi = -\frac{1}{zf}\ln\frac{D_1c_1(h)+D_2c_2(h)+i}{D_1c_1(0)+D_2c_2(0)+i}, \tag{4.276}$$

where $i \equiv IRTh/(z^2F^2\Delta\phi)$. In the absence of electric current ($I = 0$) this reduces to

$$\Delta\phi_{\mathrm{dif}} = -\frac{1}{zf}\ln\frac{D_1c_1(h)+D_2c_2(h)}{D_1c_1(0)+D_2c_2(0)}, \tag{4.277}$$

and in the absence of concentration gradients ($\mathrm{d}c_i/\mathrm{d}x = 0$) the potential drop is

$$\Delta\phi_{\mathrm{ohm}} = -IR^{\mathrm{M}}, \tag{4.278}$$

where $R^{\mathrm{M}} \equiv h/\kappa^{\mathrm{M}}$ is the membrane electrical resistance and κ^{M} is its electrical conductivity. Note, finally, that the ionic flux densities can also be expressed as

$$j_1 = \frac{D_1D_2}{D_1-D_2}G_0 + \frac{D_1}{D_1-D_2}\frac{I}{zF}, \tag{4.279}$$

$$j_2 = \frac{D_1D_2}{D_2-D_1}G_0 + \frac{D_2}{D_2-D_1}\frac{I}{zF}, \tag{4.280}$$

where $G_0 \equiv (j_1/D_1) + (j_2/D_2) = z_{\mathrm{M}}c_{\mathrm{M}}f\,\Delta\phi/h$.

If the counterions have different charge numbers $z_1 \neq z_2$, the ionic flux densities are given by [see eqns (4.248)–(4.251)]

$$j_1 = \frac{z_2 D_1}{z_2 - z_1}(1 + z_1\Gamma)G_0, \tag{4.281}$$

$$j_2 = \frac{z_1 D_2}{z_1 - z_2}(1 + z_2\Gamma)G_0, \tag{4.282}$$

where $G_0 = (-\Delta c_T + z_M c_M f \Delta\phi)/h$ and the electric potential drop is

$$f\Delta\phi = \Gamma \ln \frac{c_T(h) + z_M c_M[\Gamma + (z_1 + z_2)/(z_1 z_2)]}{c_T(0) + z_M c_M[\Gamma + (z_1 + z_2)/(z_1 z_2)]}, \tag{4.283}$$

where Γ has to be determined as the solution of the transcendental equation

$$I = \left(\Gamma + \frac{t_1^w}{z_1} + \frac{t_2^w}{z_2}\right)\frac{z_1 z_2 F(z_1 D_1 - z_2 D_2)}{(z_1 - z_2)h}$$
$$\times \left\{\Delta c_T - z_M c_M \Gamma \ln \frac{c_T(h) + z_M c_M[\Gamma + (z_1 + z_2)/(z_1 z_2)]}{c_T(0) + z_M c_M[\Gamma + (z_1 + z_2)/(z_1 z_2)]}\right\}. \tag{4.284}$$

It is worth noting that when that the membrane separates solutions of different electrolytes, $c_{13}^\beta = c_{23}^\alpha = 0$, the flux densities and the potential drop inside the membrane are independent of the values of the external concentrations c_{13}^α and c_{23}^β. This occurs because the boundary concentrations are only determined by the fixed charge concentration, $z_1 c_1(0) = z_2 c_2(h) = -z_M c_M$, and the membrane phase cannot 'know' the external concentrations. The situation of ideal permselectivity considered here is an approximation that can only be used when the external concentrations are much lower than the fixed-charge concentration, and hence this independence of the flux densities and the potential drop inside the membrane from the external concentrations should not be surprising. Yet, the Donnan potential drops

$$f\Delta\phi_D^\alpha = -\frac{1}{z_1}\ln\frac{c_1(0)}{c_1^\alpha} = -\frac{1}{z_1}\ln\frac{-z_M c_M}{z_1 c_1^\alpha} \tag{4.285}$$

$$f\Delta\phi_D^\beta = -\frac{1}{z_2}\ln\frac{c_2(h)}{c_2^\beta} = -\frac{1}{z_2}\ln\frac{-z_M c_M}{z_2 c_2^\beta} \tag{4.286}$$

are sensitive to the external concentrations.

Continuing with the discussion of the situation $c_{13}^\beta = c_{23}^\alpha = 0$, the ionic flux densities reduce under open-circuit conditions, $I = 0$, to

$$z_1 j_1 = -z_2 j_2 = -\frac{z_1 z_2 D_1 D_2 G_0}{z_1 D_1 - z_2 D_2}$$
$$= -\frac{z_M c_M}{h}\frac{z_1 z_2 D_1 D_2}{z_1 D_1 - z_2 D_2}\left(\frac{1}{z_2} - \frac{1}{z_1} + f\Delta\phi_{dif}\right), \tag{4.287}$$

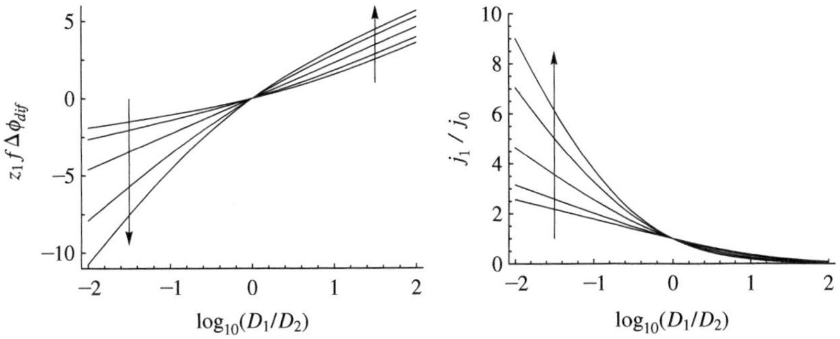

Fig. 4.37.
Diffusion potential, in RT/z_1F units, and flux density of counterion 1, in $j_0 \equiv -z_M c_M D_1/z_1 h$ units, against the counterion diffusion coefficient ratio, D_1/D_2, in the interdiffusion of counterions across an ideally permselective membrane. The counterion charge number ratios used are: $z_1/z_2 = 1/3, 1/2, 1, 2$, and 3 (increasing in the arrow direction).

where the diffusion potential

$$f\,\Delta\phi_{\text{dif}} = \frac{D_1 - D_2}{z_1 D_1 - z_2 D_2} \ln \frac{z_1 D_1}{z_2 D_2} \tag{4.288}$$

can be obtained either from eqn (4.283) or from the Henderson approximation (which is exact in this case). Figure 4.37 shows the representation of eqns (4.287) and (4.288). When $D_1 > D_2$ we observe that $z_1 f\,\Delta\phi_{\text{dif}}$ is positive, and that the flux density j_1 is smaller than $j_0 \equiv -z_M c_M D_1/z_1 h$, which can be understood as a typical value when $D_1 \approx D_2$. Contrarily, we observe that $z_1 f\,\Delta\phi_{\text{dif}}$ is negative and $j_1/j_0 > 1$ when $D_1 < D_2$.

4.3.6 Bi-ionic potential

The bi-ionic potential is the potential difference between two solutions of different binary electrolytes with a common co-ion at the same concentration that are separated by a charged membrane under open-circuit conditions, $I = 0$. We consider here that all ions are monovalent. Let us denote the counterions by indices 1 and 2, and the common co-ion by index $i = 3$. The electrolytes $A_{\nu_1}C_{\nu_{3,1}}$ and $D_{\nu_2}C_{\nu_{3,2}}$ are denoted by indices 13 and 23, respectively, and are assumed to be completely dissociated. The electrolyte concentrations are $c_{13}^{\alpha} = c_{23}^{\beta}$, and $c_{23}^{\alpha} = c_{13}^{\beta} = 0$. In the absence of chemical partition coefficients this implies that the two Donnan interfacial potential drops cancel out, and therefore the bi-ionic potential is equal to the diffusion potential drop inside the membrane.

Since the total ionic concentration, $c_T = c_1 + c_2 + c_3$, is constant throughout the membrane, the electric field is also constant, and the ionic flux densities are given by the Goldman flux equation

$$j_i = -\frac{D_i}{h} \frac{z_i f\,\Delta\phi}{e^{z_i f\,\Delta\phi} - 1} [c_i(h) e^{z_i f\,\Delta\phi} - c_i(0)]. \tag{4.289}$$

The zero current condition, $0 = j_1 + j_2 - j_3$, leads then to the Goldman equation for the diffusion potential

$$z_3 f\,\Delta\phi_{\text{dif}} = \ln \frac{D_2 c_2(h) + D_3 c_3(0)}{D_1 c_1(0) + D_3 c_3(h)}. \tag{4.290}$$

In strongly charged membranes, the bi-ionic potential is approximately given by

$$z_3 f\, \Delta\phi_{\rm dif} \approx \ln \frac{D_2}{D_1}, \tag{4.291}$$

and in very weakly charged membranes it can be evaluated as

$$z_3 f\, \Delta\phi_{\rm dif} \approx \ln \frac{D_2 + D_3}{D_1 + D_3}, \tag{4.292}$$

which agrees with eqn (4.89). This asymptotic behaviour has been represented in Fig. 4.38.

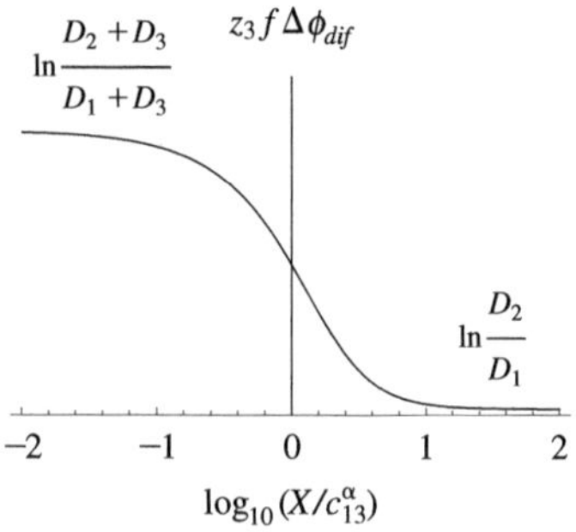

Fig. 4.38.
Schematic representation of the variation of the bi-ionic potential with the electrolyte concentration $c_{13}^{\alpha} = c_{23}^{\beta}$.

Finally, it should be mentioned that boundary-layer effects have been neglected in this simple description of the bi-ionic potential. These effects are particularly important in the case of strongly charged membranes [19] and account for the deviations of the actual measurements from eqn (4.290).

4.3.7 Transport in multi-ionic solutions

In the previous sections we have explained a solution procedure for steady-state transport equations in one-dimensional systems. This procedure is applied here to multi-ionic systems. We consider mixtures of electrolytes such that all counterions have the same charge number z_1 and all co-ions have the same charge number z_2; that is, there are only two classes of ions. It is assumed that all ionic concentrations at the (internal) membrane boundaries, $c_i(0)$ and $c_i(h)$, are known and we aim to evaluate the ionic flux densities and the potential drop across the membrane as a function of the electric current density I crossing the membrane.

The basic idea behind the method employed here to solve this problem is that the Nernst–Planck equations can be combined to yield an equation for the electric field. The solution of this equation and further integration allows us to obtain the electric-potential distribution $\phi(x)$. Then, the Nernst–Planck equation for species i can be multiplied by $e^{z_i f\,\phi(x)}$ and integrated between the membrane boundaries. This leads to the following equation for the flux density of species i

$$j_i = -D_i \frac{c_i(h)\, e^{z_i f\, \Delta\phi} - c_i(0)}{\int_0^h e^{z_i f[\phi(x)-\phi(0)]}\, dx}. \tag{4.293}$$

This equation is known as Kramer's equation and can be considered as a generalization of the Goldman flux equation, eqn (4.289).

In order to obtain the above-mentioned equation for the electric field, we introduce first a transformation of variables. The Nernst–Planck equations of the counterions can be combined in the form

$$-G_{0,1} = \frac{dC_1}{dx} + z_1 C_1 f \frac{d\phi}{dx}, \tag{4.294}$$

where

$$C_1 \equiv \sum_{\text{counterions}} c_i \tag{4.295}$$

is the total counterion concentration and

$$G_{0,1} \equiv \sum_{\text{counterions}} \frac{j_i}{D_i}. \tag{4.296}$$

After multiplication by $e^{z_1 f \phi}$, eqn (4.294) can be formally integrated between the membrane boundaries to give

$$G_{0,1} = -\frac{C_1(h)\, e^{z_1 f \Delta\phi} - C_1(0)}{\int_0^h e^{z_1 f [\phi(x) - \phi(0)]}\, dx}. \tag{4.297}$$

Combination of eqns (4.293) and (4.297) allows us to write the flux density of any counterion species as

$$j_i = D_i G_{0,1} \frac{c_i(h)\, e^{z_1 f \Delta\phi} - c_i(0)}{C_1(h)\, e^{z_1 f \Delta\phi} - C_1(0)}, \quad z_i = z_1. \tag{4.298}$$

Similarly, the flux density of any co-ion species can be written as

$$j_i = D_i G_{0,2} \frac{c_i(h)\, e^{z_2 f \Delta\phi} - c_i(0)}{C_2(h)\, e^{z_2 f \Delta\phi} - C_2(0)}, \quad z_i = z_2, \tag{4.299}$$

where

$$C_2 \equiv \sum_{\text{co-ions}} c_i \tag{4.300}$$

is the total co-ion concentration and

$$G_{0,2} \equiv \sum_{\text{co-ions}} \frac{j_i}{D_i}. \tag{4.301}$$

The ionic flux densities are coupled through the equation for the electric current density, $I = F \sum_i z_i j_i$. Equations (4.298) and (4.299) lead then to

$$\begin{aligned} \frac{I}{F} = {} & z_1 G_{0,1} \frac{\sum\limits_{z_i = z_1} D_i c_i(h)\, e^{z_1 f \Delta\phi} - \sum\limits_{z_i = z_1} D_i c_i(0)}{C_1(h)\, e^{z_1 f \Delta\phi} - C_1(0)} \\ & + z_2 G_{0,2} \frac{\sum\limits_{z_i = z_2} D_i c_i(h)\, e^{z_2 f \Delta\phi} - \sum\limits_{z_i = z_2} D_i c_i(0)}{C_2(h)\, e^{z_2 f \Delta\phi} - C_2(0)}. \end{aligned} \tag{4.302}$$

This equation plays a key role in the solution procedure but involves the variables $G_{0,1}$, $G_{0,2}$, and $\Delta\phi$ that are still to be determined.

Let us introduce the additional auxiliary variables

$$c_{\mathrm{T}} \equiv \sum_i c_i = C_1 + C_2, \tag{4.303}$$

$$S_1 \equiv z_1 C_1 + z_2 C_2 = -z_{\mathrm{M}} c_{\mathrm{M}}, \tag{4.304}$$

$$G_0 \equiv \sum_i \frac{j_i}{D_i} = G_{0,1} + G_{0,2}, \tag{4.305}$$

$$G_1 \equiv \sum_i \frac{z_i j_i}{D_i} = z_1 G_{0,1} + z_2 G_{0,2}, \tag{4.306}$$

$$\Gamma \equiv -\frac{G_1}{z_1 z_2 G_0}, \tag{4.307}$$

and combine the Nernst–Planck equations in the forms

$$-G_0 = \frac{\mathrm{d}c_{\mathrm{T}}}{\mathrm{d}x} - z_{\mathrm{M}} c_{\mathrm{M}} f \frac{\mathrm{d}\phi}{\mathrm{d}x}, \tag{4.308}$$

$$\begin{aligned} -G_1 &= (z_1^2 C_1 + z_2^2 C_2) f \frac{\mathrm{d}\phi}{\mathrm{d}x} \\ &= -z_1 z_2 [c_{\mathrm{T}} + z_{\mathrm{M}} c_{\mathrm{M}} (z_1 + z_2)/(z_1 z_2)] f \frac{\mathrm{d}\phi}{\mathrm{d}x}, \end{aligned} \tag{4.309}$$

where we have used the fact that $\mathrm{d}c_{\mathrm{M}}/\mathrm{d}x = 0$. Equation (4.308) allows us to conclude that the electric field is constant throughout the membrane when $c_{\mathrm{T}}(0) = c_{\mathrm{T}}(h)$, a case that is considered later in this section.

From eqns (4.307)–(4.309), the equation for the electric field is

$$f \frac{\mathrm{d}\phi}{\mathrm{d}x} = \frac{\Gamma}{c_{\mathrm{T}} + z_{\mathrm{M}} c_{\mathrm{M}} [\Gamma + (z_1 + z_2)/(z_1 z_2)]} \frac{\mathrm{d}c_{\mathrm{T}}}{\mathrm{d}x}, \tag{4.310}$$

and hence the electric potential drop in the membrane is

$$f\,\Delta\phi = \Gamma \ln \frac{c_{\mathrm{T}}(h) + z_{\mathrm{M}} c_{\mathrm{M}} [\Gamma + (z_1 + z_2)/(z_1 z_2)]}{c_{\mathrm{T}}(0) + z_{\mathrm{M}} c_{\mathrm{M}} [\Gamma + (z_1 + z_2)/(z_1 z_2)]}. \tag{4.311}$$

Similarly, the integration of eqn (4.308) over the membrane leads to

$$G_0 h = -\Delta c_{\mathrm{T}} + z_{\mathrm{M}} c_{\mathrm{M}} f\,\Delta\phi. \tag{4.312}$$

This equation system is solved as follows. First, we guess a value of Γ and evaluate the electric potential drop and G_0 from eqns (4.311) and (4.312). Then, we calculate

$$G_{0,1} = \frac{z_2}{z_2 - z_1}(1 + z_1 \Gamma) G_0, \tag{4.313}$$

$$G_{0,2} = \frac{z_1}{z_1 - z_2}(1 + z_2 \Gamma) G_0, \tag{4.314}$$

and substitute them in eqn (4.302). The value of Γ is iteratively modified until this equation is satisfied.

As a final comment it is worth remembering that the above equations only apply to the case of two ion classes. Although the procedure can be extended for as many ion classes as ions that are present in the system [20–22], this is obviously rather complicated, and a much simpler approximate solution gives good results in the case of multi-ionic systems. This is the so-called Goldman constant-field assumption that decouples the transport equations of the different ionic species by assuming that the electric field is a constant. The flux density of species i is then given by the Goldman equation

$$j_i = -D_i \frac{z_i f \Delta\phi}{e^{z_i f \Delta\phi} - 1} [c_i(h)\, e^{z_i f \Delta\phi} - c_i(0)]. \tag{4.315}$$

Taking eqn (4.315) to the equation for the electric current density $I = F \sum_i z_i j_i$, the electric potential drop across the membrane can be obtained as the solution of a transcendental equation. In the case of a mixture of symmetric $z : z$ electrolytes this equation is

$$zf\,\Delta\phi = \ln \frac{\sum\limits_{+} D_i c_i(0) + \sum\limits_{-} D_i c_i(h) + i}{\sum\limits_{+} D_i c_i(h) + \sum\limits_{-} D_i c_i(0) + i}, \tag{4.316}$$

where $i \equiv IRTh/(z^2F^2\Delta\phi)$ and the signs under the sums indicate that they are restricted to either cations or anions. Note also that eqns (4.315) and (4.316) become exact when $\mathrm{d}c_\mathrm{T}/\mathrm{d}x = 0$, a case to which eqns (4.310)–(4.314) cannot be applied. This means that the Goldman constant-field assumption is expected to be more accurate when $\Delta c_\mathrm{T}/c_\mathrm{T} \ll 1$ or, equivalently, when the electrical conductivity of the multi-ionic solution does not vary much with position [23].

4.3.8 Concentration polarization in ion-exchange membrane systems

Under steady-state conditions, the ionic flux density j_i is independent of position. The diffusion-conduction equation for a binary system

$$j_i = \nu_i j_{12} + \frac{t_i}{z_i} \frac{I}{F} \tag{4.317}$$

shows that the electrolyte diffusion and the conduction contributions to j_i can depend on position, but their sum cannot. That is, when the migrational transport number t_i varies with position (and $I \neq 0$), the electrolyte flux density j_{12} must vary accordingly. The fixed charge in an ion-exchange membrane increases the transport number of the counterions and decreases that of co-ions (with respect to the external solution values). This difference in the migrational transport numbers inside and outside the membrane implies that (electrolyte) concentration gradients must evolve on both sides of the membrane when an

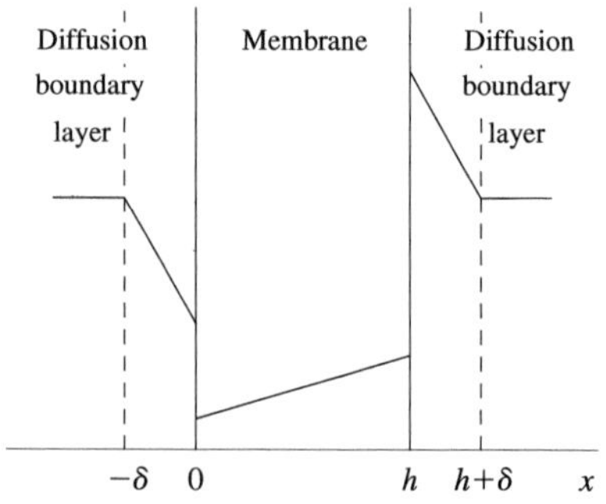

Fig. 4.39.
Schematic representation of the electrolyte concentration profile in the membrane system under consideration.

electric current is forced through it.[16] This phenomenon is known as concentration polarization. In particular, in the 'feeding' side of the membrane there is a deficit of counterions, and in the 'receiving' side there is an excess of counterions. Because of the electroneutrality condition, this also means that there is a deficit and an excess of electrolyte, respectively.

Consider that the membrane separates two solutions with the same electrolyte concentration c_{12}^{w} and compare the diffusive and ohmic contributions to the flux density of counterions ($i = 1$) in the membrane phase and in the feeding external solution. Since the flux density of the counterions is the same in both phases

$$\nu_1 j_{12}^{\mathrm{w}} + \frac{t_1^{\mathrm{w}}}{z_1}\frac{I}{F} = \nu_1 j_{12}^{\mathrm{M}} + \frac{t_1^{\mathrm{M}}}{z_1}\frac{I}{F}, \tag{4.318}$$

and

$$\nu_1 (j_{12}^{\mathrm{w}} - j_{12}^{\mathrm{M}}) = (t_1^{\mathrm{M}} - t_1^{\mathrm{w}})\frac{I}{z_1 F}. \tag{4.319}$$

The sign of j_{12}^{w} is that of I/z_1 because $t_1^{\mathrm{M}} > t_1^{\mathrm{w}}$, and this means that electrolyte diffusion in the feeding solution occurs in the same direction as the motion of counterions inside the membrane.

The flux density j_{12}^{M} takes place in the opposite direction to j_{12}^{w} if the bulk solutions have the same concentration and it is usually much smaller in magnitude. By neglecting j_{12}^{M} and replacing the local counterion transport number in the membrane by its average value, eqn (4.319) can be integrated over the external phase, extending from $x = -\delta$ to 0 (see Fig. 4.39), to give

$$\nu_1 j_{12}^{\mathrm{w}} = -\nu_1 D_{12}^{\mathrm{w}} \frac{c_{12}(0) - c_{12}^{\mathrm{w}}}{\delta} = (\overline{t_1^{\mathrm{M}}} - t_1^{\mathrm{w}})\frac{I}{z_1 F}, \tag{4.320}$$

where $c_{12}(-\delta) = c_{12}^{\mathrm{w}}$ is the bulk electrolyte concentration. An increase in the magnitude of the current density then leads to a decrease in the surface concentration $c_{12}(0)$. This concentration becomes zero when the limiting current

[16] We arrive at similar conclusions when these arguments are applied to ternary and multi-ionic systems.

density

$$I_L \equiv \frac{z_1 \nu_1 F D_{12}^{\mathrm{w}} c_{12}^{\mathrm{w}}}{\delta(\overline{t_1^{\mathrm{M}}} - t_1^{\mathrm{w}})} \tag{4.321}$$

is reached. Note that I and I_L can be either positive or negative (depending on whether the membrane is of cation- or anion-exchange type), but the ratio I/I_L is positive since the surface concentration is given by

$$c_{12}(0) = c_{12}^{\mathrm{w}}(1 - I/I_L) < c_{12}^{\mathrm{w}}. \tag{4.322}$$

In the case of an ideally selective membrane $j_{12}^{\mathrm{M}} = 0$, $\overline{t_1^{\mathrm{M}}} = 1$ and

$$I_L \equiv \frac{z_1 \nu_1 F D_{12}^{\mathrm{w}} c_{12}^{\mathrm{w}}}{\delta(1 - t_1^{\mathrm{w}})} = \frac{z_1 \nu_{12} F D_1 c_{12}^{\mathrm{w}}}{\delta}, \tag{4.323}$$

which is smaller (in magnitude) than that in eqn (4.321) because $\overline{t_1^{\mathrm{M}}} \leq 1$. Moreover, this limiting current does not depend on D_2, as should be expected from the fact that the ionic transport equations in the diffusion boundary layers reduce in this case to

$$-\frac{I}{z_1 F D_1} = \frac{\mathrm{d}c_1}{\mathrm{d}x} + z_1 c_1 f \frac{\mathrm{d}\phi}{\mathrm{d}x} = \nu_{12} \frac{\mathrm{d}c_{12}}{\mathrm{d}x}, \tag{4.324}$$

$$0 = \frac{\mathrm{d}c_2}{\mathrm{d}x} + z_2 c_2 f \frac{\mathrm{d}\phi}{\mathrm{d}x}. \tag{4.325}$$

In closing, it is interesting to observe that $\overline{t_1^{\mathrm{M}}}$ is related to the membrane permselectivity S and to $z_2\Gamma$ by eqn (4.253)

$$S \approx \frac{\overline{t_1^{\mathrm{M}}} - t_1^{\mathrm{w}}}{1 - t_1^{\mathrm{w}}} = \frac{2t_1^{\mathrm{w}}}{t_1^{\mathrm{w}} - t_2^{\mathrm{w}} - z_2\Gamma}. \tag{4.326}$$

By elimination of $\overline{t_1^{\mathrm{M}}}$, eqn (4.321) can also be written (for a symmetric electrolyte) as

$$I_L \equiv -(z_2\Gamma - t_1^{\mathrm{w}} + t_2^{\mathrm{w}}) \frac{z_1 F (D_1 + D_2) c_{12}^{\mathrm{w}}}{\delta}, \tag{4.327}$$

and the parameter $z_2\Gamma$ must be evaluated from the solution of the transcendental equation

$$I = -(z_2\Gamma - t_1^{\mathrm{w}} + t_2^{\mathrm{w}}) \frac{z_1 F (D_1 + D_2)}{h} \times \left[-\Delta c_{\mathrm{T}} + z_2\Gamma X \ln \frac{c_{\mathrm{T}}(h) + z_2\Gamma X}{c_{\mathrm{T}}(0) + z_2\Gamma X} \right], \tag{4.328}$$

where

$$c_{\rm T}(0) = \{X^2 + [c_{\rm T}^{\rm w}(0)]^2\}^{1/2}, \tag{4.329}$$

$$c_{\rm T}(h) = \{X^2 + [c_{\rm T}^{\rm w}(h)]^2\}^{1/2}, \tag{4.330}$$

$$c_{\rm T}^{\rm w}(0) = c_{\rm T}^{\rm w}(1 - I/I_L), \tag{4.331}$$

$$c_{\rm T}^{\rm w}(h) = c_{\rm T}^{\rm w}(1 + I/I_L). \tag{4.332}$$

Note that $c_{\rm T}^{\rm w} = 2c_{12}^{\rm w}$ in a symmetric electrolyte. Equation (4.321) is a very simple expression for the limiting current density I_L but it assumes that $\overline{t_1^{\rm M}}$ is known. Equations (4.327)–(4.332) look much more complicated, but they provide both I_L and $z_2\Gamma$ or, equivalently, I_L and $\overline{t_1^{\rm M}}$.

4.3.9 Influence of the diffusion boundary layers on the permselectivity

In the previous sections we have considered both the cases of ideally permselective membranes that completely exclude the co-ions from their interior, and real membranes containing both co-ions and counterions. By comparing eqns (4.321) and (4.323) it becomes apparent that the permselectivity of a membrane in relation to its performance in electrically driven separation processes can be characterized by means of the coefficient

$$S = \frac{\overline{t_1^{\rm M}} - t_1^{\rm w}}{1 - t_1^{\rm w}}. \tag{4.333}$$

The permselectivity of a membrane system, however, is not only determined by the number of co-ions in the membrane. When an electric current density I crosses the system, concentration polarization develops as explained in the previous section. Even in the limiting case $I \to 0$, this concentration polarization affects the value of the counterion flux density. To account for the influence of concentration polarization, the counterion integral transport number

$$T_1 \equiv \frac{z_1 F j_1}{I} \tag{4.334}$$

is used instead of $\overline{t_1^{\rm M}}$ in eqn (4.333). We aim at determining the value of T_1 for a membrane system composed of a membrane of thickness h flanked by two diffusion boundary layers of the same thickness δ. For the sake of simplicity, we consider that the membrane separates two identical solutions of a symmetric binary electrolyte.

In eqn (4.181), the electrolyte flux density was written in terms of the gradient of the stoichiometric concentration of the Donnan electrolyte as $\vec{j}_{12} = -D_{12}^{\rm M}\vec{\nabla}c_{12}$, where the electrolyte diffusion coefficient is

$$D_{12}^{\rm M} \equiv t_2^{\rm M} D_1 + t_1^{\rm M} D_2 = \frac{D_1 D_2 (c_1 + c_2)}{D_1 c_1 + D_2 c_2}. \tag{4.335}$$

The Donnan equilibrium requires that the mean electrolyte concentration $c_{\pm,12}$ must be continuous across the membrane/external solution interfaces. This condition makes it convenient to use $c_{\pm,12}$ rather than c_{12}. The stoichiometric concentration gradient can be related to the mean concentration gradient by

$$\frac{2}{c_{\pm,12}}\vec{\nabla}c_{\pm,12} = \frac{1}{c_1}\vec{\nabla}c_1 + \frac{1}{c_2}\vec{\nabla}c_2 = \frac{c_1+c_2}{c_1c_2}\vec{\nabla}c_{12}$$
$$= \frac{c_1+c_2}{(c_{\pm,12})^2}\vec{\nabla}c_{12}. \quad (4.336)$$

Thus, the diffusion–conduction flux equation becomes

$$\vec{j}_i = -\nu_i D^{\mathrm{M}}_{\pm,12}\vec{\nabla}c_{\pm,12} + \frac{t_i^{\mathrm{M}}}{z_i}\frac{\vec{I}}{F}, \quad (4.337)$$

where

$$D^{\mathrm{M}}_{\pm,12} \equiv \frac{2D_1D_2c_{\pm,12}}{D_1c_1+D_2c_2}, \quad (4.338)$$

and

$$t_i^{\mathrm{M}} \equiv \frac{D_ic_i}{D_1c_1+D_2c_2}. \quad (4.339)$$

In order to evaluate the counterion integral transport number T_1, eqn (4.337) has to be integrated over the membrane system extending from $x = -\delta$ to $h+\delta$. Multiplying this equation by $D_1c_1+D_2c_2$, and calculating the integral, we obtain

$$j_i(D_1\overline{c_1}+D_2\overline{c_2}) = \frac{D_i\overline{c_i}}{z_i}\frac{I}{F}, \quad (4.340)$$

where the overbars denote the average value over $-\delta \le x \le h+\delta$. In this integration, we have used the fact that $c_{\pm,12}$ is continuous across the membrane/external solution interfaces and that it takes the same value $c^{\mathrm{w}}_{\pm,12}$ at $x = -\delta$ and at $x = h+\delta$. The counterion integral transport number is then given by

$$T_1 = \frac{D_1\overline{c_1}}{D_1\overline{c_1}+D_2\overline{c_2}} = \frac{t_1^{\mathrm{w}}\overline{c_1}}{t_1^{\mathrm{w}}\overline{c_1}+t_2^{\mathrm{w}}\overline{c_2}}. \quad (4.341)$$

Since the concentration profiles in the diffusion boundary layers are linear and have the same gradient, the average concentrations over the whole membrane system can be related to the average concentrations inside the membrane by the simple relation

$$\overline{c_i} \equiv \frac{1}{2\delta+h}\left(\int_{-\delta}^{0} c_i\mathrm{d}x + \int_{h}^{h+\delta} c_i\mathrm{d}x + \int_{0}^{h} c_i\mathrm{d}x\right) = \frac{2\delta c_{12}^{\mathrm{w}} + h\overline{c_i^{\mathrm{M}}}}{2\delta+h}, \quad (4.342)$$

and eqn (4.341) becomes

$$T_1 = \frac{t_1^{\mathrm{w}}(2\delta c_{12}^{\mathrm{w}} + h\overline{c_1^{\mathrm{M}}})}{t_1^{\mathrm{w}}(2\delta c_{12}^{\mathrm{w}} + h\overline{c_1^{\mathrm{M}}}) + t_2^{\mathrm{w}}(2\delta c_{12}^{\mathrm{w}} + h\overline{c_2^{\mathrm{M}}})}$$

$$= t_1^{\mathrm{w}} \frac{r + \overline{c_1^{\mathrm{M}}}/X}{r + (t_1^{\mathrm{w}}\overline{c_1^{\mathrm{M}}} + t_2^{\mathrm{w}}\overline{c_2^{\mathrm{M}}})/X}, \qquad (4.343)$$

where $r \equiv 2\delta c_{12}^{\mathrm{w}}/hX$. Due to the boundary-layer effects, the integral counterion transport number is lower than unity even in ideally selective membranes and eqn (4.343) then reduces to

$$T_1 = t_1^{\mathrm{w}} \frac{r + \overline{c_1^{\mathrm{M}}}/X}{r + t_1^{\mathrm{w}}\overline{c_1^{\mathrm{M}}}/X} < 1. \qquad (4.344)$$

Indeed, substituting eqn (4.343) in eqn (4.333) it becomes evident that the membrane permselectivity is smaller than unity because of both the presence of co-ions inside the membrane and the boundary-layer effects

$$S = \frac{t_1^{\mathrm{w}}}{t_1^{\mathrm{w}} + r + \overline{c_2^{\mathrm{M}}}/X}. \qquad (4.345)$$

For small electric current densities, the average ionic concentrations do no differ much from the equilibrium values given by eqns (4.121) and (4.122)

$$\overline{c_1^{\mathrm{M}}} = (X/2) + [(X/2)^2 + (c_{12}^{\mathrm{w}})^2]^{1/2}, \qquad (4.346)$$

$$\overline{c_2^{\mathrm{M}}} = -(X/2) + [(X/2)^2 + (c_{12}^{\mathrm{w}})^2]^{1/2}, \qquad (4.347)$$

and the integral counterion transport number can be evaluated from eqn (4.343) for different values of the parameter r characterizing the importance of the diffusion boundary layers. Figure 4.40 shows the boundary-layer effects on the membrane permselectivity as evaluated from eqns (4.345) and (4.347).

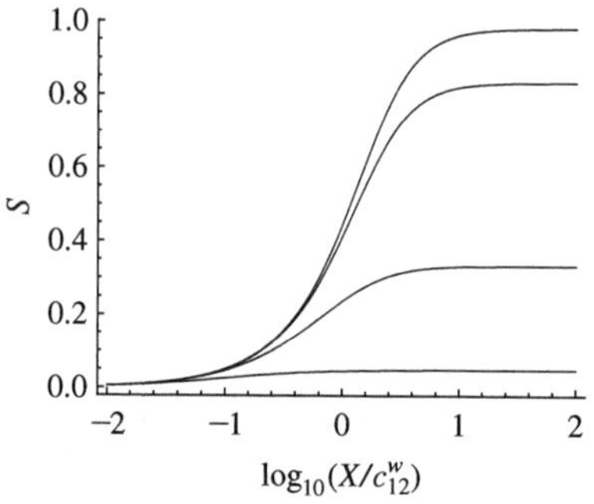

Fig. 4.40.
Variation of the membrane permselectivity with the fixed charge concentration for $t_1^{\mathrm{w}} = 0.5$ and different values of the ratio $r \equiv 2\delta c_{12}^{\mathrm{w}}/hX$ that describes the relative importance of the boundary layer effects: $r = 0.01, 0.1, 1$, and 10 (from top to bottom).

4.4 Steady-state transport across charged porous membranes

Porous membrane with charges on its pore walls are permselective. These charges may be due to dissociating groups in the membrane matrix (e.g.–COOH) or ion adsorption. For instance, a neutral membrane often exhibits cation selectivity in chloride solutions because chloride anions are strongly adsorbed on hydrophilic surfaces. In the former case, the membrane charge can be assumed to be constant along the pore, while in the latter case, it is determined through an adsorption isotherm, and therefore it depends on the concentration distribution in the pore. For the sake of simplicity, we consider throughout this section that the membrane is immersed in a symmetric (binary) electrolyte solution and that the fixed-charge groups are uniformly distributed along the transport (or axial) direction. Moreover, no pressure gradient exists inside the membrane.

4.4.1 The radial electrical double layer

a) The membrane model

Charged porous membranes are most often modelled as an array of parallel, cylindrical capillaries of radius a with a uniform surface-charge density σ on the pore walls. Since all pores are assumed to be identical, the entire membrane is analysed as if it were a single pore. The pore length h is much larger than the radius a, and hence edge (i.e. pore entrance) effects are neglected. The total surface charge in a pore is $\sigma 2\pi ah$ and the charge density per pore volume is $\sigma 2\pi ah/(\pi a^2 h)$. The charges bound to the pore walls can then be described either in terms of the surface-charge density σ or in terms of an equivalent molar concentration of fixed-charge groups

$$c_{\mathrm{M}} \equiv \frac{2\sigma}{z_{\mathrm{M}} F a}. \tag{4.348}$$

Because of the cylindrical symmetry of the membrane model, all transport magnitudes are considered to be dependent, in principle, on two spatial coordinates: the axial position x and the radial position r. The former varies between 0 at the interface with compartment α and h at the interface with compartment β. The latter varies between 0 at the pore axis and a at the pore wall.

b) The radial electrical double layer

The charges on the pore walls create an electrical double layer in the radial direction, as discussed in Section 4.2.5. The electrolyte solution filling the pores is not locally electroneutral, and the ionic concentrations $c_1(r,x)$ and $c_2(r,x)$ can be different at every location inside the pore. A global electroneutrality condition, however, must hold at every axial position

$$\sigma 2\pi a = -\int_0^a \rho_e 2\pi r \mathrm{d}r = -F\int_0^a (z_1c_1 + z_2c_2)2\pi r \mathrm{d}r$$
$$= z_2F\pi a^2(\langle c_1\rangle - \langle c_2\rangle), \tag{4.349}$$

or

$$X \equiv z_M c_M/z_2 = \langle c_1\rangle - \langle c_2\rangle\,, \tag{4.350}$$

where $\rho_e = F(z_1c_1 + z_2c_2)$ is the space-charge density associated to the ions in the pore solution and the symbol $\langle\rangle$ denotes the average value over the pore cross-section. The subscripts 1 and 2 are used for counterion and co-ion, respectively. The average ionic concentrations $\langle c_i\rangle\,(x)(i = 1, 2)$ may still vary along the pore axis.

The electric potential $\phi(r, x)$ is conveniently decomposed into two contributions, $V(x)$ and $\psi(r, x)$. The potential contribution $\psi(r, x)$ is defined from the equilibrium condition along the radial direction

$$\frac{\partial\tilde{\mu}_i}{\partial r} = RT\frac{\partial}{\partial r}(\ln c_i + z_i f\phi) = 0\ ,\quad i = 1, 2 \tag{4.351}$$

as

$$\psi \equiv -\frac{1}{z_1 f}\ln\frac{c_1}{c_{\pm,12}} = -\frac{1}{z_2 f}\ln\frac{c_2}{c_{\pm,12}}, \tag{4.352}$$

where $c_{\pm,12}(x) = [c_1(r, x)c_2(r, x)]^{1/2}$ is the mean electrolyte concentration. The contribution $V(x)$ then becomes defined as

$$V(x) \equiv \phi(r, x) - \psi(r, x). \tag{4.353}$$

Equation (4.351) simply states that the radial component of the ionic flux densities must vanish because the pore wall is impenetrable to the ions.

The variation of $\psi(r, x)$ along the radial direction is described by the Poisson equation $\nabla^2\phi = -\rho_e/\varepsilon$, where ε is the electrical permittivity of the solution. The left-hand side of this equation is

$$\nabla^2\phi = \nabla^2 V + \nabla^2\psi = \frac{\mathrm{d}^2V}{\mathrm{d}x^2} + \frac{\partial^2\psi}{\partial x^2} + \frac{\partial^2\psi}{\partial r^2} + \frac{1}{r}\frac{\partial\psi}{\partial r}, \tag{4.354}$$

and the space-charge density can be written as

$$\rho_e = F(z_1c_1 + z_2c_2) = -z_2Fc_{\pm,12}(\mathrm{e}^{-z_1f\psi} - \mathrm{e}^{-z_2f\psi})$$
$$= -2z_2Fc_{\pm,12}\sinh(z_2f\psi). \tag{4.355}$$

The Poisson equation becomes the Poisson–Boltzmann equation when the equilibrium ionic distributions are used to evaluate the space-charge density in this way. The boundary conditions for the Poisson–Boltzmann equation are

$$\left.\frac{\partial\psi}{\partial r}\right|_{r=0} = 0, \tag{4.356}$$

$$\left.\frac{\partial\psi}{\partial r}\right|_{r=a} = \frac{\sigma}{\varepsilon}, \tag{4.357}$$

and the integration of the Poisson equation over any pore cross-section leads to

$$\frac{d^2V}{dx^2} + \frac{d^2\langle\psi\rangle}{dx^2} = 0 \tag{4.358}$$

which can be considered as a straightforward consequence of the global electroneutrality condition, eqn (4.350). Although the potential $\psi(r,x)$ is not identical to its pore average value $\langle\psi\rangle(x)$, eqn (4.358) implies that the first two terms in the right-hand side of eqn (4.354) partially cancel out. Moreover, since the pore length is much larger than the pore radius, the second derivative of the potential along the axial direction is expected to be much smaller than its second derivative in the radial direction. Thus, we conclude that the Poisson–Boltzmann equation can be approximated by

$$z_2 f\left(\frac{\partial^2\psi}{\partial r^2} + \frac{1}{r}\frac{\partial\psi}{\partial r}\right) = \frac{z_2 f}{r}\frac{\partial}{\partial r}\left(r\frac{\partial\psi}{\partial r}\right) \approx (\kappa_D^{w})^2 \sinh(z_2 f\psi), \tag{4.359}$$

where $\kappa_D^{w}(x) \equiv [2z_1^2F^2c_{\pm,12}(x)/\varepsilon RT]^{1/2}$ is the Debye parameter at position x. In general, this equation must be solved numerically, but analytical solutions can also be obtained in some cases. As a preliminary step, it is interesting to write down the boundary condition at the pore wall, eqn (4.357), as

$$2z_2 f\left.\frac{\partial\psi}{\partial r}\right|_{r=a} = (\kappa_D^{X})^2 a, \tag{4.360}$$

where $\kappa_D^{X} \equiv (z_1^2F^2X/\varepsilon RT)^{1/2}$ is the Debye parameter referred to the membrane fixed-charge concentration. Moreover, by multiplication of both sides of eqn (4.359) by $2\pi r$ and further integration along the pore radius, with the boundary conditions in eqns (4.356) and (4.357), it is obtained that

$$\langle\sinh(z_2 f\psi)\rangle = \left(\frac{\kappa_D^{X}}{\kappa_D^{w}}\right)^2 = \frac{X}{2c_{\pm,12}}. \tag{4.361}$$

Equations (4.360) and (4.361) state that the electric field at the pore wall and the average value of the potential[17] are determined by $\kappa_D^X a$ and κ_D^X/κ_D^w, respectively. These conclusions help in the analysis of the approximate solutions of eqn (4.359).

c) Linear approximation

When the surface-charge density is so low that $z_2 f \psi(r,x) < 1$, the Poisson–Boltzmann equation can be linearized to

$$\frac{\partial^2 \psi}{\partial r^2} + \frac{1}{r}\frac{\partial \psi}{\partial r} \approx (\kappa_D^w)^2 \psi. \tag{4.362}$$

The solution of this equation is

$$\psi(r,x) = \frac{\sigma}{\varepsilon \kappa_D^w}\frac{I_0(\kappa_D^w r)}{I_1(\kappa_D^w a)} = \frac{X}{4 z_2 f c_{\pm,12}}\frac{\kappa_D^w a I_0(\kappa_D^w r)}{I_1(\kappa_D^w a)}$$

$$= \psi(0,x) I_0(\kappa_D^w r), \tag{4.363}$$

where $I_0(\xi)$ and $I_1(\xi)$ are the modified Bessel functions of orders 0 and 1, respectively, and argument $\xi \equiv \kappa_D^w r$. It is interesting to note that eqn (4.361) implies that the average value of this potential contribution is $\langle \psi \rangle \approx X/(2 z_2 f c_{\pm,12})$.[18] Moreover, since this approximate solution of the Poisson–Boltzmann equation is expected to be valid when the pore radius is much larger than the Debye length (i.e. the reciprocal of the Debye parameter κ_D^w) and the difference between $I_0(\kappa_D^w a)$ and $I_1(\kappa_D^w a)$ is smaller than 10% when ca. $\kappa_D^w a > 6$, we can write down the condition of validity, $z_2 f \psi(a,x) \ll 1$, of this approximation as $X \ll 4 c_{\pm,12}/(\kappa_D^w a)$.

Figure 4.41 shows a graphical representation of the potential distribution described by eqn (4.363) that evidences that the thickness of the electrical double layer adjacent to the pore wall decreases with increasing $\kappa_D^w a$. In particular, it is observed that the electrical double layer extends to ca. half of the pore radius when $\kappa_D^w a = 10$ and that it is confined to the close vicinity of the pore wall when $\kappa_D^w a$ is larger than this value. Figure 4.41 (b) shows that the value potential gradient at the pore wall is determined by $\kappa_D^X a$, as imposed by eqn (4.360). Note, finally, that the condition of validity of this linear approximation can also be stated as $(\kappa_D^X a)^2/(2\kappa_D^w a) \ll 1$.

---

[17] Expressing the potential as $\psi = \langle \psi \rangle + \tilde{\psi}$, it is easy to see that $\langle \sinh(z_2 f \psi) \rangle \approx \sinh(z_2 f \langle \psi \rangle)$ when $z_2 f \tilde{\psi} \ll 1$. It is possible, however, that $z_2 f \tilde{\psi} > 1$ when $\kappa_D^X a \gg 1$, but this fact does not significantly modify our conclusions.

[18] This conclusion can also be derived from eqn (4.360) as $\langle \psi \rangle = \sigma \langle I_0(\kappa_D^w r) \rangle / [\varepsilon \kappa_D^w I_1(\kappa_D^w a)]$ and evaluating $\langle I_0(\kappa_D^w r) \rangle = (2/\kappa_D^w a) I_1(\kappa_D^w a)$ from the relation $\xi I_0 = d(\xi I_1)/d\xi$.

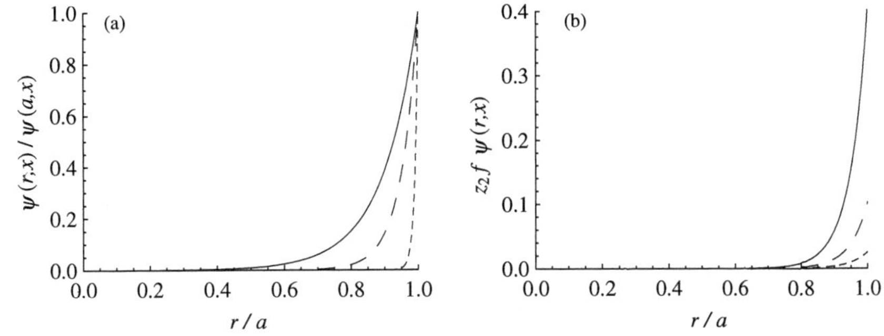

Fig. 4.41.
Radial distribution of the electric potential inside a charged cylindrical pore in the linear approximation. (a) Potential relative to its maximum value at the pore wall evaluated for $\kappa_D^W a = 10$ (solid line), 20 (long dashed line), and 100 (short dashed line). For a given value of the average potential, the potential at the pore centre decreases and that at the pore wall increases with increasing $\kappa_D^W a$. (b) Potential distribution for $\kappa_D^W a = 20$ and $\kappa_D^X a = 4$ (solid line), 2 (long dashed), and 1 (short dashed).

d) Total co-ion exclusion approximation

When the surface charge density is so high that $z_2 f\psi(r,x) \gg 1$, the Poisson–Boltzmann equation can be approximated by

$$z_2 f\left(\frac{\partial^2\psi}{\partial r^2} + \frac{1}{r}\frac{\partial\psi}{\partial r}\right) = (\kappa_D^X)^2\frac{\sinh(z_2 f\psi)}{\langle\sinh(z_2 f\psi)\rangle} \approx (\kappa_D^X)^2\frac{\exp(z_2 f\psi)}{\langle\exp(z_2 f\psi)\rangle}, \tag{4.364}$$

or, in short,

$$\frac{\partial^2\varphi}{\partial r^2} + \frac{1}{r}\frac{\partial\varphi}{\partial r} \approx (\kappa_D^X)^2\frac{e^\varphi}{\langle e^\varphi\rangle} = (\kappa_D^0)^2 e^\varphi, \tag{4.365}$$

where $\varphi \equiv z_2 f[\psi(r,x) - \psi(0,x)]$ and $\kappa_D^0 \equiv [z_1^2 F^2 c_1(0,x)/\varepsilon RT]^{1/2}$ is the Debye parameter referred to the counterion concentration at the pore axis, $c_1(0,x) = c_{\pm,12}\exp[z_2 f\psi(0,x)]$.[19] Equation (4.365) can also be obtained by noting that the co-ion concentration can be neglected inside the pore solution when $z_2 f\psi(r,x) \gg 1$. The solution of eqn (4.365) is

$$\varphi(r,x) = -2\ln[1 - (\kappa_D^0 r)^2/8]. \tag{4.366}$$

Since the (dimensionless) average space-charge density is

$$\langle e^\varphi\rangle = \left\langle [1 - (\kappa_D^0 r)^2/8]^{-2}\right\rangle = [1 - (\kappa_D^0 a)^2/8]^{-1}, \tag{4.367}$$

[19] Note that these different Debye parameters are conveniently introduced for the sake of simplicity of our notation. It is shown below, when describing the 'flat' distribution approximation, that the potential distribution inside the pore is actually determined by $\kappa_D^M \equiv [z_1^2 F^2(X + 2c_{\pm,12})/\varepsilon RT]^{1/2}$, which closely resembles that defined in eqn (4.158) and reduces to either κ_D^W or κ_D^X in the limits of weakly and strongly charged membranes, respectively.

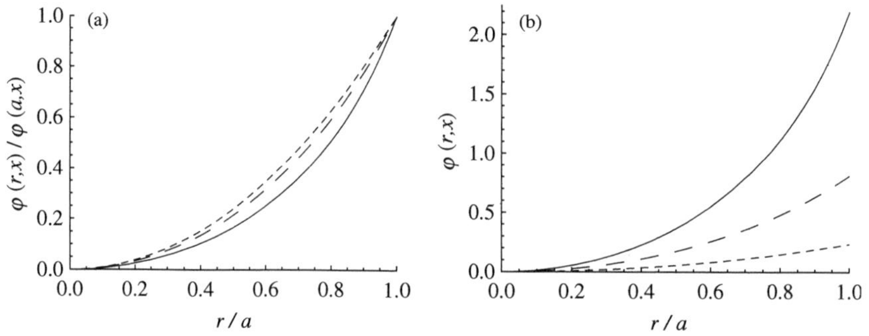

Fig. 4.42.
Radial distribution of the electric potential inside a charged cylindrical pore in the total co-ion exclusion approximation. (a) Difference between the local potential and the potential at the pore axis relative to the potential drop along the pore radius. (b) Difference between the local potential and the potential at the pore axis, in RT/z_2F units. The distributions have been evaluated for $\kappa_D^X a = 4$ (solid line), 2 (long dashed), and 1 (short dashed).

and eqn (4.361) becomes $\langle \exp(z_2 f \psi) \rangle \approx 2 \langle \sinh(z_2 f \psi) \rangle = X/c_{\pm,12}$, the counterion concentration at the pore axis can be evaluated as

$$c_1(0, x) = \frac{X}{1 + (\kappa_D^X a)^2/8}. \tag{4.368}$$

This is equivalent to the relation $(\kappa_D^0)^2 = (\kappa_D^X)^2/[1 + (\kappa_D^X a)^2/8]$, which prevents the argument of the logarithm in eqn (4.366) to take negative values. The potential drop in the radial direction can thus be evaluated as

$$\varphi(a, x) = -2 \ln[1 - (\kappa_D^0 a)^2/8] = 2 \ln[1 + (\kappa_D^X a)^2/8]. \tag{4.369}$$

The potential distribution in eqn (4.366) has been represented in Fig. 4.42 for different values of $\kappa_D^X a$. It is clear from the above equations that $\kappa_D^X a$ determines the potential gradient at the pore wall and the potential drop along the pore radius, and both increase with $\kappa_D^X a$.

Finally, since the potential at the pore axis is

$$z_2 f \psi(0, x) = \ln \frac{c_1(0, x)}{c_{\pm,12}} = \ln \frac{2(\kappa_D^0)^2}{(\kappa_D^w)^2} = \ln \frac{2(\kappa_D^X)^2}{(\kappa_D^w)^2[1 + (\kappa_D^X a)^2/8]}, \tag{4.370}$$

the condition of validity of this total co-ion exclusion approximation, $z_2 f \psi(0, x) \gg 1$, can be established as $X/[1 + (\kappa_D^X a)^2/8] \gg c_{\pm,12}$.

e) 'Flat' distribution approximation

When the deviation $\tilde{\psi} = \psi - \langle \psi \rangle$ of the local potential from the average value is small, $z_2 f \tilde{\psi} \ll 1$, it is possible to derive a third approximation solution of the Poisson–Boltzmann equation that can be used in the intermediate case

$z_2 f\,\langle\psi\rangle \approx 1$. By substituting the decomposition $\psi = \langle\psi\rangle + \tilde{\psi}$ in the Poisson–Boltzmann equation, this can be transformed to

$$z_2 f\left(\frac{\partial^2\tilde{\psi}}{\partial r^2}+\frac{1}{r}\frac{\partial\tilde{\psi}}{\partial r}\right)\approx(\kappa_{\mathrm{D}}^{\mathrm{w}})^2[\sinh(z_2 f\,\langle\psi\rangle)+\cosh(z_2 f\,\langle\psi\rangle)z_2 f\,\tilde{\psi}]$$
$$\approx(\kappa_{\mathrm{D}}^{X})^2+(\kappa_{\mathrm{D}}^{\mathrm{M}})^2 z_2 f\,\tilde{\psi}, \tag{4.371}$$

where $\kappa_{\mathrm{D}}^{\mathrm{M}} \equiv [(\kappa_{\mathrm{D}}^{\mathrm{w}})^4+(\kappa_{\mathrm{D}}^{X})^4]^{1/4} = [z_1^2F^2(2c_{\pm,12}+X)/\varepsilon RT]^{1/2}$ is the Debye parameter inside the pore and we have used the fact that $\langle\sinh(z_2 f\psi)\rangle = (\kappa_{\mathrm{D}}^{X}/\kappa_{\mathrm{D}}^{\mathrm{w}})^2 \approx \sinh(z_2 f\,\langle\psi\rangle)$. The solution of eqn (4.371) is

$$z_2 f\,\tilde{\psi}(r,x)=\frac{z_2 f\sigma}{\varepsilon\kappa_{\mathrm{D}}^{\mathrm{M}}}\frac{I_0(\kappa_{\mathrm{D}}^{\mathrm{M}}r)}{I_1(\kappa_{\mathrm{D}}^{\mathrm{M}}a)}-\frac{(\kappa_{\mathrm{D}}^{X})^2}{(\kappa_{\mathrm{D}}^{\mathrm{M}})^2}=\frac{(\kappa_{\mathrm{D}}^{X})^2}{(\kappa_{\mathrm{D}}^{\mathrm{M}})^2}\left[\frac{\kappa_{\mathrm{D}}^{\mathrm{M}}a\,I_0(\kappa_{\mathrm{D}}^{\mathrm{M}}r)}{2I_1(\kappa_{\mathrm{D}}^{\mathrm{M}}a)}-1\right]. \tag{4.372}$$

Since $\langle I_0(\kappa_{\mathrm{D}}^{\mathrm{M}}r)\rangle = (2/\kappa_{\mathrm{D}}^{\mathrm{M}}a)I_1(\kappa_{\mathrm{D}}^{\mathrm{M}}a)$, it is satisfied that $\left\langle\tilde{\psi}\right\rangle = 0$, as required by its definition. Figure 4.43 shows that this 'flat' distribution approximation is most useful when $\kappa_{\mathrm{D}}^{X}a \approx \kappa_{\mathrm{D}}^{\mathrm{w}}a \approx 1$. Finally, Fig. 4.44 shows several potential distributions obtained with the appropriate approximations. Some cases, like $\kappa_{\mathrm{D}}^{X}a = 10$ and $\kappa_{\mathrm{D}}^{\mathrm{w}}a = 10$ (not shown), cannot be described with any of the above approximations and would require a numerical solution of the Poisson–Boltzmann equation.

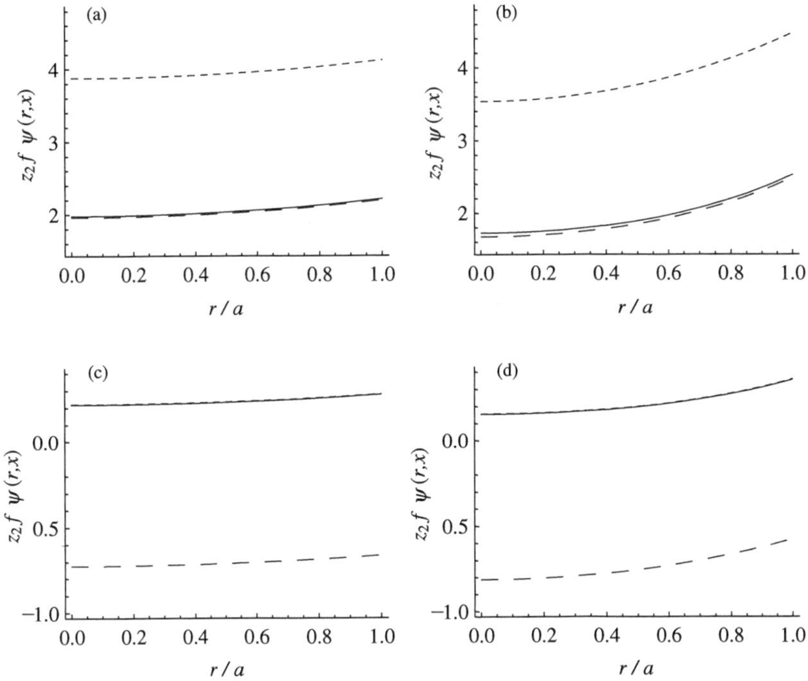

Fig. 4.43.
Radial distribution of the electric potential inside a charged cylindrical pore for: (a) $\kappa_{\mathrm{D}}^{X}a = 1, \kappa_{\mathrm{D}}^{\mathrm{w}}a = 0.5$, (b) $\kappa_{\mathrm{D}}^{X}a = 2, \kappa_{\mathrm{D}}^{\mathrm{w}}a = 1$, (c) $\kappa_{\mathrm{D}}^{X}a = 0.5, \kappa_{\mathrm{D}}^{\mathrm{w}}a = 1$, and (d) $\kappa_{\mathrm{D}}^{X}a = 1, \kappa_{\mathrm{D}}^{\mathrm{w}}a = 2$. The line styles correspond to the different approximations: (solid) 'flat' distribution, (long dashed) total co-ion exclusion, and (short dashed) linear approximation. In cases (a) and (b) the linear approximation is not valid. In cases (c) and (d) the total co-ion exclusion approximation is not valid. Note that $z_2\psi \geq 0$ because subscript 2 denotes the co-ion. In the case $\kappa_{\mathrm{D}}^{X}a = \kappa_{\mathrm{D}}^{\mathrm{w}}a = 1$ (not shown), only the 'flat' distribution approximation is valid.

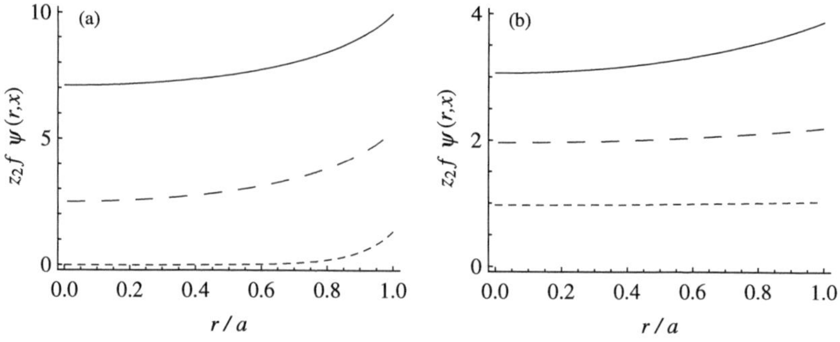

Fig. 4.44.
Radial distribution of the electric potential inside a charged cylindrical pore for: (a) $\kappa_D^X a = 5$ and $\kappa_D^w a = 0.1$ (solid line), 1 (long dashed), and 10 (short dashed), and (b) $\kappa_D^w a = 0.5$ and $\kappa_D^X a = 2$ (solid line), 1 (long dashed), and 0.5 (short dashed). Note that the electrical double layer extends over the whole pore cross section in all cases except for $\kappa_D^X a = 5$ and $\kappa_D^w a = 10$.

4.4.2 Electro-osmotic convection

In the previous section we have described the most important issue for understanding the ion transport in charged porous membranes: the radial electrical double layer. We now study another key characteristic: the occurrence of convective flow. Consider a charged porous membrane that separates two solutions with the same concentration of a symmetric electrolyte. Although we should expect neither a solution flow nor a solvent flow through the membrane under these conditions, it is experimentally observed that convective flow is established when an electric potential difference is applied between the two external solutions to drive an electric current through the pores under steady-state conditions. The origin of this solution flow is related to the radial electrical double layer. The solution inside the pores is not locally electroneutral. The applied electric field acts on the charge in every solution volume element and, in combination with the effect of viscosity, causes its steady motion in the direction of the field in cation-exchange membranes and in the opposite direction in anion-exchange membranes. In other words, the counterions are the majority in the pore solution and, due to their interaction with the field, impart more momentum to the solution than the co-ions do. Since this convective flow takes place in the same direction of motion as that of the counterions, they move faster than they would do in a stationary liquid. The opposite is true for the co-ions. As a consequence, this convection increases the current efficiency of the membrane.

a) Electro-osmotic velocity

In the situation under consideration, the ionic flux densities are

$$j_i = -z_i D_i c_i f \frac{\mathrm{d}V}{\mathrm{d}x} + c_i v, \tag{4.373}$$

and the solution velocity $v(r)$ is given by the axial component of the Navier–Stokes equation, eqn (1.86),

$$\eta \frac{1}{r}\frac{\mathrm{d}}{\mathrm{d}r}\left(r\frac{\mathrm{d}v}{\mathrm{d}r}\right) = \rho_e \frac{\mathrm{d}V}{\mathrm{d}x}. \tag{4.374}$$

By elimination of the space-charge density $\rho_e = F(z_1c_1 + z_2c_2)$ between the Navier–Stokes equation and the Poisson equation

$$\frac{1}{r}\frac{\mathrm{d}}{\mathrm{d}r}\left(r\frac{\mathrm{d}\psi}{\mathrm{d}r}\right) = -\frac{\rho_e}{\varepsilon}, \tag{4.375}$$

and after a first integration with respect to r (using the symmetry boundary conditions at the pore axis), we obtain

$$\eta\frac{\mathrm{d}v}{\mathrm{d}r} = -\varepsilon\frac{\mathrm{d}V}{\mathrm{d}x}\frac{\mathrm{d}\psi}{\mathrm{d}r}. \tag{4.376}$$

Therefore, the solution velocity is related to the electric potential contribution ψ by

$$v(r) = \frac{\varepsilon}{\eta}\frac{\mathrm{d}V}{\mathrm{d}x}[\psi(a) - \psi(r)], \tag{4.377}$$

where the non-slip boundary condition, $v(a) = 0$, has been used at the pore wall. The average velocity is then

$$\langle v\rangle = \frac{\varepsilon}{\eta}\frac{\mathrm{d}V}{\mathrm{d}x}[\psi(a) - \langle\psi\rangle] = \frac{\varepsilon}{\eta}\tilde{\psi}(a)\frac{\mathrm{d}V}{\mathrm{d}x}, \tag{4.378}$$

where $\tilde{\psi} \equiv \psi - \langle\psi\rangle$ denotes the deviation from the average value. Equation (4.378) is known as the *Helmholtz–Smoluchowski formula* in electrokinetics.

The magnitude $\tilde{\psi}(a)$ can be evaluated from the approximate solutions of the Poisson–Boltzmann equation derived above. Thus, when the radial electrical double layer is described in terms of the total co-ion exclusion approximation

$$\begin{aligned} z_2 f\,\tilde{\psi}(a) &= \varphi(a) - \langle\varphi\rangle \\ &= 2\ln[1 + (\kappa_D^X a)^2/8] - 2\left\{1 - \frac{\ln[1 + (\kappa_D^X a)^2/8]}{(\kappa_D^X a)^2/8}\right\} \\ &= 2[1 + 8/(\kappa_D^X a)^2]\ln[1 + (\kappa_D^X a)^2/8] - 2, \end{aligned} \tag{4.379}$$

where we have used eqn (4.369) and calculated the average of φ in eqn (4.366). Similarly, when the radial electrical double layer is described in terms of the 'flat' distribution approximation

$$z_2 f\,\tilde{\psi}(a) = \frac{(\kappa_D^X)^2}{(\kappa_D^M)^2}\left[\frac{\kappa_D^M a\, I_0(\kappa_D^M a)}{2I_1(\kappa_D^M a)} - 1\right]. \tag{4.380}$$

This expression can also be used when the membrane is weakly charged, $X \ll c_{\pm,12}$, and the linear approximation is then valid, because $\kappa_D^M \approx \kappa_D^w$. These expressions are represented in Fig. 4.45. As expected, $\tilde{\psi}(a)$ can take larger

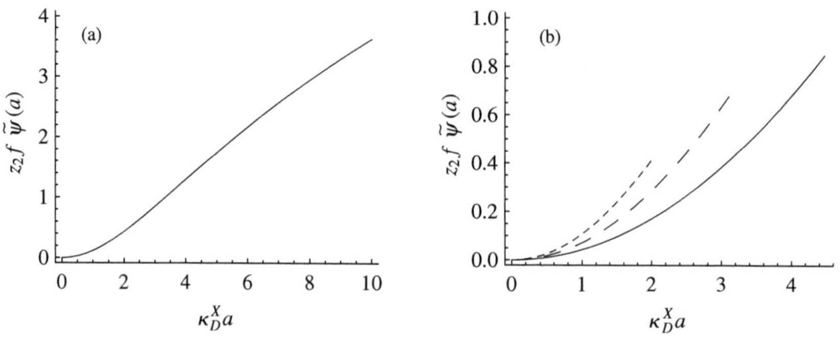

Fig. 4.45.
Difference between the electric potential at the pore wall and its average value (in RT/z_2F units) inside a charged cylindrical pore against $\kappa_{\mathrm{D}}^{X}a$: (a) total co-ion exclusion approximation and (b) 'flat' distribution approximation with $\kappa_{\mathrm{D}}^{\mathrm{W}}a = 10$ (solid line), 5 (long dashed), and 2 (short dashed). In case (a), $\kappa_{\mathrm{D}}^{\mathrm{W}}a$ should be small enough to ensure the validity of the total co-ion exclusion approximation. In case (b), the curves have been drawn only within the range of validity of the 'flat' distribution approximation.

values when the total co-ion exclusion is valid and, particularly, when $\kappa_{\mathrm{D}}^{X}a$ is large. This implies that the electro-osmotic velocity is more important in membranes with high surface-charge density and wide pores (because of their large hydraulic permeability).

Before the development of the charged capillary model by Dresner [24] and Osterle and co-workers [25–27], Schlögl clearly explained the importance of convection in transport processes across charged porous membranes. He found that it was not only responsible for the phenomenon of anomalous osmosis but also caused an increase in the electrical conductivity between ca. 10 and 45% [28–30]. Since Schlögl did not describe the variation of the electrostatic potential in the radial direction inside the pores, his expression for the average velocity[20] can be obtained after replacing ρ_{e} by $\langle\rho_{\mathrm{e}}\rangle = -z_{\mathrm{M}}Fc_{\mathrm{M}} = -z_2FX$ in the Navier–Stokes equation, eqn (4.374), and integrating it subject to the non-slip condition at the pore wall and the symmetry condition at the pore axis. Thus, the average electro-osmotic velocity can be estimated in this approach as

$$\langle v\rangle \approx \frac{z_2FXa^2}{8\eta}\frac{\mathrm{d}V}{\mathrm{d}x} = \frac{\varepsilon}{\eta}\frac{(\kappa_{\mathrm{D}}^{X}a)^2}{8z_2f}\frac{\mathrm{d}V}{\mathrm{d}x}. \tag{4.381}$$

It is noteworthy that eqn (4.381) leads to the same conclusion as the charged capillary model, i.e. that the electro-osmotic velocity is larger in membranes with large pore radius and large surface-charge density. In fact, eqn (4.381) can be obtained from eqn (4.380) when $\kappa_{\mathrm{D}}^{\mathrm{M}}a < 1$, and from eqn (4.379) when $(\kappa_{\mathrm{D}}^{X}a)^2/8 < 1$ after introducing the approximation $\ln(1+x) \approx x - x^2/2$. As a final comment, we mention that convection was neglected in the classical Teorell–Meyer–Sievers theory of membrane transport.

[20] For instance, in Ref. [29], Schlögl obtained the (average) velocity from the balance of hydrodynamical forces $K\langle v\rangle + \langle\rho_{\mathrm{e}}\rangle\,\mathrm{d}V/\mathrm{d}x = 0$, where K is the hydraulic flow resistance of the membrane (i.e. the reciprocal of the hydraulic permeability).

b) Convective conductivity

The electric current density that flows through the pores is

$$\begin{aligned} I = F(z_1 j_1 + z_2 j_2) &= -\kappa \frac{\mathrm{d}V}{\mathrm{d}x} + \rho_e v \\ &= -\left\{\kappa - \frac{\varepsilon}{\eta}\rho_e[\psi(a) - \psi(r)]\right\}\frac{\mathrm{d}V}{\mathrm{d}x}, \end{aligned} \tag{4.382}$$

where $\kappa \equiv F^2(z_1^2 D_1 c_1 + z_2^2 D_2 c_2)/RT$ is the electrical conductivity. The terms $-\kappa \mathrm{d}V/\mathrm{d}x$ and $\rho_e v$ are the conduction and the electro-osmotic current densities, respectively, and both are proportional to the applied electric field in the axial direction.[21]

The average value of the electro-osmotic current density is

$$\langle I_c \rangle = \langle \rho_e v \rangle = -\langle \kappa_c \rangle \frac{\mathrm{d}V}{\mathrm{d}x}, \tag{4.383}$$

where

$$\langle \kappa_c \rangle \equiv \frac{z_2 \varepsilon F}{\eta}[X\psi(a) - \langle (c_1 - c_2)\psi \rangle] \tag{4.384}$$

is the 'convective electrical conductivity' of the symmetric electrolyte, and we have used the global electroneutrality condition, $\langle c_1 - c_2 \rangle = X$. Since $z_2\psi(r) \leq z_2\psi(a)$, it should be clear from the right-hand side of eqn (4.384) that $\langle \kappa_c \rangle > 0$, and therefore the solution flow induced by the applied current enhances the effective electrical conductivity. In other words, the conductive and the convective contributions to the electric current density have the same direction. The equations describing electric conduction are then

$$\langle I \rangle = -(\langle \kappa \rangle + \langle \kappa_c \rangle)\frac{\mathrm{d}V}{\mathrm{d}x}, \tag{4.385}$$

$$\frac{\mathrm{d}V}{\mathrm{d}x} = -\frac{\langle I \rangle - \langle I_c \rangle}{\langle \kappa \rangle}, \tag{4.386}$$

where $\langle \kappa \rangle$ is the average electrical conductivity of the pore solution and $\langle I \rangle$ the average current density.

c) Weakly charged membranes

In weakly charged membranes the deviations of the transport magnitudes with respect to their pore average values are relatively small and the average value of a product of two magnitudes can be approximated by the product of the average values. For instance, the convective contribution to the electric current density is then

$$\langle I_c \rangle \approx \langle \rho_e \rangle \langle v \rangle = -z_2 FX \langle v \rangle, \tag{4.387}$$

[21] In fact, the proportionality between the convective velocity and the axial field can be used to define the mechanical mobility u_c of the pore solution from the relation $\langle v \rangle = z_2 F u_c \mathrm{d}V/\mathrm{d}x$ [31].

and, from eqn (4.378), the convective conductivity can be approximated by

$$\langle \kappa_c \rangle \approx z_2 F X \frac{\varepsilon}{\eta} \tilde{\psi}(a). \tag{4.388}$$

Similarly, the ionic flux densities can be approximated in weakly charged membranes by

$$\langle j_i \rangle \approx -z_i D_i \langle c_i \rangle f \frac{\mathrm{d}V}{\mathrm{d}x} + \langle c_i \rangle \langle v \rangle, \tag{4.389}$$

and the current efficiency, i.e. the integral transport number of the counterion, can be approximated by

$$T_1 \equiv \frac{z_1 F \langle j_1 \rangle}{\langle I \rangle} \approx \frac{\langle \kappa_1 \rangle + (\langle c_1 \rangle / X) \langle \kappa_c \rangle}{\langle \kappa \rangle + \langle \kappa_c \rangle}, \tag{4.390}$$

where $\kappa_1 \equiv z_1^2 F^2 D_1 c_1 / RT$ is the counterion contribution to the electrical conductivity. Note that the current efficiency increases with increasing electro-osmotic convection [32] because $\langle \kappa_1 \rangle / \langle \kappa \rangle < 1 < \langle c_1 \rangle / X$.

d) "Barycentric" reference frame

In a reference frame moving with respect to the membrane with the average solution velocity $\langle v \rangle$ (which can be understood as a kind of barycentric reference frame), the corresponding equations would be

$$\left\langle j_i^m \right\rangle \approx -z_i D_i \langle c_i \rangle f \frac{\mathrm{d}V}{\mathrm{d}x} \tag{4.391}$$

$$\left\langle I^m \right\rangle = F(z_1 \left\langle j_1^m \right\rangle + z_2 \left\langle j_2^m \right\rangle) = -\langle \kappa \rangle \frac{\mathrm{d}V}{\mathrm{d}x}, \tag{4.392}$$

where the superscript m identifies the magnitudes relative to this reference frame. Since the fixed charges bound to the membrane would move with velocity $-\langle v \rangle$ in this reference frame, they would carry an electric current

$$\left\langle I_{\mathrm{M}}^m \right\rangle = -z_{\mathrm{M}} F c_{\mathrm{M}} \langle v \rangle, \tag{4.393}$$

which, obviously, coincides with that shown in eqn (4.387), i.e. $\left\langle I_{\mathrm{M}}^m \right\rangle = \langle I_c \rangle$ and $\langle I^m \rangle = \langle I \rangle - \langle I_c \rangle$. In fact, the flux density of the fixed-charge groups in this reference frame can be written as

$$\left\langle j_{\mathrm{M}}^m \right\rangle = -c_{\mathrm{M}} \langle v \rangle = \frac{\left\langle I_{\mathrm{M}}^m \right\rangle}{z_{\mathrm{M}} F} = -z_{\mathrm{M}} D_{\mathrm{M}}^m c_{\mathrm{M}} f \frac{\mathrm{d}V}{\mathrm{d}x}, \tag{4.394}$$

where

$$D_{\mathrm{M}}^m \equiv \frac{RT}{z_{\mathrm{M}}^2 F^2 c_{\mathrm{M}}} \langle \kappa_c \rangle \tag{4.395}$$

is the effective diffusion coefficient of the fixed groups in this reference frame.

In the laboratory reference frame, the fixed groups are immobile and their flux density is obviously zero, $j_M = 0$. The average ionic flux densities can be written as

$$\langle j_i \rangle \approx \langle j_i^m \rangle + \langle c_i \rangle \langle v \rangle = -(z_i D_i - z_M D_M^m) \langle c_i \rangle f \frac{dV}{dx} \qquad (4.396)$$

Note that, as already explained above, the conductive and convective contributions are in the same direction for the counterion and in opposite directions for the co-ion.

4.4.3 Transport mechanisms in charged porous membranes

The transport of a binary electrolyte across a charged porous membrane separating two solutions at different concentrations is characterized by the ratio $a/L_D^w = \kappa_D^w a$ between the pore radius a and the thickness $L_D^w \equiv 1/\kappa_D^w$ of the radial double layer, where $\kappa_D^w \equiv [2z_1^2 F^2 c_{\pm,12}/\varepsilon RT]^{1/2}$ is the Debye parameter. Somewhat arbitrarily, we can assume that the transport mechanisms are different depending on the value of this ratio. When $a/L_D^w \ll 1$, the double layer fills the entire pore, co-ions are excluded and counterion transport takes place through a migrational hopping mechanism. That is, the counterions hope from a fixed-charge group to another. Electro-osmotic convection might also be important if the hydraulic permeability of the membrane is not too low. When $a/L_D^w \gg 1$, both diffusion and migration take place inside the pore, but the electrical double layer plays an important role in the analysis. Finally, diffusion, migration, and convection come into play when $a/L_D^w \approx 1$. We write below the corresponding transport equations in these three cases. For the sake of clarity, we avoid the use of the average symbol $\langle\rangle$ and understand implicitly that all magnitudes involved are average values over the pore cross-section.

a) Narrow pores $a \ll L_D^w$

When the double layer fills the entire pore, the membrane actually is an ideally selective membrane in which co-ions are completely excluded and electrolyte diffusion cannot take place (see Section 4.3.5). The counterion transport number is one, and its flux density (in the membrane-fixed reference system) is

$$j_1 = \frac{t_1}{z_1}\frac{I - I_c}{F} + c_1 v = \frac{I - I_c}{z_1 F} + Xv = \frac{I}{z_1 F}. \qquad (4.397)$$

Moreover, the electrical conductivity of the pore solution is $\kappa = z_1^2 F^2 D_1 X / RT$.

b) Wide pores $a \gg L_D^w$

The condition $a \gg L_D^w$ also implies that the membrane is weakly charged and that the space-charge density is small. The electro-osmotic velocity is proportional to the electrical force acting on the solution, which in turn is proportional to the space-charge density, and therefore convection is expected to be negligible in the case $a \gg L_D^M$. The charge on the pore walls, however, affects the transport phenomena because the ionic concentrations must satisfy the electroneutrality condition, eqn (4.350), $c_1 = c_2 + X$.

The flux density of species i is[22]

$$j_i = -D_i \left(\frac{\mathrm{d}c_i}{\mathrm{d}x} + z_i c_i f \frac{\mathrm{d}\phi}{\mathrm{d}x} \right). \tag{4.398}$$

The equivalent diffusion–conduction form [see eqn (4.187)] is

$$j_i = -D_{12}^{\mathrm{M}} \frac{\mathrm{d}c_i}{\mathrm{d}x} + \frac{t_i^{\mathrm{M}}}{z_i} \frac{I}{F}, \tag{4.399}$$

where, for a symmetric electrolyte, $t_i^{\mathrm{M}} = D_i c_i/(D_1 c_1 + D_2 c_2)$ and $D_{12}^{\mathrm{M}} \equiv t_2^{\mathrm{M}} D_1 + t_1^{\mathrm{M}} D_2$. Using the electroneutrality condition $c_1 = c_2 + X$, it can be easily shown that

$$\frac{1}{t_1^{\mathrm{M}}} = 1 + \frac{D_2 c_2}{D_1 c_1} = 1 + \frac{D_2}{D_1} - \frac{D_2 X}{D_1 c_1} = \frac{1}{t_1^{\mathrm{w}}} - \frac{D_2 X}{D_1 c_1}, \tag{4.400}$$

and hence we conclude that

$$\theta \equiv \frac{X}{c_1} = \frac{D_1}{D_2} \left(\frac{1}{t_1^{\mathrm{w}}} - \frac{1}{t_1^{\mathrm{M}}} \right) \tag{4.401}$$

is always a positive magnitude. Moreover, θ is measurable because the transport numbers t_1^{w} and t_1^{M} can be experimentally determined (and the diffusion coefficients D_1 and D_2 are known with good accuracy). If species 1 is the counterion, θ varies between 0 and 1, depending on the ion-exchange capacity of the membrane.

The condition $a \gg L_{\mathrm{D}}^{\mathrm{w}}$ also implies that the ionic concentrations vary very smoothly with the radial position coordinate. Then, it can be assumed that the average value of the product of ionic concentrations is equal to the product of average concentrations. The Donnan equilibrium condition then states that

$$(c_{12}^{\mathrm{w}})^2 = (c_{\pm,12}^{\mathrm{w}})^2 = (c_{\pm,12}^{\mathrm{M}})^2 \approx c_1 c_2, \tag{4.402}$$

where the right-hand side should be read as the product of the average concentrations, and it has been assumed that the activity coefficients are the same in the two phases. The combination of eqn (4.402) and the global electroneutrality condition, $c_1 = c_2 + X$, leads to

$$(c_{12}^{\mathrm{w}})^2 \approx c_1 c_2 = c_1(c_1 - X) = c_1^2(1 - \theta). \tag{4.403}$$

[22] Remember that the average symbols have been suppressed but the concentration in eqn (4.398) is the pore-average concentration and this justifies the use of the total derivative symbols.

Since θ can be measured by means of eqn (4.401), eqn (4.403) can be used to evaluate the average ionic concentrations in the membrane phase as

$$c_1 = c_{12}^{\mathrm{w}}(1-\theta)^{-1/2}, \tag{4.404}$$

$$c_2 = c_{12}^{\mathrm{w}}(1-\theta)^{1/2}. \tag{4.405}$$

c) Intermediate pore sizes, $a \approx L_{\mathrm{D}}^{\mathrm{w}}$

Ionic transport results in this case from a combination of three mechanisms: diffusion, migration, and convection. The flux density of species i (in the membrane-fixed reference frame) is

$$j_i = -D_i\left(\frac{\mathrm{d}c_i}{\mathrm{d}x} + z_i c_i f\frac{\mathrm{d}\phi}{\mathrm{d}x}\right) + c_i v, \tag{4.406}$$

and the electric current density is

$$I = -\kappa\left(\frac{\mathrm{d}\phi}{\mathrm{d}x} - \frac{\mathrm{d}\phi_{\mathrm{dif}}}{\mathrm{d}x}\right) + z_1 F X v, \tag{4.407}$$

where

$$f\frac{\mathrm{d}\phi_{\mathrm{dif}}}{\mathrm{d}x} = -\left(\frac{t_1^{\mathrm{M}}}{z_1} + \frac{t_2^{\mathrm{M}}}{z_2}\right)\frac{\mathrm{d}c_i}{\mathrm{d}x} \tag{4.408}$$

is the diffusion potential gradient and $t_i^{\mathrm{M}} = D_i c_i/(D_1 c_1 + D_2 c_2)$ for a symmetric electrolyte. The flux density of species i can also be written as

$$j_i = -D_{12}^{\mathrm{M}}\frac{\mathrm{d}c_i}{\mathrm{d}x} + \frac{t_i^{\mathrm{M}}}{z_1}\frac{I}{F} + [c_i - (z_1/z_i)t_i^{\mathrm{M}}X]v, \tag{4.409}$$

where $D_{12}^{\mathrm{M}} \equiv t_2^{\mathrm{M}}D_1 + t_1^{\mathrm{M}}D_2$.

4.4.4 Theoretical approaches for describing transport in porous membranes

In the previous sections we have discussed the basic ideas for the description of mass-transport processes in charged porous membranes. We briefly outline here some alternative theoretical approaches in order to clarify their differences.

The simplest possibility for describing transport processes in charged porous membranes makes use of a single axial position co-ordinate, and neglects the radial dependence of all transport magnitudes. This is known as the homogeneous membrane or homogeneous potential model. In contrast, the space-charge or charged-capillary model takes into account both the radial and axial position co-ordinates, and solves the Nernst–Planck (including convection) and Navier–Stokes equations, together with the Poisson–Boltzmann equation in the radial direction. This latter model is considered to be the most appropriate for the description of transport across charged porous membranes, but its solution is rather demanding from the point of view of numerical computation. Naturally, there have been several studies aiming at determining under which conditions the space-charge model leads to results significantly different from those obtained from the much simpler homogeneous potential model. Thus, for instance, one of the conclusions of these studies is that, as we explained in Section 4.2.5, the radial dependence of the electric potential inside a charged porous membrane implies that the co-ion exclusion is poorer. This is more noticeable when the radial dependence of the potential is stronger (that is, when $\kappa_D^X a \gg 1$ and $\kappa_D^w a \ll 1$; see Fig. 4.44), and implies that $(t_1^M - t_2^M)$ is smaller in the space-charge model than in the homogeneous potential model due to the poorer co-ion exclusion. Therefore, the diffusion potential inside the membrane, see eqn (4.408),

$$z_2 f \Delta\phi_{\text{dif}} = \int_{c_1(0)}^{c_1(h)} (t_1^M - t_2^M) \mathrm{d}c_1 \tag{4.410}$$

is also smaller than predicted by the homogeneous potential model [33]. Note, however, that both models give similar results when the fixed-charge concentration is small. A complete comparison of transport magnitudes evaluated from these two models can be found in Ref. [33].

In the homogeneous potential model, symbols like ϕ, c_i or t_i^M denote average values over the pore cross-section. This is the approach followed in Section 4.4.3 and closely resembles that used in Section 4.3. The fundamental set of transport equations (for a symmetric binary electrolyte) is

$$c_1 = c_2 + X, \tag{4.411}$$

$$j_i = -D_i \left(\frac{\mathrm{d}c_i}{\mathrm{d}x} + z_i c_i f \frac{\mathrm{d}\phi}{\mathrm{d}x} \right) + c_i v, \tag{4.412}$$

$$I = F(z_1 j_1 + z_2 j_2), \tag{4.413}$$

where the convective velocity v is either determined experimentally or estimated from

$$v = d_h z_2 F X \frac{\mathrm{d}\phi}{\mathrm{d}x}, \tag{4.414}$$

where d_h is the hydraulic permeability of the membrane.

The integration of the transport equations over the membrane thickness is done as follows [34]. First, we note that $\mathrm{d}X/\mathrm{d}x = 0$ because the fixed-charge concentration is independent of the axial position x, and eliminate the concentration gradient from the flux-density equations for the two ionic species, eqn (4.412), to obtain the electric potential gradient as

$$z_2 f \frac{\mathrm{d}\phi}{\mathrm{d}x} = \frac{2}{D_{12}^w (c_1 + c_2)} \left[t_2^w j_1 - t_1^w j_2 - (t_2^w c_1 - t_1^w c_2) v \right], \tag{4.415}$$

where $t_1^{\rm w} = D_1/(D_1+D_2) = 1 - t_2^{\rm w}$ and $D_{12}^{\rm w} = 2D_1D_2/(D_1+D_2)$. Similarly, we can eliminate the electric potential gradient from the flux-density equations for the two ionic species and obtain the concentration gradient

$$\frac{{\rm d}c_1}{{\rm d}x} = \frac{{\rm d}c_2}{{\rm d}x} = \frac{2}{D_{12}^{\rm w}(c_1+c_2)}\left[c_1c_2v - (c_1t_1^{\rm w}j_2 + c_2t_2^{\rm w}j_1)\right]. \tag{4.416}$$

This equation can be written as

$$\frac{2v}{D_{12}^{\rm w}}{\rm d}x = \frac{(2c_2+X){\rm d}c_2}{(c_2-c_+)(c_2-c_-)} = \left(\frac{X+2c_+}{c_2-c_+} - \frac{X+2c_-}{c_2-c_-}\right)\frac{{\rm d}c_2}{c_+-c_-}, \tag{4.417}$$

where c_+ and c_- are the two roots of the equation $(c_2+X)c_2v = (c_2+X)t_1^{\rm w}j_2 + c_2t_2^{\rm w}j_1$ (solved with respect to c_2). Equation (4.417) can be integrated over the membrane thickness to give

$$\begin{aligned}\frac{2vh}{D_{12}^{\rm w}}(c_+-c_-) = (X+2c_+)\ln\frac{c_2(h)-c_+}{c_2(0)-c_+}\\ -(X+2c_-)\ln\frac{c_2(h)-c_-}{c_2(0)-c_-},\end{aligned} \tag{4.418}$$

and the Donnan equilibrium conditions at the membrane boundaries, together with eqn (4.350), then allow determination of the flux densities. Similarly, eliminating the position co-ordinate from eqns (4.415) and (4.416), we obtain

$$\begin{aligned}z_2f{\rm d}\phi &= \frac{t_2^{\rm w}j_1 - t_1^{\rm w}j_2 - (t_2^{\rm w}c_1 - t_1^{\rm w}c_2)v}{c_1c_2v - (c_1t_1^{\rm w}j_2 + c_2t_2^{\rm w}j_1)}{\rm d}c_2\\ &= \left[\frac{(t_1^{\rm w}-t_2^{\rm w})c_+ + (t_2^{\rm w}j_1 - t_1^{\rm w}j_2)/v - t_2^{\rm w}X}{c_2-c_+}\right.\\ &\quad\left. - \frac{(t_1^{\rm w}-t_2^{\rm w})c_- + (t_2^{\rm w}j_1 - t_1^{\rm w}j_2)/v - t_2^{\rm w}X}{c_2-c_-}\right]\frac{{\rm d}c_2}{c_+-c_-},\end{aligned} \tag{4.419}$$

which can be integrated to obtain the potential drop in the membrane.

Exercises

4.1 Employing the Goldman constant-field assumption, ${\rm d}\phi/{\rm d}x = \Delta\phi/h$, solve the steady-state Nernst–Planck equation

$$j = -D\left(\frac{{\rm d}c}{{\rm d}x} + zcf\frac{{\rm d}\phi}{{\rm d}x}\right)$$

under the boundary conditions $c(0) = c^\alpha$ and $c(h) = c^\beta$, and show that the concentration profile is

$$c = c^\alpha + (c^\beta - c^\alpha)\frac{e^{-zf\,\Delta\phi\, x/h} - 1}{e^{-zf\,\Delta\phi} - 1}$$
$$= c^\beta + (c^\alpha - c^\beta)\frac{e^{-zf\,\Delta\phi} - e^{-zf\,\Delta\phi\, x/h}}{e^{-zf\,\Delta\phi} - 1}$$

and that the flux density is given by the Goldman equation

$$j = -\frac{D}{h}\frac{zf\,\Delta\phi}{e^{zf\,\Delta\phi} - 1}(c^\beta e^{zf\,\Delta\phi} - c^\alpha).$$

4.2 The lag time in diffusion processes can be evaluated solving Fick's second law

$$\frac{\partial c}{\partial t} = D\frac{\partial^2 c}{\partial x^2}$$

by the Laplace transform method.

(a) Transform this equation and its boundary conditions (see Section 4.1.2) and show that the Laplace-transformed concentration is

$$\tilde{c} = \frac{c^b}{s}\frac{\sinh q(h - x)}{\sinh qh}.$$

(b) Using the inverse transform formula

$$\mathcal{L}^{-1}\left(\frac{e^{-qx}}{s}\right) = \operatorname{erfc}\left(\frac{x}{2\sqrt{Dt}}\right) = 1 - \operatorname{erf}\left(\frac{x}{2\sqrt{Dt}}\right)$$

and a series expansion of $1/\sinh qh$ [see eqn (3.160)], show that the concentration is

$$c = \mathcal{L}^{-1}(\tilde{c}) = c^b \sum_{n=0}^{\infty}\left[\operatorname{erf}\left(\frac{nh + h - x/2}{\sqrt{Dt}}\right) - \operatorname{erf}\left(\frac{nh + x/2}{\sqrt{Dt}}\right)\right],$$

and reproduce Fig. 4.3 using this equation.

4.3 In the experimental set-up of Fig. 4.12 the membrane constant is given by

$$\frac{A}{h} = \frac{c_i^\alpha \dot{V}^\alpha + t_i IA/z_i F}{D_{12}(c_i^\beta - c_i^\alpha)},$$

and does not depend on the volume of compartment β. Explain the influence of the compartment volumes in this experimental set-up and compare it with the experimental set-up of Fig. 4.1.

4.4 In the experimental set-up considered in Fig. 4.12 the electric current density and the volume flow rate are independent parameters that can take any value. How can the membrane constant

$$\frac{A}{h} = \frac{c_i^\alpha \dot{V}^\alpha + t_i IA/z_i F}{D_{12}(c_i^\beta - c_i^\alpha)}$$

be determined when $I = -z_i F c_i^\alpha \dot{V}^\alpha / t_i A$?

4.5 It has been shown in Section 4.1.7 that when a membrane that bears no fixed-charge groups separates a mixture of two 1:1 electrolytes, the diffusion potential is

$$\Delta\phi_{\mathrm{dif}} = \frac{\Gamma}{f}\ln\frac{c_{\mathrm{T}}^{\beta}}{c_{\mathrm{T}}^{\alpha}},$$

where $c_{\mathrm{T}} \equiv \sum_i c_i$ is the total ionic concentration and Γ is a constant that must be determined from the solution of the following transcendental equation

$$\frac{(c_3^{\beta}/c_3^{\alpha})^{1+\Gamma}(r_{13}^{\beta}+r_{23}^{\beta})-(r_{13}^{\alpha}+r_{23}^{\alpha})}{(c_3^{\beta}/c_3^{\alpha})^{1+\Gamma}-1} = \frac{1-\Gamma}{1+\Gamma},$$

where $r_{13}^{\alpha} \equiv D_1c_1^{\alpha}/D_3c_3^{\alpha}$, $r_{13}^{\beta} \equiv D_1c_1^{\beta}/D_3c_3^{\beta}$, $r_{23}^{\alpha} \equiv D_2c_2^{\alpha}/D_3c_3^{\alpha}$, and $r_{23}^{\beta} \equiv D_2c_2^{\beta}/D_3c_3^{\beta}$.

(a) Using the above equations, evaluate the liquid junction potential $\Delta\phi_{\mathrm{dif}} = \phi^{\beta}-\phi^{\alpha}$ when a saturated 4.2 M KCl solution (compartment α) is in contact (through a neutral membrane) with a 0.1 M HCl (compartment β).

(b) Evaluate this liquid junction potential using Henderson's equation

$$f\,\Delta\phi_{\mathrm{dif}} = -\frac{(r_{13}^{\beta}+r_{23}^{\beta}-1)(c_3^{\beta}/c_3^{\alpha})-(r_{13}^{\alpha}+r_{23}^{\alpha}-1)}{(r_{13}^{\beta}+r_{23}^{\beta}+1)(c_3^{\beta}/c_3^{\alpha})-(r_{13}^{\alpha}+r_{23}^{\alpha}+1)} \ln\left(\frac{c_3^{\beta}}{c_3^{\alpha}}\frac{r_{13}^{\beta}+r_{23}^{\beta}+1}{r_{13}^{\alpha}+r_{23}^{\alpha}+1}\right).$$

(c) Evaluate this liquid-junction potential using Goldman's equation

$$f\,\Delta\phi_{\mathrm{dif}} = -\ln\frac{(r_{13}^{\beta}+r_{23}^{\beta})(c_3^{\beta}/c_3^{\alpha})+1}{r_{13}^{\alpha}+r_{23}^{\alpha}+(c_3^{\beta}/c_3^{\alpha})}.$$

In all cases, denote the ions H^+, K^+, Cl^- as species 1, 2, and 3, respectively, and take $D_{\mathrm{Cl^-}} = 1.037D_{\mathrm{K^+}} = 0.218D_{\mathrm{H^+}}$ and $1/f = 26$ mV.

4.6 It has been shown in Section 4.1.7 that when a membrane that bears no fixed-charge groups separates a mixture of symmetric $z:z$ electrolytes under strong stirring conditions, the exact value of the diffusion potential $\Delta\phi_{\mathrm{dif}} = \phi^{\beta}-\phi^{\alpha}$ is

$$\Delta\phi_{\mathrm{dif}} = \frac{\Gamma}{f}\ln\frac{c_{\mathrm{T}}^{\beta}}{c_{\mathrm{T}}^{\alpha}},$$

where $c_{\mathrm{T}} \equiv \sum_i c_i$ is the total ionic concentration and Γ is a constant that must be determined from the solution of the following transcendental equation

$$\frac{(c_{\mathrm{T}}^{\beta}/c_{\mathrm{T}}^{\alpha})^{z\Gamma}\sum_{+}D_ic_i^{\beta}-\sum_{+}D_ic_i^{\alpha}}{(c_{\mathrm{T}}^{\beta}/c_{\mathrm{T}}^{\alpha})^{-z\Gamma}\sum_{-}D_ic_i^{\beta}-\sum_{-}D_ic_i^{\alpha}} = \frac{1-z\Gamma}{1+z\Gamma}\,\frac{(c_{\mathrm{T}}^{\beta}/c_{\mathrm{T}}^{\alpha})^{1+z\Gamma}-1}{(c_{\mathrm{T}}^{\beta}/c_{\mathrm{T}}^{\alpha})^{1-z\Gamma}-1},$$

where the + and − signs under the sums indicate that they are restricted to cations and anions, respectively. Prove that when the solution contains only a 1:1 binary electrolyte these equations reduce to

$$f\,\Delta\phi_{\text{dif}} = \frac{D_2 - D_1}{D_1 + D_2}\ln\frac{c_{12}^{\beta}}{c_{12}^{\alpha}},$$

where species 1 is the cation and species 2 is the anion.

4.7 Using Henderson's equation

$$f\,\Delta\phi_{\text{dif}} \approx -\frac{\sum_i z_i D_i \Delta c_i}{\sum_j z_j^2 D_j \Delta c_j}\ln\frac{\sum_k z_k^2 D_k c_k^{\beta}}{\sum_l z_l^2 D_l c_l^{\alpha}}$$

evaluate the liquid-junction potential $\Delta\phi_{\text{dif}} = \phi^{\beta} - \phi^{\alpha}$ when a x M NaCl solution (compartment α) is in contact (through a neutral membrane) with a 1 M Na_2SO_4 solution (compartment β). Take $D_{Na^+} = 0.656\,D_{Cl^-} = 0.314\,D_{SO_4^{2-}}$ and $1/f = 26$ mV, and plot your result as a function of x.

4.8 In cell biophysics, the resting potential is usually estimated using Goldman's equation for the liquid-junction potential

$$\Delta\phi_{\text{dif}} = \frac{1}{f}\ln\frac{\sum\limits_{+} D_i c_i^{\alpha} + \sum\limits_{-} D_i c_i^{\beta}}{\sum\limits_{+} D_i c_i^{\beta} + \sum\limits_{-} D_i c_i^{\alpha}}.$$

Denote the outside of the cell as compartment α and the inside as compartment β, and consider that they both contain NaCl+KCl mixtures with the following concentrations $c_{Na^+}^{\alpha}$ = 145 mM, $c_{K^+}^{\alpha}$ = 4 mM, $c_{Cl^-}^{\alpha}$ = 149 mM, $c_{Na^+}^{\beta}$ = 12 mM, $c_{K^+}^{\beta}$ = 150 mM, and $c_{Cl^-}^{\beta}$ = 162 mM.

(a) Using the values $D_{Na^+} = 1.556\,D_{Cl^-} = 0.007\,D_{K^+}$ and $1/f = 26.7$ mV, show that the Goldman approximation provides a value very close to the actual liquid-junction potential. Evaluate the latter from

$$\Delta\phi_{\text{dif}} = \frac{\Gamma}{f}\ln\frac{c_T^{\beta}}{c_T^{\alpha}},$$

where Γ must be obtained first by solving numerically the equation

$$\frac{(c_3^{\beta}/c_3^{\alpha})^{1+\Gamma}(r_{13}^{\beta} + r_{23}^{\beta}) - (r_{13}^{\alpha} + r_{23}^{\alpha})}{(c_3^{\beta}/c_3^{\alpha})^{1+\Gamma} - 1} = \frac{1-\Gamma}{1+\Gamma},$$

where $r_{13}^{\alpha} \equiv D_1 c_1^{\alpha}/D_3 c_3^{\alpha}$, $r_{13}^{\beta} \equiv D_1 c_1^{\beta}/D_3 c_3^{\beta}$, $r_{23}^{\alpha} \equiv D_2 c_2^{\alpha}/D_3 c_3^{\alpha}$, and $r_{23}^{\beta} \equiv D_2 c_2^{\beta}/D_3 c_3^{\beta}$.

(b) Estimate also the liquid-junction potential under the above conditions from the Henderson approximation.

4.9 It can be considered that the ions that determine the resting potential in non-excitable cells are sodium, potassium and chloride ions, because they are in larger concentration in the outer (α) and inner (β) solutions. The chloride ions, however, do not really need to be taken into account because their distribution is very close to equilibrium and their flux density can be neglected in the open-circuit equation $I/F = j_{\mathrm{Na}^+} + j_{\mathrm{K}^+} - j_{\mathrm{Cl}^-} = 0$. This means that Goldman's equation for the liquid-junction potential can be used taking into account only the Na^+ and K^+ ions

$$\Delta\phi_{\mathrm{dif}} \approx \frac{1}{f} \ln \frac{D_{\mathrm{Na}^+} c^{\alpha}_{\mathrm{Na}^+} + D_{\mathrm{K}^+} c^{\alpha}_{\mathrm{K}^+}}{D_{\mathrm{Na}^+} c^{\beta}_{\mathrm{Na}^+} + D_{\mathrm{K}^+} c^{\beta}_{\mathrm{K}^+}}.$$

(a) Estimate the diffusion coefficient ratio $D_{\mathrm{Na}^+}/D_{\mathrm{K}^+}$ from the measured values of the resting potential $\Delta\phi_{\mathrm{dif}} = -90$ mV and ionic concentrations $c^{\alpha}_{\mathrm{Na}^+} =$ 145 mM, $c^{\alpha}_{\mathrm{K}^+} = 4$ mM, $c^{\beta}_{\mathrm{Na}^+} = 12$ mM, and $c^{\beta}_{\mathrm{K}^+} = 150$ mM. Take $1/f = 26.7$ mV.

(b) Show that the above equation for $\Delta\phi_{\mathrm{dif}}$ can be deduced from $j_{\mathrm{Na}^+} + j_{\mathrm{K}^+} \approx 0$ and the formal integration of the Nernst–Planck equations without employing Goldman's constant-field approximation.

4.10 P. Fatt and B.L. Ginsborg [*Journal of Physiology (London)*, 142 (1958) 516] studied the action potential in crustacean muscle fibres and concluded that the resting potential (i.e. the diffusion potential across the cell membrane) was determined by the exchange of Ca^{2+} and K^+ ions. Denoting the outside of the cell as compartment α and the inside as compartment β, the ionic concentrations are $c^{\alpha}_{\mathrm{Ca}^{2+}} = 1.5$ mM, $c^{\alpha}_{\mathrm{K}^+} = 4$ mM, $c^{\beta}_{\mathrm{Ca}^{2+}} = 0.1$ μM, and $c^{\beta}_{\mathrm{K}^+} = 155$ mM.

(a) From the Goldman flux equation

$$\frac{j_i}{D_i} = -\frac{z_i f\, \Delta\phi_{\mathrm{dif}}}{\mathrm{e}^{z_i f\, \Delta\phi_{\mathrm{dif}}} - 1} \frac{c_i^{\beta}\, \mathrm{e}^{z_i f\, \Delta\phi_{\mathrm{dif}}} - c_i^{\alpha}}{h}$$

and the open-circuit equation, $I/F = 2j_{\mathrm{Ca}^{2+}} + j_{\mathrm{K}^+} = 0$, show that the resting potential (under the conditions described here) is approximately given by

$$\Delta\phi_{\mathrm{dif}} = \phi^{\beta} - \phi^{\alpha} = \frac{1}{2f} \ln \frac{4D_{\mathrm{Ca}^{2+}} c^{\beta}_{\mathrm{Ca}^{2+}}}{D_{\mathrm{K}^+} c^{\alpha}_{\mathrm{K}^+}}.$$

The ionic permeability ratio in these fibres is $P_{\mathrm{Ca}^{2+}}/P_{\mathrm{K}^+} = 3000$, and we can simulate it in our formalism (without introducing partition coefficients or other magnitudes) with a diffusion coefficient ratio $D_{\mathrm{Ca}^{2+}}/D_{\mathrm{K}^+} = 3000$. Evaluate the resting potential using $1/f = 26.7$ mV.

4.11 A neutral membrane separates two compartments containing NaCl+KCl mixtures with the following concentrations $c^{\alpha}_{\mathrm{Na}^+} = 100$ mM, $c^{\alpha}_{\mathrm{K}^+} = 1$ mM, $c^{\alpha}_{\mathrm{Cl}^-} =$ 101 mM, $c^{\beta}_{\mathrm{Na}^+} = 1$ mM, $c^{\beta}_{\mathrm{K}^+} = 1.2$ mM, and $c^{\beta}_{\mathrm{Cl}^-} = 2.2$ mM. The ionic diffusion coefficients inside the membrane satisfy the ratios $D_{\mathrm{K}^+} = D_{\mathrm{Cl}^-} = 1.5D_{\mathrm{Na}^+}$.

(a) Evaluate the diffusion potential $\Delta\phi_{\mathrm{dif}} = \phi^{\beta} - \phi^{\alpha}$. (Take $1/f = 26$ mV.)

(b) Evaluate the ionic flux densities and comment on their direction in relation to the direction of the concentration gradient.

(c) Evaluate the Gibbs potential change associated to the transfer of one mole of sodium, potassium and chloride ions from compartment α to β.

4.12 Derive the expression for the Donnan potential in the case of a multi-ionic system with two ion classes.

4.13 Extend the diffuse layer model in Section 4.2.4 to the case of different dielectric permittivities in the two phases and show that the surface potential is then given by

$$\varphi_S = z_2 f(\phi_S - \phi^w) = z_2 f \Delta\phi_D - \tanh(z_2 f \Delta\phi_D/2) - \frac{2\sinh^2(\varphi_S/2)}{\sinh(z_2 f \Delta\phi_D)}\left[\left(\frac{\varepsilon^w}{\varepsilon^M}\right)^2 - 1\right].$$

4.14 Extend the diffuse layer model in Section 4.2.4 to include the effect of the ionic chemical partition coefficients $K_{c,i}$ and show that the Donnan and surface potentials are now given by

$$z_2 f \Delta\phi_D = \operatorname{arcsinh}\left(\frac{X}{2\alpha c_{12}^w}\right) + \beta,$$

$$\varphi_S = \varphi_D - \tanh[(\varphi_D - \beta)/2] + \frac{2}{\sinh(\varphi_D - \beta)} \times \left\{\sinh^2[(\varphi_S - \beta)/2] - \frac{1}{\alpha}\sinh^2(\varphi_S/2)\right\}$$

where $\alpha \equiv (K_{c,1}K_{c,2})^{1/2}$ and $\beta \equiv (1/2)\ln(K_{c,2}/K_{c,1})$.

4.15 Extend the diffuse layer model in Section 4.2.4 to the case of weakly dissociating fixed-charge groups. Consider that the dissociation reaction

$$-\mathrm{RA} \rightleftarrows -\mathrm{R}^{z_M} + \mathrm{A}^{-z_M}$$

has an equilibrium constant K, that the total molar concentration of fixed-charge groups (dissociated or not) is c_M, and that the membrane is equilibrated with a solution of a symmetric binary electrolyte of the same counterion A^{-z_M}.

4.16 Consider the interfacial region between a charged membrane occupying the region $x<0$ and a 1:1 binary electrolyte solution in the region $x>0$. Far from the interface, the electric potential in the membrane phase is ϕ^M, the electric field is zero, and the space-charge density is zero. The membrane has charged mobile groups with charge number z_M that can distribute according to the local electric potential. The concentration of this charged mobile groups in the bulk membrane phase is c_M. In the bulk of the external solution, the electrolyte concentration is c_{12}^w, the electric potential is ϕ^w, the electric field is zero, and the space-charge density is zero. Both phases are assumed to have the same dielectric permittivity ε. Calculate the equilibrium electrical-potential distribution in this interfacial region by solving the Poisson–Boltzmann equation

$$\frac{d^2\phi}{dx^2} = \frac{F}{\varepsilon}\begin{cases} 2c_{12}^w \sinh\varphi & x > 0 \\ 2c_{12}^w \sinh\varphi - z_M c_M e^{-z_M\varphi}, & x < 0 \end{cases},$$

where $\varphi(x) \equiv f[\phi(x) - \phi^w]$ is a dimensionless electric potential variable that varies continuously from zero in the bulk external phase to

$$\varphi_D \equiv f\,\Delta\phi_D = \operatorname{arcsinh}\frac{z_M c_M}{2c_{12}^w}$$

in the bulk membrane phase.

4.17 The Poisson–Boltzmann equation cannot be integrated in exact form in the membrane phase. The approximate analytical solution given in eqn (4.157) has been based on the linearization of the Poisson–Boltzmann equation in the membrane phase. An alternative approximate analytical solution can be obtained by assuming that the region $-L_{\mathrm{d}}^{\mathrm{M}} < x < 0$ (where the value of $L_{\mathrm{d}}^{\mathrm{M}}$ is yet to be determined) is depleted of mobile ions in the membrane side of the interfacial region. This is the so-called depleted layer model (DLM). The Poisson–Boltzmann equation in this region then becomes

$$\frac{\mathrm{d}^2\varphi}{\mathrm{d}\xi^2} = -\frac{c_{\mathrm{M}}}{c_{12}^{\mathrm{w}}},$$

where $\xi \equiv \kappa_{\mathrm{D}}^{\mathrm{w}} x$.

(a) Solve this equation under the boundary conditions $\varphi(\xi = -\kappa_{\mathrm{D}}^{\mathrm{w}} L_{\mathrm{d}}^{\mathrm{M}}) = z_2 f \Delta\phi_{\mathrm{D}}$, $(\mathrm{d}\varphi/\mathrm{d}\xi)_{\xi=-\kappa_{\mathrm{D}}^{\mathrm{w}} L_{\mathrm{d}}^{\mathrm{M}}} = 0$, and $\varphi(0) = \varphi_{\mathrm{s}}$ to obtain the potential distribution in the region $-L_{\mathrm{d}}^{\mathrm{M}} < x < 0$. Note that the Donnan potential is given by $\sinh(z_2 f \Delta\phi_{\mathrm{D}}) = c_{\mathrm{M}}/c_{12}^{\mathrm{w}}$.

(b) Make use of the continuity of the electric displacement at the interface as well as the electric-potential distribution in the external phase given by eqn (4.154) to show that the depleted layer model overestimates the electric potential drop in the external phase, that is, $(\varphi_{\mathrm{s}})^{\text{depleted}} > (\varphi_{\mathrm{s}})^{\text{diffuse}} > 0$.

(c) Find the value of $L_{\mathrm{d}}^{\mathrm{M}}$.

4.18 (a) Derive the expression of the diffusion potential drop inside an ion-exchange membrane for the case of a 1:1 electrolyte and singly charged fixed groups.

(b) Take the limit of a highly charged membrane and show that substitution of the resulting equation into eqn (4.203) leads to the expression $j_{12} = -D_2 \Delta c_{12}/h$.

(c) Take the limit of a weakly charged membrane and show that substitution of the resulting equation into eqn (4.203) leads to the expression $j_{12} = -D_{12}^{\mathrm{w}} \Delta_{\alpha}^{\beta} c_{12}/h$.

4.19 Under steady-state conditions and in the absence of homogeneous chemical reactions, the ionic flux densities and the electric current density have zero divergence, $\vec{\nabla} \cdot \vec{j}_i = 0$ and $\vec{\nabla} \cdot \vec{I} = 0$. In a one-dimensional membrane system this implies that these fluxes are independent of position and do not change when crossing the membrane boundaries. By analysing the diffusion-conduction flux equation

$$\vec{j}_i = \nu_i \vec{j}_{12} + \frac{t_i}{z_i}\frac{\vec{I}}{F},$$

what can you say about the position dependence of the Donnan electrolyte flux density $\vec{j}_{12}$?

4.20 Using a phenomenological approach with cross-coefficients, show that the interdiffusion of two counterions across an ideally selective ion-exchange membrane under open-circuit conditions only involves one diffusion coefficient even when the Nernst–Planck approximation is not employed.

4.21 Apply the equations derived in Section 4.3.7 to a much simpler situation in which all cations have the same charge number z and all anions have the same charge number $-z$, the membrane is neutral, the electric current is zero, and both sides of the membrane have similar ionic strength, $\sum_i c_i(h) \approx \sum_i c_i(0)$. Show that the diffusion potential is then given by the Goldman–Hodgkin–Katz (GHK) equation

$$\Delta\phi_{\text{dif}} = \frac{1}{zf} \ln \frac{\sum_{+} P_i c_i(0) + \sum_{-} P_i c_i(h)}{\sum_{+} P_i c_i(h) + \sum_{-} P_i c_i(0)},$$

where P_i denotes the permeability of the membrane to ionic species i.

4.22 From the equations derived in Section 4.3.7, determine the expression for the potential drop in a membrane that is so strongly charged that co-ions are excluded from the membrane and do not contribute to the current transport.

4.23 In Section 4.3.4 it was obtained that the ohmic potential drop, when a symmetric binary electrolyte is transported across an ion-exchange membrane, is given by

$$z_2 f\, \Delta\phi_{\text{ohm}} = z_2 \Gamma \ln \frac{c_T(h) + z_2 \Gamma X}{c_T(0) + z_2 \Gamma X} - (t_1^{\text{w}} - t_2^{\text{w}}) \ln \frac{c_T(h) + (t_1^{\text{w}} - t_2^{\text{w}})X}{c_T(0) + (t_1^{\text{w}} - t_2^{\text{w}})X}.$$

(a) Show that this expression can also be derived from the evaluation of the membrane resistance as

$$\Delta\phi_{\text{ohm}} = -I R^{\text{M}} \equiv -I \int_0^h \frac{\mathrm{d}x}{\kappa^{\text{M}}} = -\frac{IRT}{z_1^2 F^2} \int_0^h \frac{\mathrm{d}x}{D_1 c_1 + D_2 c_2}.$$

(b) Find the expression for the ohmic potential drop in the limit $c_{12}^{\alpha} \approx c_{12}^{\beta}$, i.e. when $|c_{12}^{\alpha} - c_{12}^{\beta}| \ll \overline{c_{12}^{\text{w}}}$.

Hint: Remember that $c_1 = (c_T + X)/2$ and $c_2 = (c_T - X)/2$, and evaluate $\mathrm{d}c_T/\mathrm{d}x$ from

$$xG_0 = c_T(0) - c_T(x) + z_2 \Gamma X \ln \frac{c_T(x) + z_2 \Gamma X}{c_T(0) + z_2 \Gamma X}.$$

4.24 In Section 4.4.2 we have obtained that the electric conduction equation in charged porous membranes, in the absence of concentration gradients, is

$$\langle I \rangle = -(\langle \kappa \rangle + \langle \kappa_c \rangle) \frac{\mathrm{d}V}{\mathrm{d}x}.$$

This equation can also be written as

$$\langle I \rangle = -\left(\frac{2\pi a}{\pi a^2} K^{\text{s}} + \kappa^{\text{b}} \right) \frac{\mathrm{d}V}{\mathrm{d}x}$$

where κ^{b} is the bulk conductivity (i.e. the electrical conductivity of the external solution) and K^{s} is the surface conductivity (note that it does not have the dimensions of a conductivity because of the geometrical factor). Thus, the surface conductivity accounts for the differences between electric conduction inside the membrane pores and in the external solution, which involve mainly two aspects. First, the (total) ionic concentration is larger than in the external solution because there is a need to counterbalance the fixed charge on the pore walls. And second, electro-osmotic convection influences the ionic motion.

Knowing that the convective conductivity in the linear approximation is

$$\langle \kappa_c \rangle \approx X \frac{\varepsilon RT}{\eta} z_2 f \tilde{\psi}(a) \approx X \frac{\varepsilon RT}{\eta} \frac{(\kappa_D^X)^2}{(\kappa_D^W)^2} \left[\frac{\kappa_D^W a I_0(\kappa_D^W a)}{2I_1(\kappa_D^W a)} - 1 \right]$$

obtain an expression for K^s.

4.25 In Section 4.4.4 we have obtained that the potential drop when a symmetric binary electrolyte is transported across a charged porous membrane can be evaluated in the homogeneous potential approach by integration of the equation

$$z_2 f\,\mathrm{d}\phi = \left[\frac{(t_1^W - t_2^W)c_+ + (t_2^W j_1 - t_1^W j_2)/v - t_2^W X}{c_2 - c_+} - \frac{(t_1^W - t_2^W)c_- + (t_2^W j_1 - t_1^W j_2)/v - t_2^W X}{c_2 - c_-} \right] \frac{\mathrm{d}c_2}{c_+ - c_-}$$

where c_+ and c_- are the two roots of the equation $(c_2 + X)c_2 v = (c_2 + X)t_1^W j_2 + c_2 t_2^W j_1$ (solved with respect to c_2). Show that this equation reduces to

$$z_2 f\,\mathrm{d}\phi = \frac{t_1^W - t_2^W}{c_2 + X t_1^W} \mathrm{d}c_2,$$

when convection is negligible and no electric current is transported across the membrane, $v = 0$ and $I = 0$, and compare the expression for the (diffusion) potential drop thus obtained with the expression

$$z_2 f\,\Delta\phi_{\mathrm{dif}} = (t_1^W - t_2^W) \ln \frac{c_1(h) + c_2(h) + X(t_1^W - t_2^W)}{c_1(0) + c_2(0) + X(t_1^W - t_2^W)}$$

that was obtained in Section 4.3.3.

References

[1] J.C. Keister and G.B. Kasting, 'Ionic mass transport through a homogeneous membrane in the presence of a uniform electric field', *J. Membrane Sci.*, 29 (1986) 155–167.

[2] V.M. Aguilella, S. Mafé, and J. Pellicer, 'Ionic transport through a homogeneous membrane in the presence of simultaneous diffusion, conduction and convection', *J. Chem. Soc. Faraday Trans. 1*, 85 (1989) 223–235.

[3] P. Henderson, 'Zur Thermodynamic der Flüssigkeitsketten', *Z. Phys. Chem.*, 59 (1907) 118–128.

[4] P. Henderson, 'Zur Thermodynamic der Flüssigkeitsketten', *Z. Phys. Chem.*, 63 (1908) 325–345.

[5] P. Ramírez, A. Alcaraz, and S. Mafé, 'Model calculations of ion transport against its concentration gradient when the driving force is a pH difference across a charged membrane', *J. Membrane Sci.*, 135 (1997) 135–144.

[6] P. Ramírez, A. Alcaraz, and S. Mafé, 'Uphill transport of amino acids through fixed charge membranes', *Encyclopedia of Surface and Colloid Science*, vol. 1 (1) pp. 1–12, Marcel Dekker, New York, 2005.

[7] N. Lakshminarayanaiah, *Transport Phenomena in Membranes*, Academic Press, New York, 1969.

[8] H. Reiss and I.C. Bassignana, 'Critique of the mechanism of superselectivity in ion exchange membranes', *J. Membrane Sci.*, 11 (1982) 219–229.
[9] J.A. Manzanares, S. Mafé, and J. Pellicer, 'Current efficiency enhancement in membranes with macroscopic inhomogeneities in the fixed charge distribution', *J. Chem. Soc. Faraday Trans.*, 88 (1992) 2355–2364.
[10] J.P. Hsu and K.C. Ting, 'Current efficiency of an ion-selective membrane: Effect of fixed charge distribution', *J. Electrochem. Soc.*, 145 (1998) 1088–1092.
[11] E. Hawkins Cwirko and R.G. Carbonell, 'Transport of electrolytes in charged pores: analysis using the method of spatial averaging', *J. Colloid Interface Sci.*, 129 (1989) 513–531.
[12] E. Glueckauf, 'A new approach to ion exchange polymers', *Proc. Roy. Soc. London A*, 268 (1962) 350–370.
[13] J.M. Crabtree and E. Glueckauf, 'Structural analysis of ion semi-permeable membranes by co-ion uptake and diffusion studies', *Trans. Faraday Soc.*, 59 (1963) 2639–2654.
[14] J.H. Petropoulos, 'Membrane transport properties in relation to microscopic and macroscopic structural inhomogeneity', *J. Membrane Sci.*, 52 (1990) 305–323.
[15] A.V. Sokirko, J.A. Manzanares, and J. Pellicer, 'The permeability of membrane systems with an inhomogeneous distribution of fixed charged groups', *J. Colloid Interface Sci.*, 168 (1994) 32–39.
[16] F. Helfferich, *Ion Exchange*, Dover, New York, 1995, p. 303 ff.
[17] J.A. Manzanares and K. Kontturi, 'Transport numbers of ions in charged membrane systems' in *Surface Chemistry and Electrochemistry of Membranes*, T.S. Sørensen (ed.), Marcel Dekker, New York, 1999, Ch.11.
[18] N. Lakshminarayanaiah, *Equations of Membrane Biophysics*, Academic Press, Orlando, FL, 1984.
[19] A. Guirao, S. Mafé, J.A. Manzanares, and J.A. Ibáñez, 'Bi-ionic potential of charged membranes: effects of the diffusion boundary layers', *J. Phys. Chem.*, 99 (1995) 3387–3393.
[20] J.A. Manzanares and K. Kontturi, 'Diffusion and migration' in *Encyclopedia of Electrochemistry, Vol. 2 Interfacial Kinetics and Mass Transport*, E.J. Calvo (ed.), Wiley-VCH, Weinheim, 2003.
[21] M.Kh. Urtenov, *Methods of Solution of the Nernst–Planck–Poisson Equation System* [in Russian], Kuban State University, Krasnodar, Russia, 1998.
[22] V.I. Zabolotsky and V.V. Nikonenko, *Ion Transport in Membranes* [in Russian], Nauka, Moscow, 1996.
[23] J. Pellicer, S. Mafé, and V.M. Aguilella, 'Ionic transport across porous charged membranes and the Goldman constant field assumption', *Ber. Bunsenges. Phys. Chem.*, 90 (1986) 867–872.
[24] L. Dresner, 'Electrokinetic phenomena in charged microcapillaries', *J. Phys. Chem.*, 67 (1963) 1635–1641.
[25] F.A. Morrison Jr. and J.F. Osterle, 'Electrokinetic energy conversion in ultrafine capillaries', *J. Chem. Phys.*, 43 (1965) 2111–2115.
[26] R.J. Gross and J.F. Osterle, 'Membrane transport characteristics in ultrafine capillaries', *J. Chem. Phys.*, 49 (1968) 228–234.
[27] J.C. Fair and J.F. Osterle, 'Reverse electrodialysis in charged capillary membranes', *J. Chem. Phys.*, 54 (1971) 3307–3316.
[28] R. Schlögl, 'Zur theorie der anomalen osmose', *Z. physik. Chem. N.F.*, 3 (1955) 73–102.

[29] R. Schlögl, 'The significance of convection in transport processes across porous membranes', *Discuss. Faraday Soc.*, 21 (1956) 46–52.

[30] R. Schlögl, *Stofftransport durch Membranen*, D. Steinkopff, Darmstadt, 1964.

[31] F. Helfferich, *Ion Exchange*, Dover, New York, 1995, p.330.

[32] J. Cervera, J.A. Manzanares, and S. Mafé, 'Ion size effects on the current efficiency of narrow charged pores', *J. Membrane Sci.*, 191 (2001) 179–187.

[33] G.B. Westermann-Clark and C.C. Christoforou, 'The exclusion-diffusion potential in charged porous membranes', *J. Electroanal. Chem.*, 198 (1986) 213–231.

[34] R. Telaranta, J.A. Manzanares, and K. Kontturi, 'Convective electrodiffusion processes through graft-modified charged porous membranes', *J. Electroanal. Chem.*, 464 (1999) 222–229.

5 Transport through liquid membranes

5.1 Distribution equilibria in liquid membrane systems

5.1.1 Liquid membranes

Liquid membranes form an interesting group of membranes, which can be used to selectively separate or extract solutes from one phase to another [1]. They can also be used as a crude model for a biomembrane, although they have no organized structure on the molecular level like a phospholipid bilayer. Liquid membranes can either be emulsion-like or supported. In an emulsion liquid membrane (or surfactant liquid membrane) the solutes to be separated are enriched in the stripping phase inside the micelle created by the membrane-forming surfactant. This type of liquid membrane is common in practical extraction processes, but they are not considered here.

A supported liquid membrane is usually a porous hydrophobic membrane[1] where an organic solvent is impregnated. The solvent is held inside the membrane by capillary forces. Extraction by liquid membranes takes place in one stage only, i.e. separate extraction and stripping stages are not required. Extraction is often based on some selective carrier molecule in the membrane phase, whereby solutes can be transferred across the membrane against their concentration gradient. In this case, the chemical energy needed for the process is taken from the transfer of some other species, e.g. a proton. As the volume of the membrane phase is very small, expensive and/or even toxic carriers can be utilized in extraction because the amount of material is minimized. By the appropriate choice of the carrier molecule and the solvent, extraction can be made very selective.

Also, electric fields can be used to run extraction processes of metal cations, for example. There exist, however, a few difficulties. Firstly, the membrane phase must remain electroneutral. If an ion enters the membrane at one interface, either another ion must leave the membrane at the opposing interface, or an ion with an opposite charge has to enter the membrane. Secondly, because the membrane presumably is thin, of the order of 100 μm, the potential difference across the individual aqueous/organic interfaces cannot be monitored

[1] Although other geometries might be preferable for some applications [2], in this chapter we consider only planar membranes.

or controlled. Only the total potential difference across the membrane can be controlled, and as a consequence of the coupling of charges, the total potential difference distributes itself asymmetrically in the system. It may create initially, for instance, such a high potential difference on the stripping side, as no ions are available yet, that the interface becomes unstable, or the base electrolytes of the aqueous or membrane phase cross the interface. Next, we are considering the distribution equilibrium of different species between the liquid membrane and one of the aqueous solutions in contact with it.

5.1.2 Equilibrium partitioning at the aqueous/organic solution interface

When a liquid membrane is clamped by two aqueous phases, i.e. the feed and the strip solutions, two aqueous/organic solution interfaces are formed. We consider here one of those interfaces and describe the equilibrium partitioning $i(\mathrm{w}) \rightleftarrows i(\mathrm{o})$ of a solute i. The ability of neutral solute to cross the interface depends only on the solute and solvent properties. The equilibrium partition ratio $a_i^{\mathrm{o}}/a_i^{\mathrm{w}}$ is then equal to its chemical partition coefficient K_i, and this is determined by the Gibbs free energy of transfer of the solute i from the organic to the aqueous solution, $\Delta_{\mathrm{o}}^{\mathrm{w}} G_i^{\circ} \equiv \mu_i^{\mathrm{o,w}} - \mu_i^{\mathrm{o,o}}$, through the thermodynamic relation $K_i \equiv \mathrm{e}^{\Delta_{\mathrm{o}}^{\mathrm{w}} G_i^{\circ}/RT}$. For hydrophobic neutral solutes, $\Delta_{\mathrm{o}}^{\mathrm{w}} G_i^{\circ} > 0$ and $a_{\mathrm{i}}^{\mathrm{o}}/a_i^{\mathrm{w}} = K_i > 1$ so that they have a higher activity in the organic phase. On the contrary, for hydrophilic neutral solutes $\Delta_{\mathrm{o}}^{\mathrm{w}} G_i^{\circ} < 0$ and $a_i^{\mathrm{o}}/a_{\mathrm{i}}^{\mathrm{w}} = K_i < 1$.

In the case of an ionic solute i, the distribution equilibrium requires that the electrochemical potential of this solute takes the same value in both phases, $\tilde{\mu}_i^{\mathrm{w}} = \tilde{\mu}_i^{\mathrm{o}}$, and therefore the equilibrium condition involves the electric potentials in these phases

$$\mu_i^{\mathrm{o,w}} + RT \ln a_i^{\mathrm{w}} + z_i F \phi^{\mathrm{w}} = \mu_i^{\mathrm{o,o}} + RT \ln a_i^{\mathrm{o}} + z_i F \phi^{\mathrm{o}}. \tag{5.1}$$

This implies that ionic solutes can be 'pushed' across the interface by adjusting the interfacial potential difference $\Delta_{\mathrm{o}}^{\mathrm{w}}\phi \equiv \phi^{\mathrm{w}} - \phi^{\mathrm{o}}$. In fact, their partitioning is often explained in terms of potential differences only by writing the Gibbs free energy of transfer as $\Delta_{\mathrm{o}}^{\mathrm{w}} G_i^{\circ} \equiv -z_i F \Delta_{\mathrm{o}}^{\mathrm{w}} \phi_i^{\circ}$. This expression constitutes the definition of the standard transfer potential of species i, $\Delta_{\mathrm{o}}^{\mathrm{w}} \phi_i^{\circ}$. The transfer potential is positive for hydrophilic cations and hydrophobic anions, and negative otherwise. A few values are given in Table 5.1. The chemical affinity of a hydrophilic cation for the aqueous phase is then interpreted in terms of a positive transfer potential $\Delta_{\mathrm{o}}^{\mathrm{w}} \phi_i^{\circ}$ so that the interfacial potential drop $\Delta_{\mathrm{o}}^{\mathrm{w}}\phi$ must be increased above $\Delta_{\mathrm{o}}^{\mathrm{w}} \phi_i^{\circ}$ to 'push' this hydrophilic cation from the aqueous to the organic phase. The equilibrium partition ratio for ionic solutes is then

$$\frac{a_i^{\mathrm{o}}}{a_i^{\mathrm{w}}} = K_i \mathrm{e}^{z_i f \Delta_{\mathrm{o}}^{\mathrm{w}}\phi} = \mathrm{e}^{z_i f (\Delta_{\mathrm{o}}^{\mathrm{w}}\phi - \Delta_{\mathrm{o}}^{\mathrm{w}}\phi_i^{\circ})}, \tag{5.2}$$

which is equivalent to eqn (5.1). Thus, for instance, increasing the electrostatic energy of a hydrophilic solute in the aqueous phase so much that the difference

Table 5.1. Ionic standard transfer potentials $\Delta_o^w \phi_i^o$ (mV) for mutually saturated water–organic solvent systems at 25°C. (Extracted from Ref. [3] with permission.)

Ion	nitrobenzene	1,2-dichloroethane	dichloromethane
Li^+	298	493	
Na^+	355	490	
H^+	337		
NH_4^+	284		
K^+	241	499	
Rb^+	201	445	
Cs^+	159	360	
acetylcholine	52		
$(CH_3)_4N^+$	37	182	195
$(C_2H_5)_4N^+$	−63	44	44
$(C_3H_7)_4N^+$	−160	−91	−91
$(C_4H_9)_4N^+$	−270	−225	−230
$(C_5H_{11})_4N^+$		−360	−377
$(C_6H_5)_4As^+$	−372	−364	
crystal violet	−410		
$(C_6H_{13})_4N^+$	−472	−494	−455
Mg^{2+}	370		
Ca^{2+}	354		
Sr^{2+}	348		
Ba^{2+}	328		
Cl^-	−395	−481	−481
Br^-	−335	−408	−408
NO_3^-	−270		
I^-	−195	−273	−273
SCN^-	−161		
BF_4^-	−91		
ClO_4^-	−91	−178	−221
2,4-dinitrophenolate	−77		
PF_6^-	12		
picrate	47	−69	
$(C_6H_5)_4B^-$	372	364	
dipicrylaminate	414		
dicarbolylcobaltate	520		

in standard chemical potentials is overcome, $z_i \Delta_o^w \phi > z_i \Delta_o^w \phi_i^o$, it can be forced into the organic phase.[2]

The activity coefficients of species i in the two phases are often assumed to be equal to each other, $\gamma_i^o \approx \gamma_i^w$, and the partitioning equilibrium equation is simplified to

$$\frac{c_i^o}{c_i^w} \approx K_i e^{z_i f \Delta_o^w \phi}, \tag{5.3}$$

which can also be applied to neutral solutes ($z_i = 0$).

[2] This actually means that the activity of the hydrophobic solute in the aqueous phase can be raised above the activity in the organic phase.

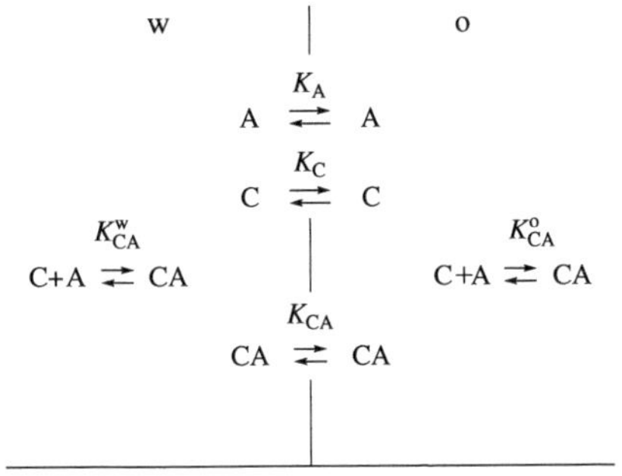

Fig. 5.1. Equilibrium partitioning of a neutral solute A and a ligand C that can associate to form a complex CA.

5.1.3 Finite-volume effects on the partitioning of a neutral solute

As the volume of the liquid membrane is much smaller than those of the bathing aqueous solutions, partitioning of solutes between an aqueous solution and the membrane has some interesting characteristics. Let us consider first a case where the aqueous solution contains a hydrophilic solute A that can partition to the membrane and the membrane contains a hydrophobic ligand C that can also partition to the aqueous solution (Fig. 5.1). Initially, i.e. before the equilibrium distribution is established, the solute A is only present in the aqueous phase with a concentration $c_A^{w,0}$ and the ligand C is only present in the liquid membrane with a concentration $c_C^{o,0}$. After the equilibrium has been established, their concentrations in both phases are[3]

$$c_A^o = K_A c_A^w = c_A^{w,0} \frac{K_A r}{K_A + r}, \tag{5.4}$$

$$c_C^o = K_C c_C^w = c_C^{o,0} \frac{K_C}{K_C + r}, \tag{5.5}$$

where $r = V^w/V^o$ is the volume ratio of the aqueous and membrane phase, and K_A, K_C are the partition coefficients of A and C, respectively. Thus, for instance, in order not to lose ligand from the membrane, it must be satisfied that $r \ll K_C$.

Consider now that A and C can form a hydrophobic ligand–solute complex CA that can also be present in both phases. The chemical equilibrium condition for this complexation reaction in the aqueous phase is

$$K_{CA}^w = \frac{c_{CA}^w}{c_C^w c_A^w}, \tag{5.6}$$

and a similar mass-action law can be written for the membrane phase, although this is not an independent relation because the corresponding equilibrium

[3] These equations are derived from mass balances as explained below.

constant K^{o}_{CA} is determined by K^{w}_{CA} and the partition coefficients as[4] $K^{\text{o}}_{\text{CA}} = K^{\text{w}}_{\text{CA}} K_{\text{CA}}/K_{\text{C}}K_{\text{A}}$. The equilibrium concentrations of the different species in the aqueous and membrane phases are no longer given by eqns (5.4) and (5.5) and they must be found out from the mass balances

$$c^{\text{w},0}_{\text{A}} V^{\text{w}} = (c^{\text{w}}_{\text{A}} + c^{\text{w}}_{\text{CA}})V^{\text{w}} + (c^{\text{o}}_{\text{A}} + c^{\text{o}}_{\text{CA}})V^{\text{o}} \quad \text{solute balance,} \tag{5.7}$$

$$c^{\text{o},0}_{\text{C}} V^{\text{o}} = (c^{\text{w}}_{\text{C}} + c^{\text{w}}_{\text{CA}})V^{\text{w}} + (c^{\text{o}}_{\text{C}} + c^{\text{o}}_{\text{CA}})V^{\text{o}} \quad \text{ligand balance.} \tag{5.8}$$

Introducing the volume ratio r, the partition coefficients, and the reaction equilibrium constant, the mass balances can be transformed to

$$c^{\text{w},0}_{\text{A}} r = [K_{\text{A}} + r + (K_{\text{CA}} + r)K^{\text{w}}_{\text{CA}} c^{\text{w}}_{\text{C}}]c^{\text{w}}_{\text{A}}, \tag{5.9}$$

$$c^{\text{o},0}_{\text{C}} = [K_{\text{C}} + r + (K_{\text{CA}} + r)K^{\text{w}}_{\text{CA}} c^{\text{w}}_{\text{A}}]c^{\text{w}}_{\text{C}}, \tag{5.10}$$

which is a simple equation system that can be solved for the variables c^{w}_{A} and c^{w}_{C}. Some typical results are shown in Fig. 5.2 for the parameter values: $K_{\text{A}} = 10^{-3}$, $K_{\text{C}} = 10^4$, $K_{\text{CA}} = 10^3$, $K^{\text{w}}_{\text{CA}} = 3 \times 10^3\,\text{M}$, $c^{\text{w},0}_{\text{A}} = 0.01\,\text{M}$, and $c^{\text{o},0}_{\text{C}} = 0.1\,\text{M}$.

Figure 5.2 shows that the total ligand concentration in the membrane, $c^{\text{o}}_{\text{C}} + c^{\text{o}}_{\text{CA}}$, is practically equal to the initial value $c^{\text{o},0}_{\text{C}}$ for all values of the volume ratio in the range, $r \ll K_{\text{C}}$, as was also concluded from eqn (5.5) in the absence of a complexation reaction. The solute concentration in the aqueous phase c^{w}_{A} drops significantly from the initial value, $c^{\text{w},0}_{\text{A}}$, due to the complexation reaction when $r \approx K_{\text{CA}} K^{\text{w}}_{\text{CA}} c^{\text{o,w}}_{\text{C}}/K_{\text{C}}$, which corresponds to the condition $c^{\text{w}}_{\text{A}} V^{\text{w}} \approx c^{\text{o}}_{\text{CA}} V^{\text{o}}$, i.e. when the amount of solute in free form in the aqueous solution is of the same order of magnitude as the amount of solute in complexed form inside the membrane.

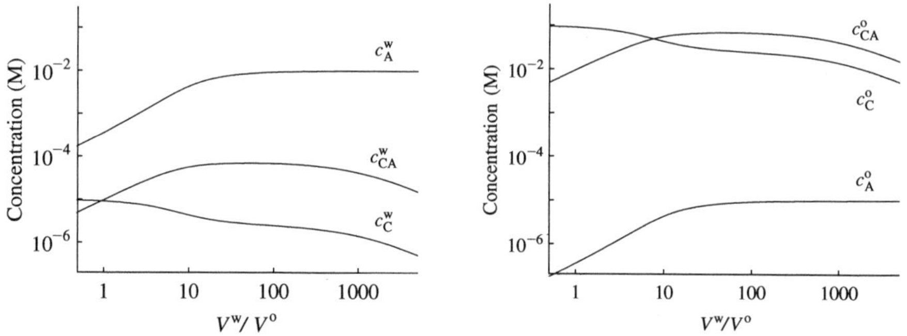

Fig. 5.2.
Equilibrium concentrations (aqueous left, membrane right) as a function of the volume ratio; parameter values given in the text.

[4] Note that K_{CA} without superscript denotes here the partition coefficient of the ligand–solute complex.

5.1.4 Finite-volume effects on the partitioning of an ionic solute

Consider next the case where the bathing solution contains a strong acid AB, say HCl, and the organic phase contains a hydrophobic ligand C (e.g. an amine R–NH_2) that can extract the acid from the aqueous to the organic phase by forming an acid–amine complex. This partitioning process can be illustrated by either of the two schemes in Fig. 5.3. In the aqueous solution the strong acid is dissociated into the ions A^+ and B^-(e.g. H^+ and Cl^-), and in the organic solution it is more likely in the form of ion pairs AB due to its lower relative electrical permittivity. Similarly, the acid–amine complex CAB can be considered to be dissociated into the ions CA^+ and B^- (e.g. $R-NH_3Cl$ or $R-NH_3^+$ and Cl^-). The important point, however, is that the partition constant of species i is now the product of two terms, a chemical partition coefficient K_i and an electrostatic coefficient $e^{z_i f \Delta_o^w \phi}$, and that the latter is not a property of the ion and the solvents, but it is rather determined by the concentrations of all the ions dissolved in both phases. Let us then introduce the variable $K_e = e^{f \Delta_o^w \phi}$ and write the partition constant of the cations as $K_i K_e$ and that of the anion as K_i/K_e. The partition coefficient of the ion pairs is the product of the chemical partition coefficients of the corresponding ions, e.g. $K_{R-NH_3Cl} = K_{R-NH_3^+} K_{Cl^-}$ and $K_{HCl} = K_{H^+} K_{Cl^-}$.

Initially, i.e. before the equilibrium distribution is established, the acid AB is only present in the aqueous phase with a concentration $c_{AB}^{w,0}$ and the amine C is only present in the liquid membrane with a concentration $c_C^{o,0}$. The equilibrium concentrations of the different species in the aqueous and organic phases must be found from the mass balances

$$c_{AB}^{w,0} V^w = (c_{A^+}^w + c_{CA^+}^w) V^w + (c_{A^+}^o + c_{CA^+}^o) V^o \quad \text{ionic solute balance,} \tag{5.11}$$

$$c_C^{o,0} V^o = (c_C^w + c_{CA^+}^w) V^w + (c_C^o + c_{CA^+}^o) V^o \quad \text{amine balance.} \tag{5.12}$$

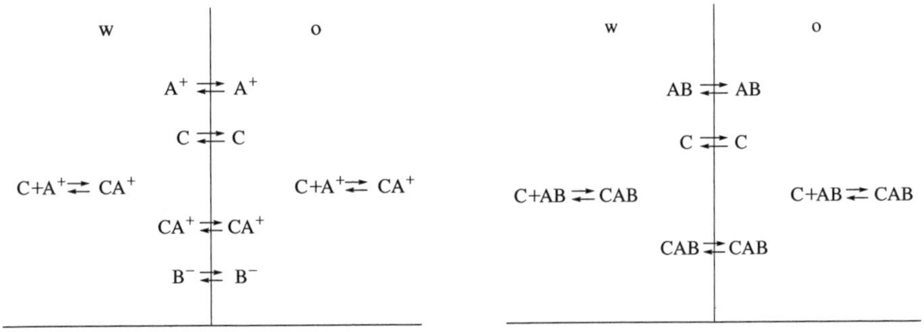

Fig. 5.3.
Equilibrium partitioning of an acid AB and an amine C that can associate to form an acid–amine complex CAB. The equilibrium distribution can be described considering either ions or ion pairs.

The chemical equilibrium condition for the complexation reaction in the aqueous phase is

$$K^{\mathrm{w}}_{\mathrm{CA^+}} = \frac{c^{\mathrm{w}}_{\mathrm{CA^+}}}{c^{\mathrm{w}}_{\mathrm{C}} c^{\mathrm{w}}_{\mathrm{A^+}}}, \tag{5.13}$$

and a similar mass-action law can be written for the organic phase with an equilibrium constant $K^{\mathrm{o}}_{\mathrm{CA^+}} = K^{\mathrm{w}}_{\mathrm{CA^+}} K_{\mathrm{CA^+}}/K_{\mathrm{C}} K_{\mathrm{A^+}}$.

The electrostatic contribution to the partition coefficient of the cations, $K_{\mathrm{e}} = \mathrm{e}^{f\Delta^{\mathrm{w}}_{\mathrm{o}}\phi}$, can be determined from the electroneutrality condition in both phases, $c^{\mathrm{w}}_{\mathrm{B^-}} = c^{\mathrm{w}}_{\mathrm{A^+}} + c^{\mathrm{w}}_{\mathrm{CA^+}}$ and $c^{\mathrm{o}}_{\mathrm{B^-}} = c^{\mathrm{o}}_{\mathrm{A^+}} + c^{\mathrm{o}}_{\mathrm{CA^+}}$. The latter can be transformed to

$$\begin{aligned}(K_{\mathrm{B^-}}/K_{\mathrm{e}}) c^{\mathrm{w}}_{\mathrm{B^-}} &= (K_{\mathrm{B^-}}/K_{\mathrm{e}})(c^{\mathrm{w}}_{\mathrm{A^+}} + c^{\mathrm{w}}_{\mathrm{CA^+}}) \\ &= K_{\mathrm{A^+}} K_{\mathrm{e}} c^{\mathrm{w}}_{\mathrm{A^+}} + K_{\mathrm{CA^+}} K_{\mathrm{e}} c^{\mathrm{w}}_{\mathrm{CA^+}},\end{aligned} \tag{5.14}$$

and, therefore,

$$K_{\mathrm{e}} = \left[\frac{K_{\mathrm{B^-}}(1 + K^{\mathrm{w}}_{\mathrm{CA^+}} c^{\mathrm{w}}_{\mathrm{C}})}{K_{\mathrm{A^+}} + K_{\mathrm{CA^+}} K^{\mathrm{w}}_{\mathrm{CA^+}} c^{\mathrm{w}}_{\mathrm{C}}}\right]^{1/2}, \tag{5.15}$$

where we have introduced the complexation constant defined in eqn (5.13).

Similarly, the mass balances can be transformed to

$$c^{\mathrm{w},0}_{\mathrm{AB}} r = [K_{\mathrm{A^+}} K_{\mathrm{e}} + r + (K_{\mathrm{CA^+}} K_{\mathrm{e}} + r) K^{\mathrm{w}}_{\mathrm{CA^+}} c^{\mathrm{w}}_{\mathrm{C}})] c^{\mathrm{w}}_{\mathrm{A^+}}, \tag{5.16}$$

$$c^{\mathrm{o},0}_{\mathrm{C}} = [K_{\mathrm{C}} + r + (K_{\mathrm{CA^+}} K_{\mathrm{e}} + r) K^{\mathrm{w}}_{\mathrm{CA^+}} c^{\mathrm{w}}_{\mathrm{A^+}}] c^{\mathrm{w}}_{\mathrm{C}}, \tag{5.17}$$

where $r = V^{\mathrm{w}}/V^{\mathrm{o}}$ is the volume ratio of the aqueous and membrane phase. Note that these equations only differ from eqns (5.9) and (5.10) in the factor K_{e}, which is given by eqn (5.15). This equation system can be solved for the variables $c^{\mathrm{w}}_{\mathrm{A}}$ and $c^{\mathrm{w}}_{\mathrm{C}}$ in terms of r,[5] and some typical results are shown in Figs. 5.4 and 5.5 for the parameter values: $K_{\mathrm{A^+}} = 10^{-5}$, $K_{\mathrm{B^-}} = 2 \times 10^{-5}$, $K_{\mathrm{C}} = 10^4$, $K_{\mathrm{CA^+}} = 10^{-2}$, $K^{\mathrm{w}}_{\mathrm{CA^+}} = 10^6\,\mathrm{M}$, $c^{\mathrm{w},0}_{\mathrm{AB}} = 0.01\,\mathrm{M}$, and $c^{\mathrm{o},0}_{\mathrm{C}} = 0.1\,\mathrm{M}$.

Figure 5.5 shows that, even in the absence of an external electric circuit, a substantial galvanic potential difference $\Delta^{\mathrm{w}}_{\mathrm{o}}\phi \equiv \phi^{\mathrm{w}} - \phi^{\mathrm{o}}$ is developed across the phase boundary, which changes the partition equilibrium accordingly. This has a significant impact in the partitioning of ionic drugs into biomembranes, for example. The sign of the potential drop, negative in the aqueous phase, is due to the fact that we have considered an acid–amine complex that is much less hydrophilic than the acid ions $\mathrm{A^+}$ and $\mathrm{B^-}$, see eqn (5.15).

Although the exact form of the curves depends strongly on the given parameter values, Fig. 5.4 shows that the total ligand concentration in the membrane, $c^{\mathrm{o}}_{\mathrm{C}} + c^{\mathrm{o}}_{\mathrm{CA^+}}$, is significantly smaller than the initial value $c^{\mathrm{o},0}_{\mathrm{C}}$ for all values of the

[5] In fact, it is simpler to solve for the variables r and $c^{\mathrm{w}}_{\mathrm{A}}$ in terms of $c^{\mathrm{w}}_{\mathrm{C}}$, and the graphical representation of the solution is then made as a parametric plot.

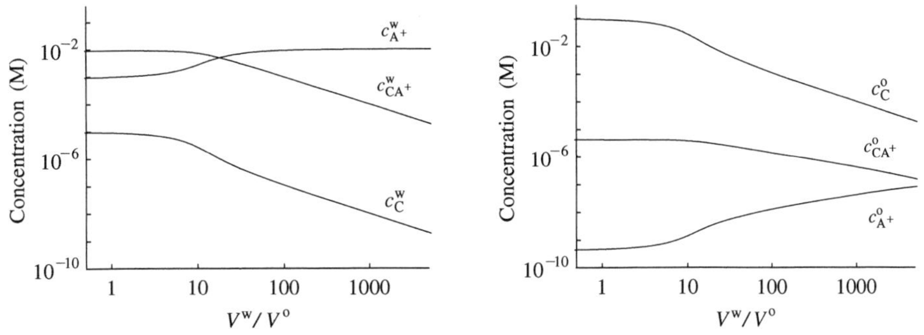

Fig. 5.4.
Equilibrium concentrations (aqueous left, membrane right) as a function of the volume ratio; parameter values given in the text.

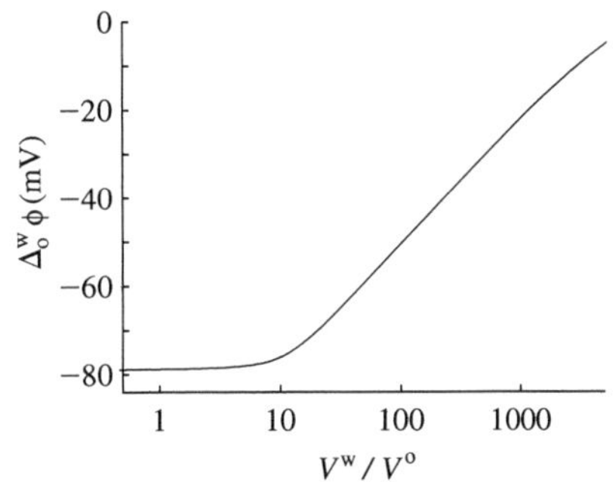

Fig. 5.5.
Distribution potential (at 25°C) as a function of the volume ratio r for the partitioning equilibrium considered in Fig. 5.4.

volume ratio in the range, $r > 10$. This is due in this case to the complexation reaction and that fact that the acid–amine complex is hydrophilic and partitions preferably to the aqueous phase. Similarly, the solute concentration in the aqueous phase $c^{\mathrm{w}}_{\mathrm{A}+}$ drops significantly from the initial value, $c^{\mathrm{w},0}_{\mathrm{AB}}$, due to the complexation reaction except for such high values of r that most of the ligand is in complex form.

A detailed analysis of the multi-ionic equilibria at the liquid/liquid interface, considering the effect of the volume ratio and pH was first given by Hung [4, 5] and later by Kakiuchi [6], emphasizing the effect on the ion-selective electrodes, in particular.

5.2 Ion transfer across a liquid membrane

To illustrate the description of steady-state ion transfer across a liquid membrane, we consider in this section the electrically driven transport of a trace ion between two aqueous solutions, α and β, of identical composition. Such an ion (identified by subscript 1) could typically be tetraethylammonium $(C_2H_5)_4N^+$, or TEA^+, and its transfer from compartment α to β across the membrane is driven by applying a potential difference $\Delta^{\beta}_{\alpha}\phi \equiv \phi^{\beta} - \phi^{\alpha} < 0$. Moreover, this is the only electroactive species in the system, and its flux density is proportional to the electric current density, $j_1 = I/z_1F$. Our aim is to find the equation describing the current–voltage curve of this membrane system and to analyse the influence of the different transport parameters, such as the concentrations of the transferring ion in the aqueous solutions, c_b^w, and in the membrane phase, c_b^o, and the thickness of the membrane, h, and of the diffusion boundary layers in the aqueous phases, δ. The most interesting characteristic of this system is that the transferring ion encounters some resistance to its transfer from both the membrane and the aqueous diffusion boundary layers. The current density can then be limited by either the membrane or the diffusion boundary layers.

Supporting electrolyte is added in the aqueous phases, such that transfer of TEA^+ is under diffusion control. Under steady-state conditions, the flux density of this ion is constant throughout the system and in the aqueous diffusion boundary layers it is

$$j_1 = -D_1^\mathrm{w}\frac{\mathrm{d}c_1}{\mathrm{d}x} = \frac{I}{z_1 F} \qquad (-\delta < x < 0,\ h < x < h+\delta). \tag{5.18}$$

The concentration profiles are then linear and their values at the (external) membrane boundaries are

$$c_1^\mathrm{w}(0) = c_\mathrm{b}^\mathrm{w}(1 - I/I_\mathrm{L}^\mathrm{w}), \tag{5.19}$$

$$c_1^\mathrm{w}(h) = c_\mathrm{b}^\mathrm{w}(1 + I/I_\mathrm{L}^\mathrm{w}), \tag{5.20}$$

where

$$I_\mathrm{L}^\mathrm{w} \equiv \frac{z_1 F D_1^\mathrm{w} c_\mathrm{b}^\mathrm{w}}{\delta} \tag{5.21}$$

is the limiting current density in the aqueous phases[6] and the boundary conditions $c_1^\mathrm{w}(-\delta) = c_1^\mathrm{w}(h+\delta) = c_\mathrm{b}^\mathrm{w}$ have been used.

The electrolyte solution inside the membrane is binary (i.e. there is no supporting electrolyte and migration is not negligible), but the other ion, e.g. tetraphenylborate TPB^-, is not capable of crossing the interface. Since only species 1 can cross the membrane boundaries, the transport equations in the membrane are

$$j_1 = -D_1^\mathrm{o}\left(\frac{\mathrm{d}c_1}{\mathrm{d}x} + z_1 c_1 f\frac{\mathrm{d}\phi}{\mathrm{d}x}\right) = \frac{I}{z_1 F} \qquad (0 < x < h), \tag{5.22}$$

$$j_2 = -D_2^\mathrm{o}\left(\frac{\mathrm{d}c_2}{\mathrm{d}x} + z_2 c_2 f\frac{\mathrm{d}\phi}{\mathrm{d}x}\right) = 0 \qquad (0 < x < h). \tag{5.23}$$

The local electroneutrality condition $z_1 c_1 + z_2 c_2 = 0$ then implies that the ionic concentration distributions are also linear in the organic phase. The average

[6] For the sake of convenience, we have chosen $I = I_\mathrm{L}^\mathrm{w}$ when $c_1^\mathrm{w}(0) = 0$. If we had chosen $I = I_\mathrm{L}^\mathrm{w}$ when $c_1^\mathrm{w}(h) = 0$, a negative sign would appear and eqn (5.21) would then resemble eqn (3.22).

concentration in the membrane phase is bound to be its 'bulk' value

$$c_{\mathrm{b}}^{\mathrm{o}} \equiv \frac{1}{h}\int_0^h c_1\,\mathrm{d}x, \tag{5.24}$$

and, therefore, at the membrane centre the transferring ion concentration is $c_1^{\mathrm{o}}(h/2) = c_{\mathrm{b}}^{\mathrm{o}}$. The concentration drop inside the membrane then depends on the electric current density, and its maximum is $\Delta c_1^{\mathrm{o}} = c_1^{\mathrm{o}}(h) - c_1^{\mathrm{o}}(0) = -2c_{\mathrm{b}}^{\mathrm{o}}$ that corresponds to the limiting current in the organic phase, $I_{\mathrm{L}}^{\mathrm{o}}$. Thus, at the (internal) membrane boundaries, the concentrations of the transferring ion are

$$c_1^{\mathrm{o}}(0) = c_{\mathrm{b}}^{\mathrm{o}}(1 + I/I_{\mathrm{L}}^{\mathrm{o}}), \tag{5.25}$$

$$c_1^{\mathrm{o}}(h) = c_{\mathrm{b}}^{\mathrm{o}}(1 - I/I_{\mathrm{L}}^{\mathrm{o}}), \tag{5.26}$$

where

$$I_{\mathrm{L}}^{\mathrm{o}} \equiv \frac{2z_1(1 - z_1/z_2)FD_1^{\mathrm{o}}c_{\mathrm{b}}^{\mathrm{o}}}{h}. \tag{5.27}$$

The interfacial electric potential drops are given by eqn (5.2) (with the assumption $\gamma_1^{\mathrm{o}} \approx \gamma_1^{\mathrm{w}}$) as

$$\Delta_{\mathrm{o}}^{\mathrm{w}}\phi(0) \approx \Delta_{\mathrm{o}}^{\mathrm{w}}\phi_1^{\circ} + \frac{1}{z_1 f}\ln\frac{c_1^{\mathrm{o}}(0)}{c_1^{\mathrm{w}}(0)} = \Delta_{\mathrm{o}}^{\mathrm{w}}\phi_1^{\circ} + \frac{1}{z_1 f}\ln\frac{c_{\mathrm{b}}^{\mathrm{o}}(1 + I/I_{\mathrm{L}}^{\mathrm{o}})}{c_{\mathrm{b}}^{\mathrm{w}}(1 - I/I_{\mathrm{L}}^{\mathrm{w}})}, \tag{5.28}$$

$$\Delta_{\mathrm{o}}^{\mathrm{w}}\phi(h) \approx \Delta_{\mathrm{o}}^{\mathrm{w}}\phi_1^{\circ} + \frac{1}{z_1 f}\ln\frac{c_1^{\mathrm{o}}(h)}{c_1^{\mathrm{w}}(h)} = \Delta_{\mathrm{o}}^{\mathrm{w}}\phi_1^{\circ} + \frac{1}{z_1 f}\ln\frac{c_{\mathrm{b}}^{\mathrm{o}}(1 - I/I_{\mathrm{L}}^{\mathrm{o}})}{c_{\mathrm{b}}^{\mathrm{w}}(1 + I/I_{\mathrm{L}}^{\mathrm{w}})}, \tag{5.29}$$

and the potential drop in the membrane can be obtained from eqns (5.23)–(5.26) as

$$\begin{aligned}\Delta\phi^{\mathrm{o}} \equiv \phi^{\mathrm{o}}(h) - \phi^{\mathrm{o}}(0) &= -\frac{1}{z_2 f}\ln\frac{c_2^{\mathrm{o}}(h)}{c_2^{\mathrm{o}}(0)} = -\frac{1}{z_2 f}\ln\frac{c_1^{\mathrm{o}}(h)}{c_1^{\mathrm{o}}(0)} \\ &= -\frac{1}{z_2 f}\ln\frac{1 - I/I_{\mathrm{L}}^{\mathrm{o}}}{1 + I/I_{\mathrm{L}}^{\mathrm{o}}}.\end{aligned} \tag{5.30}$$

The cell potential $\Delta\phi_{\mathrm{cell}} \equiv \phi^{\beta} - \phi^{\alpha} = \phi^{\mathrm{w}}(h) - \phi^{\mathrm{w}}(0)$ is then

$$\begin{aligned}\Delta\phi_{\mathrm{cell}} &= \Delta_{\mathrm{o}}^{\mathrm{w}}\phi(h) + \Delta\phi^{\mathrm{o}} - \Delta_{\mathrm{o}}^{\mathrm{w}}\phi(0) \\ &= \frac{1}{z_1 f}\ln\frac{1 - I/I_{\mathrm{L}}^{\mathrm{w}}}{1 + I/I_{\mathrm{L}}^{\mathrm{w}}} + \frac{z_2 - z_1}{z_1 z_2 f}\ln\frac{1 - I/I_{\mathrm{L}}^{\mathrm{o}}}{1 + I/I_{\mathrm{L}}^{\mathrm{o}}} \\ &= -\frac{2}{z_1 f}\,\mathrm{arctanh}\frac{I}{I_{\mathrm{L}}^{\mathrm{w}}} - \frac{2(z_2 - z_1)}{z_1 z_2 f}\,\mathrm{arctanh}\frac{I}{I_{\mathrm{L}}^{\mathrm{o}}},\end{aligned} \tag{5.31}$$

where the standard potentials have cancelled out.

Equation (5.31) describes the current–voltage characteristics of the membrane system under consideration. The current–voltage relation is linear at low deviations from equilibrium (i.e. low electric current or low electric potential difference). The behaviour of the system is then ohmic, $-\Delta\phi_{cell} = IR$, where the total electrical resistance is the sum of the contributions from the three layers, $R = 2R^{w} + R^{o}$ with $R^{w} = \delta/\kappa_{eff}^{w}$, $R^{o} = h/\kappa_{eff}^{o}$, and $\kappa_{eff}^{\varphi} = z_1^2 F^2 D_1^{\varphi} c_b^{\varphi}/RT$ being the effective electrical conductivity of phase $\varphi(\varphi = \mathrm{o, w})$. At higher applied potentials, the current–voltage curve is non-linear and shows a limiting current density, which is the lower of I_L^{o} and I_L^{w}. It can also be noticed that the current–voltage characteristics are symmetric. That is, reversing the sign of the cell potential leads to a reversal of the electric current density without affecting its magnitude.

Figures 5.6 and 5.7 show some typical results corresponding to $z_1 = -z_2 = 1$ and $\Delta_o^w \phi_1^o = 44$ mV, which is roughly the value corresponding to TEA$^+$ at the water/1,2-dichloroethane interface. Figure 5.6 shows the current–voltage curve and the contributions $\Delta_o^w\phi(h)$, $\Delta\phi^{o}$, and $-\Delta_o^w\phi(0)$ to the cell potential for a case in which $c_b^{o}/c_b^{w} = 100$ and $I_L^{o}/I_L^{w} = 5$. The limiting current density is then determined by the aqueous phase (i.e. the depleted diffusion boundary layer) and the membrane does not get polarized practically (and hence, $|\Delta\phi^{o}| \ll |\Delta\phi_{cell}|$).

Figure 5.7 shows the current–voltage curve and the contributions $\Delta_o^w\phi(h)$, $\Delta\phi^{o}$, and $-\Delta_o^w\phi(0)$ to the cell potential for a case in which $c_b^{o}/c_b^{w} = 10$ and $I_L^{o}/I_L^{w} = 0.5$. The limiting current density is then determined by the membrane, but the interfacial potential drops are still relevant. In fact, it can be shown that $\Delta\phi^{o} \approx \Delta_o^w\phi(h) - \Delta_o^w\phi(0) \approx \Delta\phi_{cell}/2$. Note also that, in spite of the close similarity of the current–voltage curves in Figs. 5.6 and 5.7, these two curves are not exactly equal to each other. First, the limiting current is different. And second, the cell potential axis is also different. Note, for instance, that their initial slopes are $(5/14)fI_L^{w}$ and $(1/5)fI_L^{o}$, respectively, as can be

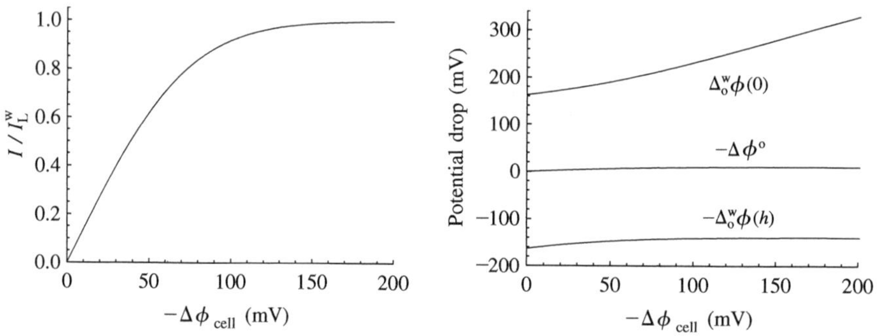

Fig. 5.6.
(Left) Current-voltage curve (at 25°C) for cation transfer across a liquid membrane when $c_b^{o}/c_b^{w} = 100$ and $I_L^{o}/I_L^{w} = 5$, a case in which the current is limited by the transport in the depleted diffusion boundary layer. (Right) The cell potential is the sum of three contributions: the potential drop in the organic phase $\Delta\phi^{o}$, and the interfacial potential drops at the boundaries $x = 0$ and h, $\Delta_o^w\phi(0)$ and $\Delta_o^w\phi(h)$. When the limiting current density is approached, the interfacial potential drop $\Delta_o^w\phi(0)$ diverges.

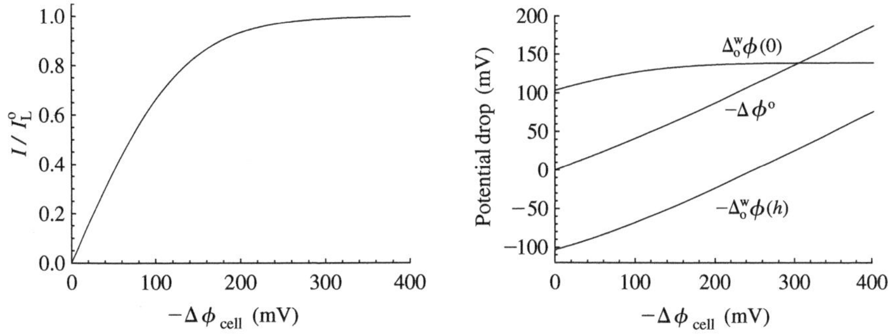

Fig. 5.7.
(Left) Current-voltage curve (at 25°C) for a cation transfer across a liquid membrane when $c_b^o/c_b^w = 10$ and $I_L^o/I_L^w = 0.5$, a case in which the current is limited by the transport in the membrane. (Right) The cell potential is the sum of three contributions: the potential drop in the organic phase $\Delta\phi^o$, and the interfacial potential drops at the boundaries $x = 0$ and h, $\Delta_o^w\phi(0)$ and $\Delta_o^w\phi(h)$. When the limiting current density is approached, both the potential drop inside the membrane $\Delta\phi^o$ and the interfacial potential drop $\Delta_o^w\phi(h)$ diverge.

deduced from

$$\left(\frac{I}{-\Delta\phi_{\text{cell}}}\right)_{I=0} = \frac{z_1 f}{2}\frac{I_L^w I_L^o}{I_L^o + (1 - z_1/z_2)I_L^w}. \tag{5.32}$$

Experimentally, ion-transfer processes in liquid membranes can be conveniently studied, e.g., in a rotating diffusion cell [7]. This cell consists of a rotating cylinder that contains one cell compartment and is limited by the membrane from below. The other solution compartment surrounds the rotating cylinder. By adjusting the rotation speed, regular convection profiles are created on both sides of the membrane and the thickness of the aqueous diffusion boundary layers varies in a controlled way with the rotation speed. The difference with the rotating-disc electrode is that only slow rotation speeds (ca. 20 rpm) are possible in this system due to the mechanical instability of the liquid membrane.

5.3 Carrier-mediated transport

5.3.1 Solute permeability in a supported liquid membrane

In this section we describe the steady-state transport of a neutral solute A across a supported liquid membrane from a concentrated to a dilute aqueous solution. The solute transport is driven by the molar concentration difference $\Delta_\beta^\alpha c_A \equiv c_A^\alpha - c_A^\beta > 0$, where compartment α contains the source or feed solution and compartment β contains the receiving, strip or sweep solution, and hence the molar flux density of this solute j_A is positive in the direction from α to β. In this transport process, the solute has to overcome different (transport) resistances: diffusional resistance to transport in the aqueous phase α, interfacial resistance to partition to the membrane phase, diffusional resistance in the membrane phase, interfacial resistance to partition to the receiving aqueous phase β, and diffusional resistance to transport in the latter. In the case of supported liquid membranes, the interfacial resistances are negligible compared to the other ones, and hence the partitioning between the aqueous and organic phase can be assumed to take place under equilibrium conditions. The solute permeability P_A in the membrane system is defined (implicitly) by the equation

$$j_A = P_A(c_A^\alpha - c_A^\beta). \tag{5.33}$$

It is a positive-definite quantity with dimensions of velocity. Our aim in the following sections is to identify the factors that determine the solute permeability.

5.3.2 Free solute transport: the solubility–diffusion mechanism

The diffusional transport resistance in the aqueous phase can be reduced by stirring, so that the rate-limiting transport resistance comes from diffusion in the liquid membrane. In this case the solute concentration difference $(c_A^\alpha - c_A^\beta)$ in eqn (5.33) is approximately equal to the concentration drop across the outer (i.e. aqueous phase) membrane boundaries, $c_A^w(0) - c_A^w(h)$, where h is the membrane thickness (Fig. 5.8). The transport can then be explained in terms of the so-called solubility–diffusion mechanism. The solute that reaches the supported liquid membrane from the concentrated aqueous solution must become soluble in the organic membrane phase before it can diffuse across it. We assume that this partitioning process is thermodynamically reversible. The membrane offers some resistance to the diffusion process, which is the only resistance under consideration in this section. At the other interface, the solute must transfer the aqueous phase, and this process releases the same amount of Gibbs free energy of transfer that was consumed when crossing the first interface.

The flux density of a neutral solute in the membrane phase is given by

$$j_A = -D_A^o \frac{dc_A^o}{dx}. \tag{5.34}$$

Under steady-state conditions, the flux density is constant and the integration of eqn (5.34) over the membrane (i.e. from $x = 0$ to h) leads to

$$j_A = (D_A^o/h)[c_A^o(0) - c_A^o(h)]. \tag{5.35}$$

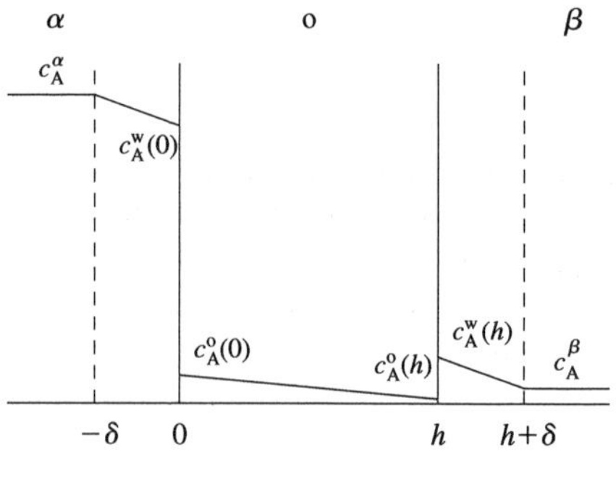

Fig. 5.8.
Schematic concentration profile for free solute transport across a supported liquid membrane flanked by two aqueous diffusion boundary layers.

Since the solute concentration ratio at the membrane/external solution interfaces is given by the solute chemical partition coefficient $K_A = c_A^o(0)/c_A^w(0) = c_A^o(h)/c_A^w(h)$, see eqn (5.2), the solute permeability across the membrane reduces in this case to

$$P_A = P_A^o \equiv \frac{K_A D_A^o}{h}. \tag{5.36}$$

This equation shows that solutes that are poorly soluble in the organic phase find a large overall resistance to their transfer across the membrane. If we want to increase the rate of solute transfer, we can only change the organic solvent (trying to maximize the product $K_A D_A^o$) and to decrease the membrane thickness h. Moreover, the selectivity of this type of solute transport is quite limited because the choice of solvent is our only degree of freedom.

It is worth observing that the chemical partition coefficient of the solute was not considered in the description of transport across the membranes considered in Chapter 4. This was justified on the basis that no significant difference in the standard chemical potential of the solute should be expected between the solutions internal and external to the membrane because they were both aqueous; this holds true unless the water content of the membrane were extremely low. However, when considering the transport across supported liquid membranes, the use of organic solvents in the membrane phase makes it necessary to take into account the difference in the standard chemical potential of the solute between the internal and external solutions at the membrane boundaries.

In the case of ions, the electrostatic solvation energy in an organic phase, as deduced for instance from the Born equation, is much larger than in an aqueous phase due to their different dielectric permittivities. The solubility of ions in the organic phase is then very small (particularly in the case of small ions) and the membrane is practically impermeable to them. There are, however, some ways to increase the solubility of charged solutes in the membrane phase. These are studied in the next sections and can be either chemical methods, for example the complexation with some appropriate ligand in the membrane phase, or electrochemical methods, which require control of the electric potential difference between the aqueous and the organic phases (and not necessarily via the use of electrodes in these phases).

5.3.3 The effect of the diffusion boundary layers on free solute transport

When the stirring of the aqueous solutions is not so efficient, we must consider that the membrane is flanked by two diffusion boundary layers of thickness δ that also offer some resistance to the solute transport and, therefore, affect the solute permeability (Fig. 5.8). Using an electrical analogy, the three layers (liquid membrane and aqueous diffusion boundary layers) behave as three resistors in series (Fig. 5.9) and, therefore, the overall resistance of the membrane system to the solute transport is the sum of the three resistances. Hence, we show below

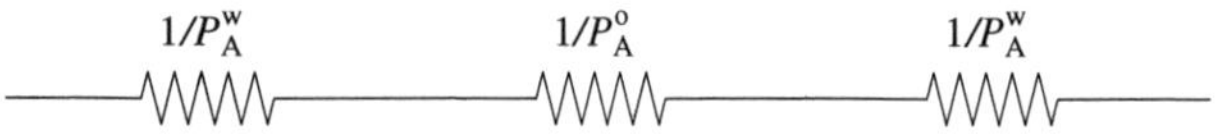

Fig. 5.9.
The total resistance to solute transport of a supported liquid membrane flanked by two aqueous diffusion boundary layers can be evaluated as the sum of three transport resistances in series. The permeability of each one of these layers to the solute is the reciprocal of its transport resistance.

that the solute permeability is

$$\frac{1}{P_A} = \frac{2}{P_A^w} + \frac{1}{P_A^o} = \frac{2\delta}{D_A^w} + \frac{h}{K_A D_A^o}. \tag{5.37}$$

Across the aqueous diffusion boundary layer in the source solution the solute concentration profile is linear and its flux density is

$$j_A = -D_A^w \frac{dc_A^w}{dx} = P_A^w [c_A^\alpha - c_A^w(0)], \tag{5.38}$$

where $P_A^w \equiv D_A^w/\delta$ is the solute permeability in this layer. Similarly, in the boundary layer of the receiving solution, we have $j_A = P_A^w[c_A^w(h) - c_A^\beta]$. Note that under steady-state conditions the solute flux density is independent of position, and this justifies the absence of a phase superscript on j_A. By writing the overall concentration drop as the sum of three concentration drops

$$c_A^\alpha - c_A^\beta = c_A^\alpha - c_A^w(0) + c_A^w(0) - c_A^w(h) + c_A^w(h) - c_A^\beta, \tag{5.39}$$

and using the above equation, this can be transformed to

$$\frac{j_A}{P_A} = \frac{j_A \delta}{D_A^w} + \frac{j_A h}{K_A D_A^o} + \frac{j_A \delta}{D_A^w}, \tag{5.40}$$

which is the same as eqn (5.37). We conclude then that the aqueous diffusion boundary layers decrease the permeability of the membrane system to the solute, and this effect is particularly important when the permeability of the organic phase is high and when the boundary layers have a thickness comparable to that of the membrane because stirring is not very strong.

5.3.4 Carrier-mediated solute transport: the facilitation factor

The study of liquid membranes is important because of industrial applications, but also because it resembles in some aspects the solute transport across biological membranes. The mechanisms of transport across biological membranes are quite diverse, but many of them are characterized by a high specificity with respect to the solutes and by a saturation of the rate of transport at high solute concentrations in the source solution. These characteristics can be explained (although not exclusively) assuming that the solute transport is mediated by a carrier molecule inside the membrane. This is a mobile molecule that binds the solute at one interface and releases it at the opposite interface. The presence of carriers in biological membranes was first proposed by Pfeffer in 1890 and received widespread interest after Pressman found that valinomycin facilitated selective transport of potassium in a factor of several thousands [8]. When its role in transport across biological membranes was understood, the use of carriers was also proposed in commercial membrane-separation processes [9, 10], and in ion-selective electrodes [11].

Consider a liquid membrane that separates two aqueous solutions of a neutral solute A at concentrations c_A^α (source or feed compartment) and c_A^β (receiving or strip compartment). These solutions are assumed to be ideally mixed, so that the aqueous diffusion boundary layers offer no transport resistance, and approximately c_A^α and c_A^β are the concentrations at the external membrane boundaries. The (free) solute can partition inside the membrane and diffuse from the source to the receiving solution. As we have studied in Section 5.3.2, the solute permeability in the membrane associated to this transport mechanism is $P_A^o = K_A D_A^o/h$, where D_A^o is the diffusion coefficient of the free solute inside the membrane and $K_A = c_A^o/c_A^w$ is its partition coefficient.

The membrane also contains a carrier C (Fig. 5.10) that is able to complex with the solute *at* the membrane aqueous solution interfaces as described by the reaction[7]

$$\mathrm{A(w) + C(o) \rightleftarrows CA(o)}, \tag{5.41}$$

with an equilibrium constant[8]

$$K_{CA} = \frac{c_{CA}}{c_C c_A^w}. \tag{5.42}$$

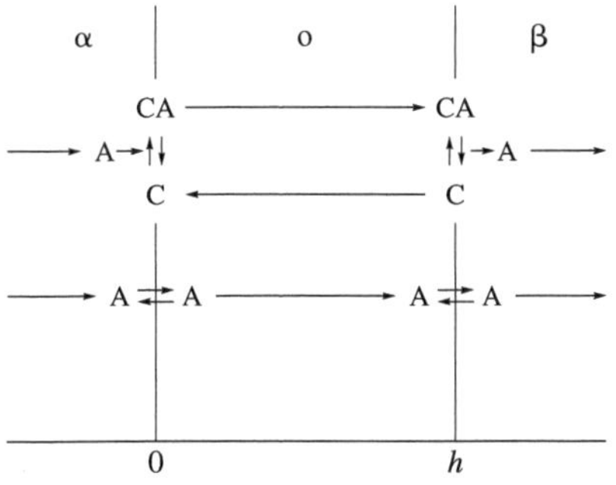

Fig. 5.10.
Carrier-mediated transport of a neutral solute A across a supported liquid membrane. A solubility–diffusion transport mechanism of the free solute runs in parallel with the carrier-mediated transport mechanism.

[7] The complexation reaction could be presented as $\mathrm{A(o) + C(o) \rightleftarrows CA(o)}$, but we recommend eqn (5.41), involving species in different phases, to emphasize that it does not take place inside the membrane (i.e. it is not a homogeneous reaction) but at the interfaces (i.e. it is a heterogeneous reaction).

[8] Not to be confused with a partition equilibrium constant for the complex CA, because it is assumed here that this complex cannot partition to the aqueous phase. This interfacial complexation constant also differs from the bulk complexation constants K_{CA}^w and K_{CA}^o defined in Sections 5.1.3 and 5.1.4.

Note that no phase superscript is used for the carrier and the complex CA because they can only be in the membrane phase. Since the supported liquid membranes are relatively thick and the limitation to solute transport arises from the diffusion inside the membrane and not from the interfacial kinetics [12], it is a very good approximation to consider that the reaction (5.41) is under conditions of thermodynamic equilibrium. Yet, there is a small, positive (i.e. in the forward direction) net rate at the source solution/liquid membrane interface, and a small, negative (i.e. in the backward direction) net rate at the liquid membrane/receiving solution interface. The carrier-mediated transport then proceeds following the steps:

1) at the source solution/liquid membrane interface, the solute A binds to the free carrier molecules C and forms the complex CA,
2) the complex CA diffuses through the membrane,
3) at the liquid membrane/receiving solution interface, the complex CA dissociates, the solute is released and partitions to the receiving aqueous solution, and the complex C remains in the liquid membrane because it has a negligible solubility in the aqueous phase,
4) the carrier C diffuses back to the source solution/liquid membrane interface, and the cycle is repeated.

The solute flux density across the membrane due to the carrier-mediated mechanism is also driven by the concentration difference $\Delta_\beta^\alpha c_\mathrm{A} \equiv c_\mathrm{A}^\alpha - c_\mathrm{A}^\beta > 0$, and we can formally write the contribution of this mechanism to the solute flux density as $j_\mathrm{CA} = P_\mathrm{CA}\Delta_\beta^\alpha c_\mathrm{A}$, which constitutes the definition of the solute permeability P_CA due to the carrier-mediated transport mechanism. Since two transport mechanisms (free and carrier-mediated transport) take place in parallel in the membrane phase, the total permeability P_A of the membrane to the solute is the sum of the permeabilities P_A^o and P_CA,

$$P_\mathrm{A} = P_\mathrm{A}^\mathrm{o} + P_\mathrm{CA}. \tag{5.43}$$

This equation simply states that the total solute flux density across the membrane is the sum of the flux density of free solute and that of the complex, $j_\mathrm{A} = j_\mathrm{A}^\mathrm{o} + j_\mathrm{CA}$. Remember that the total permeability is defined by eqn (5.33).

The facilitation factor F is defined as the relative increase in the flux across the membrane due to the carrier

$$F = \frac{j_\mathrm{A}}{j_\mathrm{A}^\mathrm{o}} = \frac{P_\mathrm{A}}{P_\mathrm{A}^\mathrm{o}} = 1 + \frac{P_\mathrm{CA}}{P_\mathrm{A}^\mathrm{o}}. \tag{5.44}$$

When the chemical partition coefficient K_A is small and the solubility–diffusion mechanism is practically inoperative, the facilitation factor is very large, thus showing the importance of the carrier-mediated transport to increase the rate of solute transport from the source to the receiving solution.

We aim at evaluating the permeability P_CA, and hence, the total permeability P_A as a function of the solute concentrations c_A^α and c_A^β, the equilibrium and partition constants, the carrier concentration and the diffusion coefficients of the free carrier and complex. Since the complexation reaction, eqn (5.41), takes place at the interfaces and not inside the membrane, the concentrations of the free and complexed forms of the carrier, and hence also its total concentration $c_{\mathrm{CT}} \equiv c_{\mathrm{C}} + c_{\mathrm{CA}}$, vary linearly with position. Introducing the average total carrier concentration, $c_{\mathrm{CT}}^{\mathrm{b}} \equiv (1/h)\int_0^h c_{\mathrm{CT}}(x)\mathrm{d}x$, the total carrier concentration at the membrane boundaries satisfies the relation

$$c_{\mathrm{CT}}(0) + c_{\mathrm{CT}}(h) = 2c_{\mathrm{CT}}^{\mathrm{b}}. \tag{5.45}$$

Note that $c_{\mathrm{CT}}^{\mathrm{b}}$ is determined by the amount of carrier dissolved in the organic solvent when the liquid membrane is prepared and is one of the key parameters in carrier-mediated transport.

The flux densities of free and complexed carrier are

$$j_{\mathrm{C}} = -D_{\mathrm{C}}\frac{\mathrm{d}c_{\mathrm{C}}}{\mathrm{d}x} = (D_{\mathrm{C}}/h)[c_{\mathrm{C}}(0) - c_{\mathrm{C}}(h)], \tag{5.46}$$

$$j_{\mathrm{CA}} = -D_{\mathrm{CA}}\frac{\mathrm{d}c_{\mathrm{CA}}}{\mathrm{d}x} = (D_{\mathrm{CA}}/h)[c_{\mathrm{CA}}(0) - c_{\mathrm{CA}}(h)], \tag{5.47}$$

and it is assumed that $j_{\mathrm{C}} + j_{\mathrm{CA}} = 0$, which states that the total (i.e. free and complexed) carrier flux density must be zero because the carrier cannot partition to the aqueous phase. Moreover, since the carrier is often a macrocyclic compound that accommodates the solute in its interior, the size and shape of the carrier is not significantly modified by the complexation with the solute and it can be assumed that $D_{\mathrm{C}} = D_{\mathrm{CA}}$. The above equations then imply that the total carrier has a uniform distribution, $c_{\mathrm{CT}}(0) = c_{\mathrm{CT}}(h) = c_{\mathrm{CT}}^{\mathrm{b}}$. We restrict the discussion to this case hereafter and eliminate the superscript b for the sake of clarity.

The fraction of complexed carrier $\theta \equiv c_{\mathrm{CA}}/c_{\mathrm{CT}}$ depends on the amount of solute available. At the membrane boundaries this is given by

$$\theta^{\alpha} = \frac{K_{\mathrm{CA}}c_{\mathrm{A}}^{\alpha}}{1 + K_{\mathrm{CA}}c_{\mathrm{A}}^{\alpha}}, \tag{5.48}$$

$$\theta^{\beta} = \frac{K_{\mathrm{CA}}c_{\mathrm{A}}^{\beta}}{1 + K_{\mathrm{CA}}c_{\mathrm{A}}^{\beta}}, \tag{5.49}$$

where eqn (5.42) has been used.[9] The solute flux density due to the carrier-mediated mechanism can then be written as

$$j_{\mathrm{CA}} = \frac{D_{\mathrm{C}}c_{\mathrm{CT}}}{h}(\theta^{\alpha} - \theta^{\beta}), \tag{5.50}$$

[9] Although eqns (5.48) and (5.49) are of the form of the Langmuir adsorption isotherm, no adsorption isotherm is assumed.

and a comparison with the expression $j_{\mathrm{CA}} = P_{\mathrm{CA}} \Delta^{\alpha}_{\beta} c_{\mathrm{A}}$ allows us to determine the permeability P_{CA} as

$$P_{\mathrm{CA}} = \frac{D_{\mathrm{C}}}{h} \frac{K_{\mathrm{CA}} c_{\mathrm{CT}}}{(1 + K_{\mathrm{CA}} c^{\alpha}_{\mathrm{A}})(1 + K_{\mathrm{CA}} c^{\beta}_{\mathrm{A}})}. \tag{5.51}$$

It is noteworthy that this permeability depends on the solute concentrations c^{α}_{A} and c^{β}_{A}, which implies that the flux density j_{CA}, and hence j_{A}, is a non-linear function of the concentration drop $\Delta^{\alpha}_{\beta} c_{\mathrm{A}}$. Only when the solute concentration is small, $c^{\beta}_{\mathrm{A}} < c^{\alpha}_{\mathrm{A}} \ll 1/K_{\mathrm{CA}}$, does the permeability P_{CA} reduce to $P_{\mathrm{CA,dil}} \approx K_{\mathrm{CA}} D_{\mathrm{C}} c_{\mathrm{CT}}/h$, and the solute flux density j_{A} become proportional to $\Delta^{\alpha}_{\beta} c_{\mathrm{A}}$.

From an experimental point of view, carrier-mediated transport can be conveniently studied by analysing the initial-time solute flux density. This corresponds to a situation where the solute is present only in the source compartment and the running time of the experiment is so short that we can assume $c^{\beta}_{\mathrm{A}} \approx 0$. In this case, the solute flux density due to the carrier-mediated mechanism reduces to

$$j_{\mathrm{CA}} = P_{\mathrm{CA}} c^{\alpha}_{\mathrm{A}} \approx j_{\mathrm{C,L}} \frac{K_{\mathrm{CA}} c^{\alpha}_{\mathrm{A}}}{1 + K_{\mathrm{CA}} c^{\alpha}_{\mathrm{A}}} \quad (c^{\beta}_{\mathrm{A}} \approx 0), \tag{5.52}$$

and saturates to its maximum value $j_{\mathrm{C,L}} \equiv D_{\mathrm{C}} c_{\mathrm{CT}}/h$ in the limit of high solute concentration, $K_{\mathrm{CA}} c^{\alpha}_{\mathrm{A}} \gg 1$. Since $j_{\mathrm{C,L}}$ is determined by the amount of carrier inside the membrane, and not by the solute concentration c^{α}_{A}, a plot of j_{A} vs. c^{α}_{A} shows a tendency to saturation at high solute concentrations that corresponds to $j_{\mathrm{A}} \approx j^{\mathrm{o}}_{\mathrm{A}} + j_{\mathrm{C,L}} = P^{\mathrm{o}}_{\mathrm{A}} c^{\alpha}_{\mathrm{A}} + j_{\mathrm{C,L}}$ (Fig. 5.11). This is one of the most distinguishing characteristics of carrier-mediated transport.

Due to this tendency to saturation in the flux density (for the case $c^{\beta}_{\mathrm{A}} \approx 0$), the solute permeability in the membrane due to carrier-mediated transport is a decreasing function of the solute concentration in the source solution (Fig. 5.12). At low concentrations, $K_{\mathrm{CA}} c^{\alpha}_{\mathrm{A}} \ll 1$, the permeability P_{CA} takes its maximum value, $P_{\mathrm{CA,dil}} = K_{\mathrm{CA}} D_{\mathrm{C}} c_{\mathrm{CT}}/h$, and at high concentrations it decreases with increasing concentration as $P_{\mathrm{CA,sat}} = D_{\mathrm{C}} c_{\mathrm{CT}}/c^{\alpha}_{\mathrm{A}} h$. Note that the complexation constant K_{CA} can be experimentally determined, e.g., from the ratio of the permeability at low concentrations $P_{\mathrm{CA,dil}}$ and the saturation flux density $j_{\mathrm{C,L}}$.

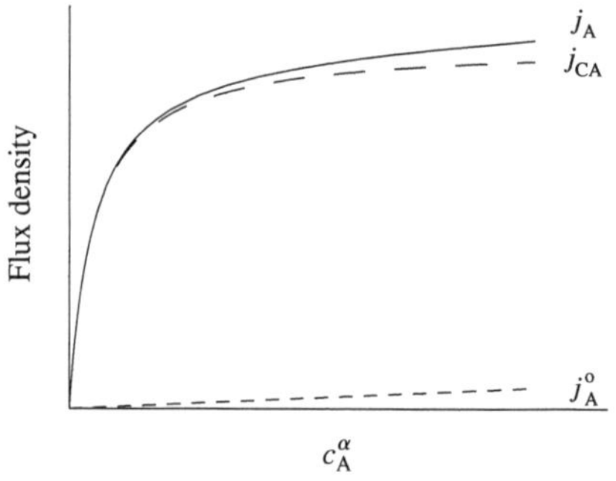

Fig. 5.11.
Schematic representation of the initial ($c^{\beta}_{\mathrm{A}} \approx 0$) solute flux density vs. the solute concentration in the source solution for carrier-mediated transport. The solute flux density is the sum of the contributions of the carrier-mediated mechanism and the solubility–diffusion mechanism. The first one saturates at high solute concentrations, while the latter shows a linear behaviour.

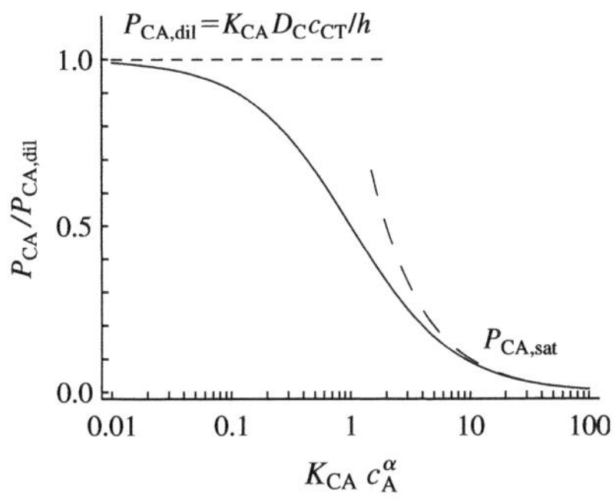

Fig. 5.12.
The solute permeability in the membrane due to carrier-mediated transport is a decreasing function of the solute concentration in the source solution (for the case $c_A^\beta \approx 0$). At low concentrations, $K_{CA}c_A^\alpha \ll 1$, the permeability P_{CA} takes its maximum value, $P_{CA,dil} = K_{CA}j_{C,L}$. At high concentrations, $K_A c_A^\alpha \gg 1$, it decreases with increasing c_A^α as $P_{CA,sat} = j_{C,L}/c_A^\alpha$.

It was mentioned at the beginning of this section that carrier-mediated transport is also characterized by a high solute selectivity. Specific carriers are available (or can be designed with the current supramolecular chemistry techniques) for a large number of solutes. This means that the equilibrium constant K_{CA} is large for the complexation of the carrier with a given solute and very small for other solutes, so that only the former is effectively transported across the liquid membrane. In other words, while in the solubility–diffusion mechanism we could only change the organic solvent to affect the solute flux across the membrane, the choice of different carriers implies a much more significant change in the solute flux.

This comment on the selectivity leads us to another interesting issue of carrier-mediated transport. In the case $c_A^\beta \approx 0$ considered above we have concluded that the solute flux density is a monotonously increasing function of the solute concentration in the source solution that saturates to the value $j_{C,L} \equiv D_C c_{CT}/h$ in the limit of high solute concentration, $K_{CA}c_A^\alpha \gg 1$. Thus, one is tempted to conclude that the facilitation factor increases with increasing complexation equilibrium constant K_{CA}. However, this is not correct and a very high value of this constant may result in a low solute flux density, and hence, on a facilitation factor of the order of one. The reason for this behaviour is the fact that the solute is not released at the membrane receiving solution interface when K_{CA} is very large. All the carrier molecules are then in the form of solute–carrier complex, and they do not contribute to solute transport. To analyse this effect, we have to consider the case $c_A^\beta \neq 0$. Figure 5.13 shows the solute flux density against $K_{CA}c_A^\alpha$ for some values of the solute concentration ratio c_A^β/c_A^α. It is observed that there is an optimum value of the complexation constant K_{CA} that leads to a maximum facilitation of the solute transport through the membrane when $K_{CA}c_A^\alpha \approx 1$. The saturation behaviour of the solute flux density mentioned above is therefore a peculiarity of the case $c_A^\beta = 0$. Moreover, since the diffusion boundary layer at the receiving side makes the solution concentration at the membrane interface larger than c_A^β, it is to be expected that this layer leads to a significant reduction in the solute flux density in the limit of high solute concentrations (in the source solution).

In conclusion, the coupling between diffusion of the different species and the complexation reactions at the membrane interfaces gives rise to a non-linear

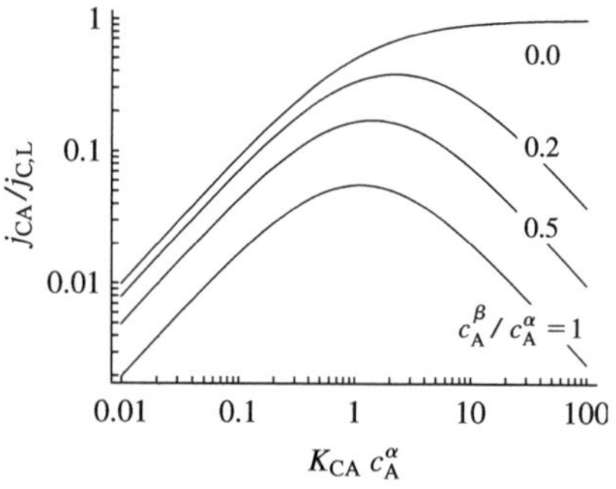

Fig. 5.13.
The solute flux density due to carrier-mediated transport shows a maximum when $K_{CA} c_A^\alpha \approx 1$ for non-zero values of the concentration ratio c_A^β / c_A^α. This implies that the complexation constant has to be optimized in order to maximize the facilitation factor.

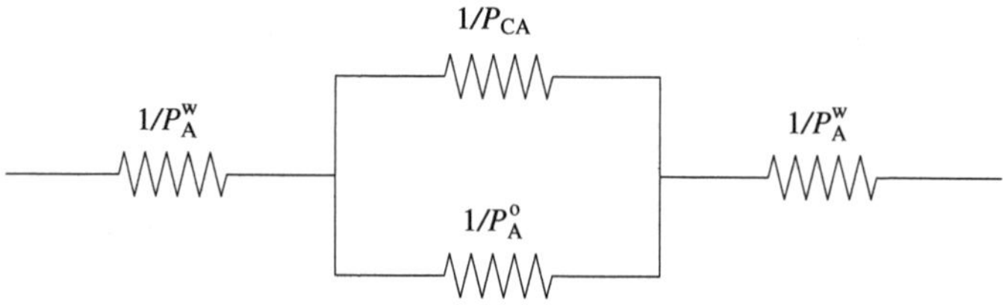

Fig. 5.14.
The total resistance to solute transport of a supported liquid membrane that incorporates a carrier and is flanked by two aqueous diffusion boundary layers can be evaluated as the sum of three transport resistances in series, the middle one representing the liquid membrane and consisting in turn in a parallel association of two transport resistances.

dependence of the solute flux density with the system variables (concentrations and equilibrium constant) and, therefore, to a rather rich system behaviour. The general equations that describe the solute flux are eqns (5.43) and (5.51). Their particular cases have limited ranges of validity that have to be clearly stated to avoid errors in the analysis of experimental data.

5.3.5 The effect of the diffusion boundary layers on carrier-mediated transport

As explained in Section 5.3.3, the liquid membrane is flanked by two diffusion boundary layers of thickness δ when the aqueous solutions are poorly stirred and offer some resistance to the solute transport. Using an electrical analogy, the three layers (liquid membrane and aqueous diffusion boundary layers) behave as three resistors in series, but the resistor representing the liquid membrane consists in turn in a parallel association of resistors representing the different transport mechanisms inside the membrane, such as free and carrier-mediated solute transport. Since the transport resistance is the reciprocal of the solute permeability in the corresponding phase, it can be shown that the solute permeability is (Fig. 5.14)

$$\frac{1}{P_A} = \frac{2}{P_A^w} + \frac{1}{P_A^o + P_{CA}}. \tag{5.53}$$

The derivation of this formula runs parallel to that of eqn (5.37) and the conclusion is similar, that is, the aqueous diffusion boundary layers decrease the permeability of the membrane system to the solute, and this effect is particularly important when the permeability of the organic phase is high. Thus, since the presence of the carrier significantly increases the permeability of the solute in the liquid membrane, we conclude that the effect of the aqueous diffusion boundary layers is more important in carrier-mediated transport than when solute is only transported by the solubility–diffusion mechanism.

In this section we describe the solute flux density across the supported liquid membrane taking into account the diffusion boundary layers, thus extending the analyses made in the previous two sections. For the sake of simplicity, we introduce the following assumptions:

1) $D_C \approx D_{CA}$, and hence c_{CT} is independent on position,
2) $c_A^\beta \approx 0$, i.e. we discuss only the initial measurements, and
3) $P_A^o \ll P_{CA}$, so that the free solute cannot get inside the membrane.

Under these conditions, the total permeability of solute A in the membrane system is $P_A = [2/P_A^w + 1/P_{CA}]^{-1}$, where $P_A^w \equiv D_A^w/\delta$ is the permeability in one boundary layer and the permeability in the membrane

$$P_{CA} = \frac{D_C}{h} \frac{K_{CA} c_{CT}}{[1 + K_{CA} c_A^w(0)][1 + K_{CA} c_A^w(h)]} \tag{5.54}$$

is a non-linear function of the solute concentration at the membrane boundaries (see below). We aim at finding the relation between the solute flux density j_A and its source concentration c_A^α.

When the liquid membrane offers a negligible transport resistance, it is said that the system operates under conditions of film control. The maximum solute flux density would then be $j_{A,L}^w \equiv D_A^w c_A^\alpha / 2\delta$. This is expected to occur at low solute concentrations, $K_{CA} c_A^\alpha \ll 1$, since P_{CA} then takes its maximum value $P_{CA,dil} = K_{CA} D_C c_{CT}/h$ (Fig. 5.12), but it also requires that $P_A^w \ll P_{CA,dil}$ (otherwise both the membrane and the boundary layers control the solute flux).

When the diffusion boundary layers were disregarded in Section 5.3.4 it was concluded that the maximum solute flux density was the saturation flux density $j_{C,L} \equiv D_C c_{CT}/h$, which corresponds to the limit of high solute concentration in the source solution, $K_{CA} c_A^\alpha \gg 1$. The question to be solved in this section is whether $j_{C,L}$ is also the maximum solute flux density when the diffusion boundary layers in the aqueous solution have a high solute permeability. In other words, we have to determine whether it is possible that $P_A^w \gg P_{CA}$ and hence $P_A \approx P_{CA}$ under any conditions. If this were the case, we would say that the system operates under conditions of membrane control. On the contrary, if P_A^w and P_{CA} are of the same order of magnitude, the system operates under conditions of mixed control, i.e. both the diffusion boundary layers and the liquid membrane offer a significant transport resistance.

The flux density of the solute across the diffusion boundary layer in the source compartment is $j_A = P_A^w[c_A^\alpha - c_A^w(0)]$. Then, the solute concentration at the membrane source solution interface is given by

$$c_A^w(0) = c_A^\alpha - \frac{j_A}{P_A^w}. \tag{5.55}$$

Similarly, at the membrane receiving solution interface the solute concentration is

$$c_A^w(h) = \frac{j_A}{P_{A,L}^w}. \tag{5.56}$$

Since we are assuming that $P_A^o \ll P_{CA}$, the transport in the membrane phase is only carrier-mediated, and hence $j_A \approx j_{CA} = P_{CA}[c_A^w(0) - c_A^w(h)]$. Denoting the concentration of the complexed carrier as $c_{CA} = c_{CT}\theta$, the solute flux density in the liquid membrane is

$$j_A \approx j_{CA} = \frac{D_C c_{CT}}{h}[\theta(0) - \theta(h)] = j_{C,L}\frac{K_{CA}[c_A^w(0) - c_A^w(h)]}{[1 + K_{CA}c_A^w(0)][1 + K_{CA}c_A^w(h)]}. \tag{5.57}$$

Introducing the auxiliary parameter

$$p \equiv \frac{P_{CA,dil}}{P_A^w} = \frac{K_{CA}D_C c_{CT}\delta}{D_A^w h} = K_{CA}\frac{j_{C,L}}{P_A^w}, \tag{5.58}$$

eqn (5.57) can be transformed to

$$\frac{j_A}{j_{C,L}}\left[1 + K_{CA}c_A^\alpha - p\frac{j_A}{j_{C,L}}\right]\left[1 + p\frac{j_A}{j_{C,L}}\right] = K_{CA}c_A^\alpha - 2p\frac{j_A}{j_{C,L}}. \tag{5.59}$$

This is the relation between j_A and c_A^α that we were looking for. This can be interpreted as a third-order algebraic equation in j_A, but it is much more convenient to consider it as a first-order equation in c_A^α. Thus, eqn (5.59) can be presented as

$$K_{CA}c_A^\alpha = \frac{j_A}{j_{C,L}}\frac{p^2(j_A/j_{C,L})^2 - 2p - 1}{p(j_A/j_{C,L})^2 + (j_A/j_{C,L}) - 1}. \tag{5.60}$$

At low concentrations, $K_{CA}c_A^\alpha \ll 1$, (which also implies $j_A/j_{C,L} \ll 1$) this reduces to

$$j_{A,dil} = j_{C,L}\frac{K_{CA}c_A^\alpha}{1 + 2p}, \tag{5.61}$$

which is equivalent to $j_A \approx j^w_{A,L}$ if $p \gg 1$. At high concentrations, $K_{CA}c^\alpha_A \gg 1$, the solute flux density saturates to

$$j_{A,sat} = j_{C,L}\frac{\sqrt{1+4p}-1}{2p}, \tag{5.62}$$

which is equivalent to $j_A \approx j_{C,L}$ only if $p \ll 1$.

Figure 5.15 shows the graphical representation of eqn (5.60) for different values of the parameter p. It is observed that the diffusion boundary layers make the saturation current smaller than $j_{C,L}$, and this effect becomes more significant as P^w_A decreases (or p increases). In the limit of low concentration c^α_A and low permeability P^w_A, the solute flux density is $j_A \approx j^w_{A,L}$, as indicated by the dashed line in Fig. 5.15. This is the only case where the diffusion boundary layers completely control the solute transport. Membrane control can be observed when $p = 0$. Otherwise, solute transport occurs under conditions of mixed control. In fact, the most significant conclusion is that even at high concentrations c^α_A, when the permeability P_{CA} decrease to low values, we cannot neglect the influence of the diffusion boundary layers on the solute transport.

5.3.6 Extraction of an acid

Facilitated diffusion was originally employed in gas separations but the range of applications has widened considerably in recent decades. We consider here as a practical example the case of acid extraction. The source (or feed) solution α contains the acid A (e.g., HCl) to be extracted, and contains an excess of supporting electrolyte (e.g., NaCl). The receiving (or strip) solution β contains an excess of base B (e.g., NaOH). The acid extracted reacts with the base and

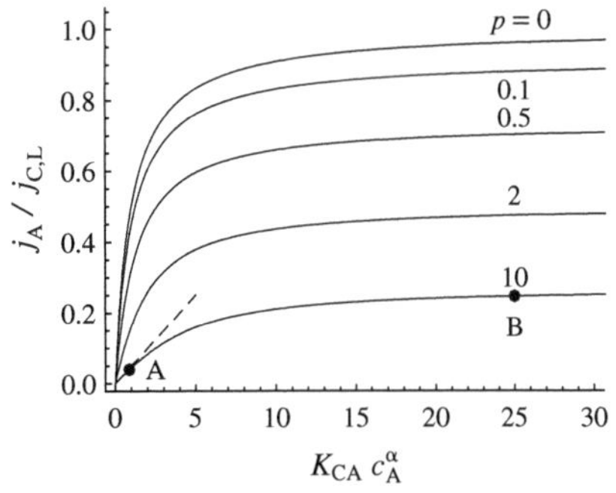

Fig. 5.15.
Solute flux density against concentration in source compartment for carrier-mediated transport across a liquid membrane flanked by two diffusion boundary layers. Parameter p is the ratio between the solute permeability in the membrane when the source solution is diluted and the permeability in the boundary layers. Thus, $p = 0$ corresponds to the absence of boundary layers. The saturation current decreases with increasing p, and therefore the effect of the boundary layers cannot be neglected even at high solute concentrations. At low concentrations and high p, the boundary layers control the solute flux and $j_A \approx j^w_{A,L}$ (dashed line). The situations marked with the labels A and B are analysed in Fig. 5.16.

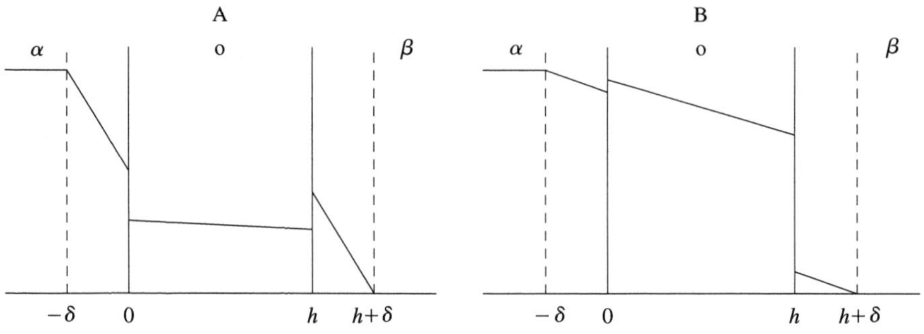

Fig. 5.16.
Concentration profiles in the membrane system corresponding to a permeability ratio $p = 10$ and the two solute concentrations marked with the labels A and B in Fig. 5.15. In the boundary layers, the magnitude represented in the ordinate axis is c_A/c_A^α, and in the membrane it is $\theta = c_{CA}/c_{CT}$; note that $j_{A,sat}/j_{C,L} = \theta(0) - \theta(h)$. At low solute concentrations and high p (case A) most of the concentration drop takes place in the diffusion boundary layers, a situation known as film control. At high concentrations (case B) the concentration drops in these layers are smaller but significant. Hence, their influence cannot be neglected. This occurs because the solute concentration at the membrane receiving solution interface determines the concentration drop (of the complexed solute) inside the membrane and it is finite, although $c_A^\beta = 0$.

produces salt BA (e.g., NaCl) in the receiving compartment. The membrane is flanked by two diffusion boundary layers of thickness δ. The heterogeneous reactions at the membrane interfaces are

$$A(\alpha) + C(o) \rightleftarrows CA(o) \quad \text{feed,} \tag{5.63}$$

$$CA(o) + B(\beta) \rightleftarrows C(o) + BA(\beta) \quad \text{strip,} \tag{5.64}$$

where the carrier C is typically an amine and CA is the amine–acid complex. That is, reaction (5.63) could be

$$HCl(w) + R - NH_2(o) \rightleftarrows R - NH_3^+Cl^-(o). \tag{5.65}$$

The five stages of this transport process are:

1) diffusion of A across the diffusion boundary layer in the source solution ($-\delta < x < 0$),
2) reaction of A with C at the source solution/membrane interface,
3) diffusion of complex CA across the membrane, accompanied by simultaneous diffusion of free C in the opposite direction,
4) dissociation of complex CA via the reaction with B at the membrane/receiving solution interface, and
5) diffusion of BA and B across the receiving diffusion layer ($h < x < h + \delta$).

Figure 5.17 illustrates the interfacial reactions involved in this acid-extraction-process. At the interface between the membrane and the source solution, the amine carrier reacts with the acid. At the opposite interface, the amine–acid complex reacts with the base, producing salt and the free carrier. This figure also shows schematically the concentration profiles in the system.

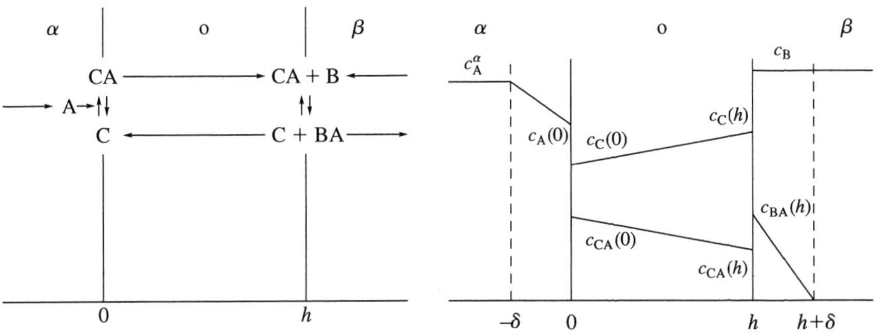

Fig. 5.17.
(Left) Schematic reaction mechanism for the extraction of an acid A using carrier-mediated transport across a supported liquid membrane. Species B at the receiving solution β is a base. (Right) Schematic concentration profiles in the membrane and the aqueous diffusion boundary layers for the situation considered in this section.

The interfacial heterogeneous reactions are not rate limiting and can be assumed to be described by the equilibria[10]

$$K_{CA} = \frac{c_{CA}}{c_C c_A}, \tag{5.66}$$

$$K_{BA} = \frac{c_C c_{BA}}{c_{CA} c_B}, \tag{5.67}$$

where no phase superscripts are needed because it is assumed that species A, B and BA can only be present in the aqueous phases and species C and CA cannot leave the membrane. Since the receiving solution contains an excess of base, it can be assumed that c_B in eqn (5.67) is a constant (equal to c_B^β). Although the acid does not exist as a solute in the receiving solution, it is convenient to defined its 'concentration' in the receiving diffusion boundary layer as

$$c_A \equiv \frac{c_{BA}}{K_{CA} K_{BA} c_B}, \quad (h < x < h + \delta). \tag{5.68}$$

Under steady-state conditions the flux density of the acid A across the diffusion boundary layer in the source solution is the same as the flux density of amine–acid complex CA in the liquid membrane, and also the same as the flux density of salt BA in the receiving diffusion boundary layer, $j_A = j_{CA} = j_{BA}$. These flux densities can be written as

$$j_A = P_A^w [c_A^\alpha - c_A(0)], \tag{5.69}$$

$$j_{CA} = (D_C/h)[c_{CA}(0) - c_{CA}(h)] = P_{CA}[c_A(0) - c_A(h)], \tag{5.70}$$

$$j_{BA} = (D_{BA}/\delta)[c_{BA}(h) - c_{BA}^\beta] = P_{BA}[c_A(h) - c_A^\beta], \tag{5.71}$$

[10] The reaction of the acid with the base at the membrane/receiving solution interface also produces water, and the equilibrium constant includes the water concentration.

where $P_A^w \equiv D_A/\delta, P_{BA} \equiv K_{CA}K_{BA}c_B D_{BA}/\delta$, and

$$P_{CA} \equiv \frac{D_C}{h} \frac{K_{CA}c_{CT}}{[1 + K_{CA}c_A(0)][1 + K_{CA}c_A(h)]}. \tag{5.72}$$

The solute permeability across the membrane system, $P_A \equiv j_A/(c_A^\alpha - c_A^\beta)$, is then

$$\frac{1}{P_A} = \frac{1}{P_A^w} + \frac{1}{P_{CA}} + \frac{1}{P_{BA}}, \tag{5.73}$$

as expected for the transport across three layers in series. In these expressions, the effective diffusion coefficients have to be determined. In the source diffusion boundary layer, the acid is completely dissociated. Since there is an excess of salt (acting as supporting electrolyte), this is the trace-ion diffusion case, see eqn (3.16), and the effective diffusion coefficient of the HCl component is then the diffusion coefficient of the H^+ ion, $D_A = D_{H^+}$. Similarly, the (H_2O-NaOH-NaCl) solution in the receiving diffusion boundary layer is essentially ternary (Na^+, OH^-, Cl^-) because the concentration of H^+ ions is negligible, and the situation is again that of trace-ion diffusion (due to the excess of base). So, the effective diffusion coefficient of the salt is $D_{BA} = D_{Cl^-}$. In the membrane, it is assumed that $D_{CA} = D_C$ as usual.

As in the previous sections, we consider the situation of practical importance of short running times (and efficient stirring) where the amount of salt AB in the bulk of compartment β is negligible, $c_{BA}^\beta = 0 = c_A^\beta$. If we consider the solute flux density j_A to be known, the above equations can be used to determine all the concentrations at the membrane boundaries as follows

$$c_A(0) = c_A^\alpha - \frac{j_A}{P_A^w}, \tag{5.74}$$

$$c_A(h) \equiv \frac{c_{BA}(h)}{K_{CA}K_{BA}c_B} = \frac{j_A}{P_{BA}}, \tag{5.75}$$

$$\frac{j_A}{j_{C,L}}[1 + K_{CA}c_A(0)][1 + K_{CA}c_A(h)] = K_{CA}[c_A(0) - c_A(h)], \tag{5.76}$$

where $j_{C,L} \equiv D_C c_{CT}/h$ is the limiting, carrier-mediated solute flux density.

Introducing the auxiliary parameters

$$p \equiv \frac{P_{CA,dil}}{P_A^w} = \frac{K_{CA}D_C c_{CT}\delta}{D_A^w h}, \tag{5.77}$$

$$q \equiv \frac{P_A^w}{P_{BA}} = \frac{D_A^w}{K_{CA}K_{BA}c_B D_{BA}}, \tag{5.78}$$

eqn (5.76) can be transformed to

$$\frac{j_A}{j_{C,L}}\left[1 + K_{CA}c_A^\alpha - p\frac{j_A}{j_{C,L}}\right]\left[1 + pq\frac{j_A}{j_{C,L}}\right] = K_{CA}c_A^\alpha - p(1+q)\frac{j_A}{j_{C,L}}. \tag{5.79}$$

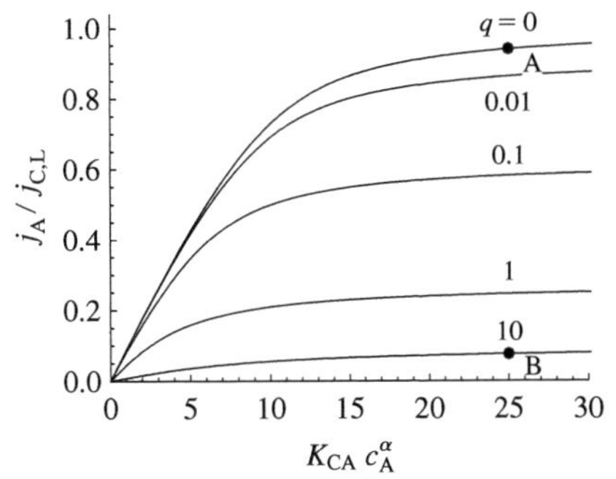

Fig. 5.18.
Solute flux density against concentration in source compartment for the acid extraction as an example of carrier-mediated transport across a liquid membrane flanked by two diffusion boundary layers. Parameter q is the ratio between the solute permeability in the source boundary layer and in the receiving boundary layer, where it is transported in the form of salt; the value $q = 1$ reproduces the situation analysed in Fig. 5.15. The ratio between the solute permeability across the liquid membrane in dilute source solutions and the solute permeability in the source boundary layer has been fixed as $p = 10$. Thus, the curve $q = 0$ corresponds to no polarization (i.e. no concentration drop) of the receiving boundary layer and saturates to $j_A = j_{C,L}$. The situations marked with the labels A and B are analysed in Fig. 5.19.

This relation between j_A and c_A^α can be interpreted as a third-order algebraic equation in j_A or, more conveniently, as a first-order equation in c_A^α that can be solved to obtain the following solute flux density–source concentration relation

$$K_{CA} c_A^\alpha = \frac{j_A}{j_{C,L}} \frac{p^2 q (j_A/j_{C,L})^2 + p(1-q)(j_A/j_{C,L}) - p(1+q) - 1}{pq(j_A/j_{C,L})^2 + (j_A/j_{C,L}) - 1}. \tag{5.80}$$

At high concentrations, $K_{CA} c_A^\alpha \gg 1$, the solute flux density saturates to

$$j_{A,\mathrm{sat}} = j_{C,L} \frac{\sqrt{1+4pq} - 1}{2pq}, \tag{5.81}$$

which is equivalent to $j_A \approx j_{C,L}$ if $p \ll 1$ or $q \ll 1$.

Figure 5.18 shows the graphical representation of eqn (5.80) for $p = 10$ and different values of the parameter q. At high solute concentrations, $K_{CA} c_A^\alpha \gg 1$, the liquid membrane is always limiting the transport because this is essential to carrier-mediated transport. If $pq \ll 1$, the diffusion boundary layers do not limit the transport at high concentrations and the maximum solute flux density is $j_{C,L}$. However, if pq is of the order of or larger than unity, mixed control (membrane and boundary layers) takes place at high concentrations. At low concentrations, $K_{CA} c_A^\alpha \ll 1$, the membrane determines the rate of solute transport if $p \ll 1$ and $pq \ll 1$, and the flux density is then $j_A \approx P_{CA,\mathrm{dil}} c_A^\alpha$. The source boundary layer determines the rate if $p \gg 1$ and $q \ll 1$, so that the flux density is $j_A \approx P_A^w c_A^\alpha$. Finally, the receiving boundary layer determines the rate if $p \gg 1$ and $q \gg 1$, so that the flux density is $j_A \approx P_{BA} c_A^\alpha$.

In conclusion, since the acid transport takes place in different form in the source and receiving boundary layers, the phenomenology is richer than in

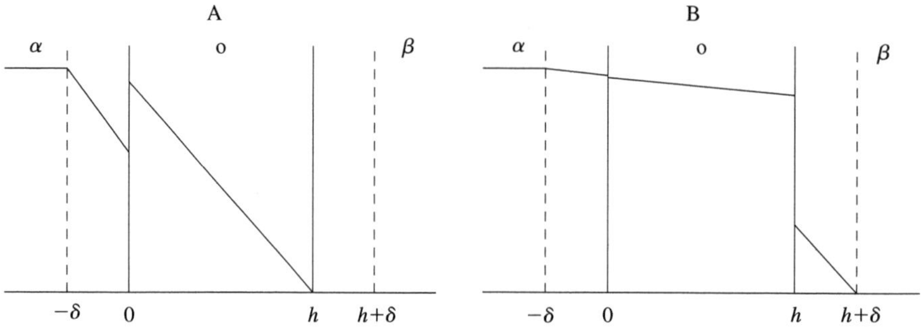

Fig. 5.19.
Concentration profiles in the membrane system corresponding to a solute concentration in the source compartment $K_{\mathrm{CA}}c_{\mathrm{A}}^{\alpha} = 25$, and permeability ratios (A) $p = 10$, $q = 0$ and (B) $p = 10$, $q = 10$, corresponding to the labels A and B in Fig. 5.18. In the boundary layers, the magnitude represented is $c_{\mathrm{A}}/c_{\mathrm{A}}^{\alpha}$, and in the liquid membrane it is $\theta = c_{\mathrm{CA}}/c_{\mathrm{CT}}$; note that $j_{\mathrm{A,sat}}/j_{\mathrm{C,L}} = \theta(0) - \theta(h)$. At low values of q (case A) the concentration drop in the receiving boundary layer vanishes, the fraction of complexed carrier vanishes at the membrane receiving solution interface and, therefore, the solute flux density can reach the saturation value $j_{\mathrm{C,L}}$ at high solute concentration in the source solution. At high values of q (case B) the concentration drop in the source boundary layer is small compared to that in the receiving layer.

Section 5.3.5. Any of the three layers (membrane and the two boundary layers) may determine the rate the acid transport, and this depends on the component concentrations in the different phases. By measuring the flux with varying bulk concentrations of A, B and C, the rate-determining step can be found experimentally.[11]

5.3.7 Additional comments on the modelling of carrier-mediated transport

The mathematical modelling of carrier-mediated transport constitutes a complicated and diverse problem [13]. Both the complexation reactions and diffusion in the aqueous phases as well as in the membrane need to be taken into account. Analysis is further complicated by the fact that solvents used in liquid membranes are usually non-polar and, as a consequence, their relative permittivities (dielectric constants) are very low, causing agglomeration of carriers. Therefore, the transport of electrolytes is usually feasible only as ion pairs. If electroassisted transport is considered, the membrane solvent must have sufficiently high relative permittivity to allow for the dissociation of electrolytes, and hence ensure some conductivity in the membrane.

Some simplifying assumptions often introduced in the theoretical modelling are widely accepted, such as:

1) the transport can be studied using the quasi-steady-state approximation,

[11] A secondary problem when analysing the experimental data is that the numerical values of the equilibrium constants and the diffusion coefficients are not always available in the literature but they can be either estimated or found from the best fit to the experimental data. The diffusion coefficients can be estimated, if no better data is available, with Walden's rule, which states that the ratio $D\eta/T$ is independent of the solvent. In this relation, D is the diffusion coefficient, η the solvent viscosity and T the absolute temperature. The Walden rule can formally be derived from the Stokes law for spherical solutes.

2) the membrane is so porous that the area of the aqueous/membrane interface is equal to the geometrical area of the membrane, and
3) both aqueous phases are well stirred, forming diffusion boundary layers of thickness δ.

Other approximations need to be tested for the experimental set-up used, such as the assumption that the carrier does not dissolve in the aqueous phases. As we have learned from Sections 5.1.3 and 5.1.4 this approximation might fail and the partitioning of the carrier must then be considered. Possibly, the approximation that is more open to discussion refers to the complexation reactions [14]. In this chapter we assume that they are heterogeneous and take place at the interfaces, not inside the membrane. This assumption has been experimentally confirmed in some systems discussed in Sections 5.4.4 and 5.4.5, and it is widely used when describing carrier-mediated transport in biomembranes. On the contrary, most theoretical analyses of carrier-mediated transport in chemical engineering assume that the complexation reactions take place inside the membrane. The theoretical description of these diffusion–reaction problems is more complicated because the flux densities are not constant with position under steady-state conditions, and the differential transport equations are non-linear [15]. However, in some cases the homogeneous reaction takes place in a region so close to the interface that it can hardly be discriminated from a true heterogeneous reaction.

5.4 Carrier-mediated coupled transport

5.4.1 Introduction

When two neutral solutes A and B are simultaneously transported across a supported liquid membrane by a carrier-mediated mechanism involving a single carrier, their transport is *coupled*. From the point of view of the transport equations, this implies that the driving force for the flux of solute A is not only the concentration gradient of this solute but also the concentration gradient of solute B, and vice versa. The coupled transport of the two solutes in the same direction is known as *co-transport* and the coupled transport in opposite directions in known as *countertransport*. Coupled transport is widely used in practical applications like liquid-phase extraction.

The transport of these two solutes is a spontaneous process towards equilibrium that leads to a decrease of Gibbs free energy and, hence, it does not require an external source of energy. In the biophysical jargon, it is said that this is a case of *passive transport*, as opposed to the case of *primary active transport* (not considered here) where the coupled transport requires an external source of metabolic energy because it drives the system to a non-equilibrium state.

Two related concepts are that of *downhill transport* and *uphill transport*. When applied to neutral solutes, the former refers to the diffusion from a more concentrated solution to a less concentration solution, and the uphill transport refers to the opposite case of diffusion from a less to a more concentrated solution. In the case of independent transport of a solute, the second law of thermodynamics requires that passive transport occurs downhill. However, when the two solutes share the same carrier and their transport is coupled, it is possible that one of them is transported uphill. Again, the second law of thermodynamics requires that the other solute must be transported downhill. From the point of view of the rate of variation of Gibbs free energy in the transport process, the uphill transport of one of the solutes requires a source of free energy. Such a source is provided by the downhill transport of the second solute. Similarly, from the point of view of entropy production, we could say then that the solute that is transported against its concentration gradient has a negative contribution to the entropy production, and the second law of thermodynamics requires that the downhill transport of the other solute produces more entropy than the uphill transport of the first one. In biophysics, the uphill transport of a solute using the free energy released by the simultaneous, coupled downhill transport of another solute is known as *secondary active transport*.

In the next sections we consider two mechanisms for carrier-mediated coupled transport of two neutral solutes A and B. We choose solute B as the one that is always transported downhill. First, we consider competitive binding to a carrier dissolved in the membrane. When the two solutes flow in the same directions, solute A can only be transported downhill. However, when the solutes flow in opposite directions, solute A can be transported uphill under some conditions. For this reason, competitive binding to a carrier is often related to countertransport. Second, we consider sequential binding to a carrier. When the two solutes flow in opposite directions, solute A can only be transported downhill. However, when the solutes flow in the same direction, solute A can be transported against its concentration gradient under some conditions. For this reason, sequential binding to a carrier is often related to co-transport.

5.4.2 Competitive binding of two neutral solutes to a carrier

Consider a liquid membrane that separates two aqueous solutions, α and β, containing two neutral solutes A and B at molar concentrations c_{A}^{α}, c_{B}^{α}, c_{A}^{β}, and c_{B}^{β}. Without loss of generality, we consider that $c_{\mathrm{B}}^{\alpha} > c_{\mathrm{B}}^{\beta}$ and that the transport of this solute proceeds from compartment α to β. These solutions are assumed to be ideally mixed, so that the aqueous diffusion boundary layers offer no transport resistance. The membrane contains a neutral carrier C that can bind competitively to either solute A or solute B (Figs. 5.20 and 5.21). For the sake of simplicity, we neglect free solute diffusion. That is, none of the solutes can partition inside the membrane without binding to the carrier. Thus, only the solute–carrier complexes CA and CB, and the free carrier C, can diffuse throughout the membrane.

The complexation reactions at the membrane aqueous solution interfaces

$$\mathrm{A(w)} + \mathrm{C(o)} \rightleftarrows \mathrm{CA(o)}, \tag{5.82}$$

$$\mathrm{B(w)} + \mathrm{C(o)} \rightleftarrows \mathrm{CB(o)} \tag{5.83}$$

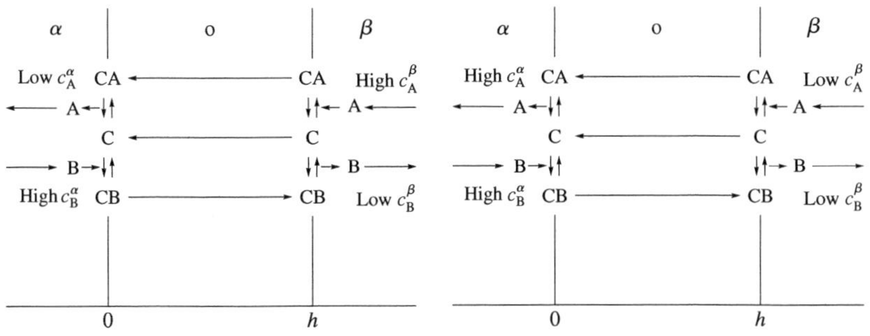

Fig. 5.20.
The carrier-mediated countertransport of two competing neutral solutes A and B across a supported liquid membrane may proceed downhill (left) or uphill (right) for one of the solutes.

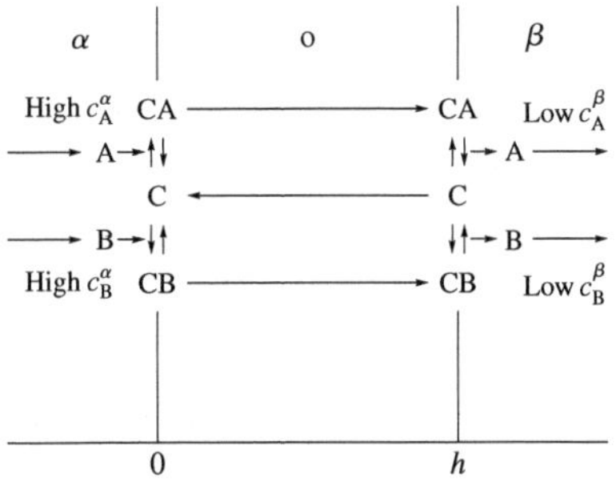

Fig. 5.21.
The carrier-mediated co-transport of two competing neutral solutes A and B across a supported liquid membrane always proceeds downhill for both solutes.

have equilibrium constants[12]

$$K_{CA} = \frac{c_{CA}}{c_C c_A^w}, \tag{5.84}$$

$$K_{CB} = \frac{c_{CB}}{c_C c_B^w}. \tag{5.85}$$

The flux densities of free and complexed carrier are

$$j_C = -D_C \frac{dc_C}{dx} = (D_C/h)[c_C(0) - c_C(h)], \tag{5.86}$$

$$j_{CA} = -D_{CA} \frac{dc_{CA}}{dx} = (D_{CA}/h)[c_{CA}(0) - c_{CA}(h)], \tag{5.87}$$

$$j_{CB} = -D_{CB} \frac{dc_{CB}}{dx} = (D_{CB}/h)[c_{CB}(0) - c_{CB}(h)], \tag{5.88}$$

and it is assumed that $j_C + j_{CA} + j_{CB} = 0$, which states that the total (i.e. free and complexed) carrier flux density must be zero because the carrier cannot partition to the aqueous phase. Furthermore, we assume that $D_C = D_{CA} = D_{CB}$, so that the above equations imply that the total carrier $c_{CT} = c_C + c_{CA} + c_{CB}$ has a

[12] Since the supported liquid membranes are relatively thick and the limitation to solute transport arises from the diffusion inside the membrane and not from the interfacial kinetics, it is a very good approximation to consider that reactions (5.82) and (5.83) are under conditions of thermodynamic equilibrium.

uniform distribution. Then, the fraction of carrier in CA form at the membrane solution α interface is

$$\theta(0) = \frac{c_{\mathrm{CA}}(0)}{c_{\mathrm{CT}}} = \frac{K_{\mathrm{CA}} c_{\mathrm{A}}^{\alpha}}{1 + K_{\mathrm{CA}} c_{\mathrm{A}}^{\alpha} + K_{\mathrm{CB}} c_{\mathrm{B}}^{\alpha}}, \tag{5.89}$$

and similar expressions can be written for CB and the other interface.

Since free solute transport is not possible, the flux density of solute A across the membrane is $j_{\mathrm{A}} = j_{\mathrm{CA}}$, and using eqns (5.87) and (5.89) it can be evaluated as

$$j_{\mathrm{A}} = \frac{D_{\mathrm{C}} c_{\mathrm{CT}}}{h} K_{\mathrm{CA}} \frac{c_{\mathrm{A}}^{\alpha}(1 + K_{\mathrm{CB}} c_{\mathrm{B}}^{\beta}) - c_{\mathrm{A}}^{\beta}(1 + K_{\mathrm{CB}} c_{\mathrm{B}}^{\alpha})}{[1 + K_{\mathrm{CA}} c_{\mathrm{A}}^{\alpha} + K_{\mathrm{CB}} c_{\mathrm{B}}^{\alpha}][1 + K_{\mathrm{CA}} c_{\mathrm{A}}^{\beta} + K_{\mathrm{CB}} c_{\mathrm{B}}^{\beta}]}. \tag{5.90}$$

The most important characteristic of this equation is that it is no longer convenient to present it in the form $j_{\mathrm{A}} = P_{\mathrm{CA}}(c_{\mathrm{A}}^{\alpha} - c_{\mathrm{A}}^{\beta})$, because the permeability P_{CA} would then be a complicated function of the concentrations c_{A}^{α}, c_{B}^{α}, c_{A}^{β}, and c_{B}^{β}, and the equilibrium constants K_{CA} and K_{CB}. In fact, the transport of solute A is not only driven by the concentration difference $\Delta_{\beta}^{\alpha} c_{\mathrm{A}} \equiv c_{\mathrm{A}}^{\alpha} - c_{\mathrm{A}}^{\beta}$ but the concentration difference of the other solute $\Delta_{\beta}^{\alpha} c_{\mathrm{B}} \equiv c_{\mathrm{B}}^{\alpha} - c_{\mathrm{B}}^{\beta} > 0$ can also act as a driving force for A, and vice versa. Indeed, the solute flux densities are given by the equations

$$j_{\mathrm{A}} = j_{\mathrm{C,L}} \frac{(1 + K_{\mathrm{CB}} \bar{c}_{\mathrm{B}}) K_{\mathrm{CA}}(c_{\mathrm{A}}^{\alpha} - c_{\mathrm{A}}^{\beta}) - K_{\mathrm{CA}} \bar{c}_{\mathrm{A}} K_{\mathrm{CB}}(c_{\mathrm{B}}^{\alpha} - c_{\mathrm{B}}^{\beta})}{[1 + K_{\mathrm{CA}} c_{\mathrm{A}}^{\alpha} + K_{\mathrm{CB}} c_{\mathrm{B}}^{\alpha}][1 + K_{\mathrm{CA}} c_{\mathrm{A}}^{\beta} + K_{\mathrm{CB}} c_{\mathrm{B}}^{\beta}]}, \tag{5.91}$$

$$j_{\mathrm{B}} = j_{\mathrm{C,L}} \frac{(1 + K_{\mathrm{CA}} \bar{c}_{\mathrm{A}}) K_{\mathrm{CB}}(c_{\mathrm{B}}^{\alpha} - c_{\mathrm{B}}^{\beta}) - K_{\mathrm{CB}} \bar{c}_{\mathrm{B}} K_{\mathrm{CA}}(c_{\mathrm{A}}^{\alpha} - c_{\mathrm{A}}^{\beta})}{[1 + K_{\mathrm{CA}} c_{\mathrm{A}}^{\alpha} + K_{\mathrm{CB}} c_{\mathrm{B}}^{\alpha}][1 + K_{\mathrm{CA}} c_{\mathrm{A}}^{\beta} + K_{\mathrm{CB}} c_{\mathrm{B}}^{\beta}]}, \tag{5.92}$$

where $\bar{c}_{\mathrm{A}} \equiv (c_{\mathrm{A}}^{\alpha} + c_{\mathrm{A}}^{\beta})/2$ and $\bar{c}_{\mathrm{B}} \equiv (c_{\mathrm{B}}^{\alpha} + c_{\mathrm{B}}^{\beta})/2$ are the average solute concentrations and $j_{\mathrm{C,L}} = D_{\mathrm{C}} c_{\mathrm{CT}}/h$ is the limiting flux density in a carrier-mediated transport mechanism.

Equation (5.90) clearly shows that the sign of j_{A}, and hence the flow direction for solute A, is equal to the sign of $[c_{\mathrm{A}}^{\alpha}(1 + K_{\mathrm{CB}} c_{\mathrm{B}}^{\beta}) - c_{\mathrm{A}}^{\beta}(1 + K_{\mathrm{CB}} c_{\mathrm{B}}^{\alpha})]$ and not to the sign of $(c_{\mathrm{A}}^{\alpha} - c_{\mathrm{A}}^{\beta})$. This opens up the possibility for uphill transport. If $c_{\mathrm{A}}^{\alpha} < c_{\mathrm{A}}^{\beta}$ then $j_{\mathrm{A}} < 0$ and the countertransport proceeds downhill for both solutes (Fig. 5.20, left). If $c_{\mathrm{A}}^{\alpha} > c_{\mathrm{A}}^{\beta}$, we can observe that $j_{\mathrm{A}} < 0$ (i.e. solute A can flow from a low concentration solution β to a high concentration solution α) provided that $1 < c_{\mathrm{A}}^{\alpha}/c_{\mathrm{A}}^{\beta} < (1 + K_{\mathrm{CB}} c_{\mathrm{B}}^{\alpha})/(1 + K_{\mathrm{CB}} c_{\mathrm{B}}^{\beta})$ (Fig. 5.20, right). Obviously, this requires that $c_{\mathrm{B}}^{\alpha} > c_{\mathrm{B}}^{\beta}$ and also that a significant fraction of the carrier is coupled to solute B, because otherwise $K_{\mathrm{CB}} c_{\mathrm{B}}^{\beta} < K_{\mathrm{CB}} c_{\mathrm{B}}^{\alpha} \ll 1$ and the above requirement could not be satisfied. These two conditions then imply that $j_{\mathrm{B}} > 0$ and, therefore, that solute B must be transported downhill if solute A is transported uphill. Finally, if the concentration difference for solute A is so high that $c_A^\alpha/c_A^\beta > (1 + K_{CB}c_B^\alpha)/(1 + K_{CB}c_B^\beta) > 1$, then the coupled transport of the two solutes proceeds in the same direction (Fig. 5.21).

Let us analyse in more detail the carrier-mediated countertransport illustrated in Fig. 5.20. If the free carrier were not able to diffuse across the membrane, the countertransport could be understood as the following sequence of steps:

1) at the interface between the liquid membrane and the solution with a low concentration of solute B, the equilibria in eqns (5.82) and (5.83) are displaced towards the dissociation of carrier complex CB and the formation of carrier complex CA, releasing solute B to and taking solute A from the aqueous solution,
2) the carrier complex CA diffuses across the membrane,
3) at the opposite interface, the equilibria in eqns (5.82) and (5.83) are displaced towards the dissociation of carrier complex CA and the formation of carrier complex CB, releasing solute A to and taking solute B from the aqueous solution,
4) the carrier complex CB diffuses back across the membrane, and the cycle is repeated.

In this case, the transport of the two solutes satisfies a 1:1 stoichiometric relation since $j_{CA} = -j_{CB}$ when $j_C = 0$. Stoichiometric countertransport is very common in biomembranes, although the coupling mechanism there is usually a channel protein or an ionic pump rather than a carrier.

When the free carrier also diffuses across the membrane, as we have considered in the above theoretical description, the countertransport could be understood as the following sequence of steps:

1) at the interface between the liquid membrane and the solution β (diluted in solute B), the equilibria in eqns (5.82) and (5.83) are displaced towards the dissociation of carrier complex CB and the formation of carrier complex CA, releasing solute B to and taking solute A from the aqueous solution. In addition, the interfacial reaction produces some free carrier C, which has a high concentration there,
2) the carrier complex CA and the free carrier C diffuse across the membrane,
3) at the opposite interface, the equilibria in eqns (5.82) and (5.83) are displaced towards the dissociation of carrier complex CA and the formation of carrier complex CB, releasing solute A to and taking solute B from the solution. In addition, the interfacial reaction consumes some free carrier C, which has a low concentration there,
4) the carrier complex CB diffuses back across the membrane, and the cycle is repeated.

The interesting thing here is that the fraction of complex CA at the interfaces with the solution α and β is not only determined by the concentrations of solute A at the respective solutions but also by the concentrations of solute B. Thus, if the concentration of solute B in solution α is very high, the fraction of carrier in CA form at this interface is very small. Conversely, the concentration of solute B in solution β is low, and the fraction of carrier in CA form at this interface is large, even if the concentration of solute A in solution β is lower than in solution α. This explains why uphill countertransport is then observed.

Figure 5.22 shows the solute flux densities evaluated from eqns (5.91) and (5.92) as a function of $K_{CA}\bar{c}_A$ for different values of the concentration ratio $r_A \equiv c_A^\alpha/c_A^\beta$ and fixed values of $r_B \equiv c_B^\alpha/c_B^\beta$ and $K_{CB}\bar{c}_B$. It is observed that the flux density j_A vanishes when $K_{CA}\bar{c}_A$ does. This is because $j_A \propto K_{CA}\bar{c}_A$ in this range. It is also observed that j_A vanishes when $K_{CA}\bar{c}_A$ takes very large values, a phenomenon already noted and illustrated in Fig. 5.13. At intermediate values of $K_{CA}\bar{c}_A$ the flux density j_A is of the order of the maximum value $j_{C,L} = D_C c_{CT}/h$. If $r_A > (1 + K_{CB}c_B^\alpha)/(1 + K_{CB}c_B^\beta)$, both solutes flow from solution α to β. If $r_A < 1$, solute A flows from solution β to α. Uphill countertransport of solute A is observed when $1 < r_A < (1 + K_{CB}c_B^\alpha)/(1 + K_{CB}c_B^\beta)$, which corresponds to $1 < r_A < 1.16$ when $K_{CB}\bar{c}_B = 0.1$ and $1 < r_A < 6.80$ when $K_{CB}\bar{c}_B = 10$. In relation to the flux density j_B, it is noticed that it is independent of r_A when $K_{CA}\bar{c}_A \ll 1$, as expected because the transport of B must be independent of solute A when the latter is present in trace amounts. Again, the flux density j_B vanishes when $K_{CA}\bar{c}_A \gg 1$ because the carrier is then in CA form and is not able to transport any of the solutes. At intermediate values of $K_{CA}\bar{c}_A$ the coupling between the transport of the two solutes is more important and it is observed that it can lead to either an enhancement of j_B or even to a situation of uphill countertransport of B driven by the concentration gradient of solute A.

Figure 5.23 clearly shows the regions when uphill countertransport is possible. Note that the asymmetry in this diagram arises from the values considered for $K_{CA}\bar{c}_A$ and $K_{CB}\bar{c}_B$. Otherwise, eqns (5.91) and (5.92) evidence the symmetry of the transport equations with respect to solutes A and B, although

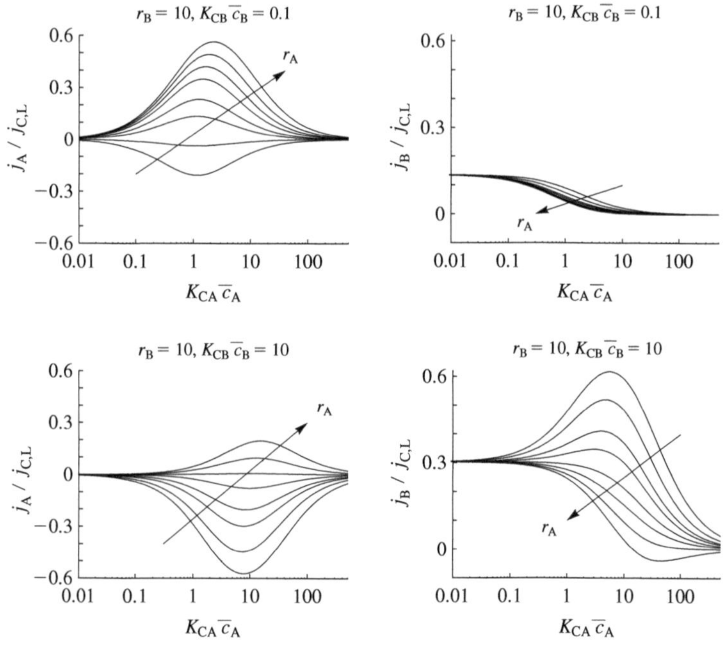

Fig. 5.22.
Solute flux densities as a function of $K_{CA}\bar{c}_A$ evaluated from eqns (5.91) and (5.92) for the following values of the concentration ratio $r_A \equiv c_A^\alpha/c_A^\beta = 0.5$, 1, 2, 3, 5, 7, 10, and 15 (increasing in the arrow direction). The concentration ratio of solute B has been fixed to $r_B \equiv c_B^\alpha/c_B^\beta = 10$ and two values of its average concentration have been considered, $K_{CB}\bar{c}_B = 0.1$ and 10.

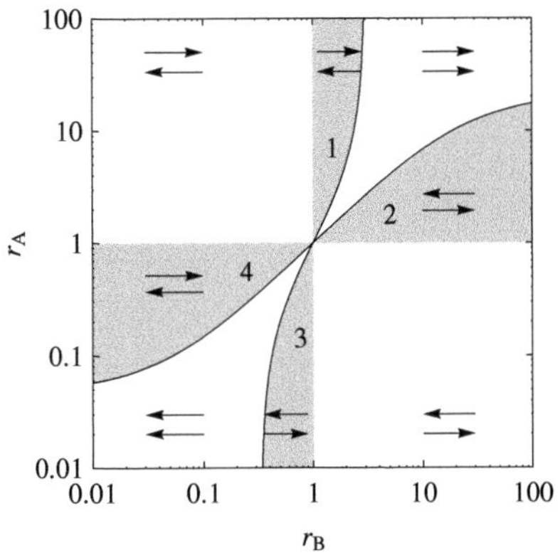

Fig. 5.23.
The arrows in every region of this diagram show the flow direction of solutes A and B (upper and lower arrows, respectively), as evaluated from eqn (5.91) and (5.92) for the values $K_{CA}\bar{c}_A = 1$ and $K_{CB}\bar{c}_B = 10$. The shaded regions correspond to uphill countertransport: in regions 1 and 3, solute B is transported uphill driven by the concentration gradient of solute A, and in regions 2 and 4, the uphill transport of solute A is driven by the concentration gradient of solute B.

we have chosen above to discuss the situations when B is transported downhill for the sake of clarity.

The energetics of uphill countertransport is analysed next. The rate of change of Gibbs free energy per unit membrane area can be understood as a sum of two contributions, $\Delta_r G = \Delta_r G_A + \Delta_r G_B$ where $\Delta_r G_A \equiv j_A(\mu_A^\beta - \mu_A^\alpha) = -RTj_A \ln r_A$, $\Delta_r G_B \equiv -RTj_B \ln r_B$ and the solute flux densities are given by eqns (5.91) and (5.92). The second law of thermodynamics requires that $\Delta_r G < 0$ but there is no constraint on the contributions $\Delta_r G_A$ and $\Delta_r G_B$. Uphill transport of solute A is characterized by the condition $\Delta_r G_A > 0$, and similarly for solute B. Figure 5.24 shows the contributions $\Delta_r G_A$ and $\Delta_r G_B$ for the transport conditions considered in Fig. 5.22. For the situations considered here, uphill transport of A is observed when r_A= 2, 3, 5 and $K_{CB}\bar{c}_B = 10$, and uphill transport of B is observed, in a limited range of values of $K_{CA}\bar{c}_A$, when r_A= 15 and $K_{CB}\bar{c}_B = 10$.

5.4.3 Sequential binding of two neutral solutes to a carrier

Consider a liquid membrane that separates two aqueous solutions, α and β, containing two neutral solutes A and B at molar concentrations c_A^α, c_B^α, c_A^β, and c_B^β. Without loss of generality, we consider that $c_B^\alpha > c_B^\beta$ and that the transport of this solute proceeds from compartment α to β. The solutions are assumed to be ideally mixed, so that the aqueous diffusion boundary layers offer no transport resistance. The membrane contains a neutral carrier C that can bind sequentially first to solute A and then to solute B (Fig. 5.25). For the sake of simplicity, we neglect free solute diffusion. That is, none of the solutes can partition inside the membrane without binding to the carrier. Thus, only the solute–carrier complexes CA and CAB, and the free carrier C, can diffuse throughout the membrane.

The complexation reactions at the membrane aqueous solution interfaces

$$A(w) + C(o) \rightleftarrows CA(o), \qquad (5.82)$$

$$CA(o) + B(w) \rightleftarrows CAB(o), \qquad (5.93)$$

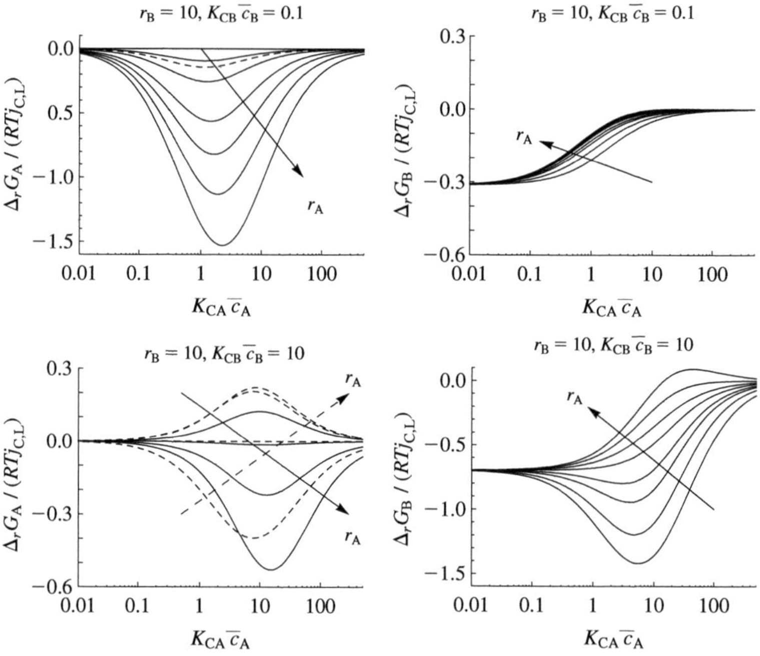

Fig. 5.24.
Rate of Gibbs free energy change in the carrier-mediated coupled transport of two neutral solutes A and B. The contributions of the two solutes are shown as a function of $K_{CA}\bar{c}_A$ for the following values of the concentration ratio $r_A \equiv c_A^\alpha / c_A^\beta = 0.5$, 1, 2, 3, 5, 7, 10, and 15 (increasing in the direction indicated by the arrows); in the first plot the dashed curve corresponds to $r_A = 0.5$ and in the third one the dashed curves correspond to $r_A = 0.5$, 1, 2, and 3. Two average concentrations of solute B have been considered, $K_{CB}\bar{c}_B = 0.1$ and 10, and its ratio has been fixed to $r_B \equiv c_B^\alpha / c_B^\beta = 10$.

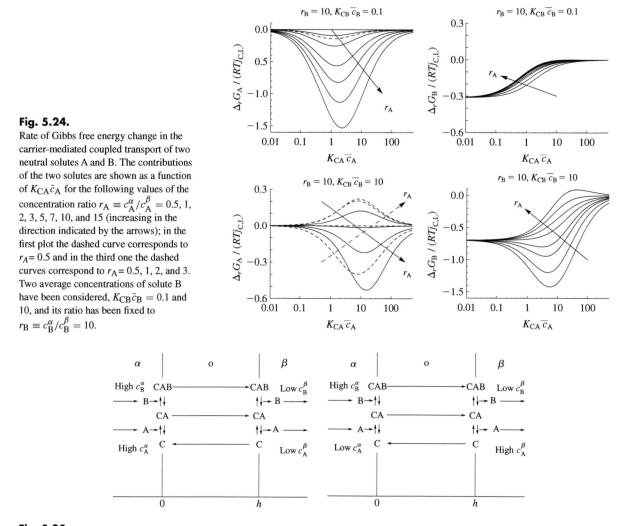

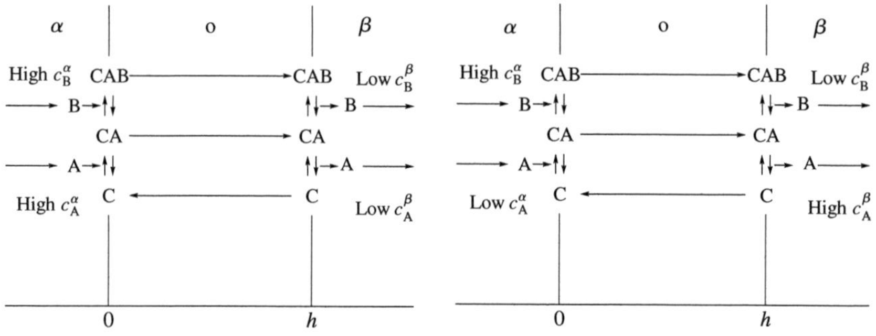

Fig. 5.25.
The carrier-mediated co-transport of two neutral solutes A and B that bind sequentially to the same carrier may proceed downhill (left) or uphill (right) for one of the solutes.

have equilibrium constants[13]

$$K_{CA} = \frac{c_{CA}}{c_C c_A^w}, \tag{5.84}$$

$$K_{CAB} = \frac{c_{CAB}}{c_{CA} c_B^w}. \tag{5.94}$$

[13] As explained before, it is a very good approximation to consider that these reactions are under conditions of thermodynamic equilibrium.

The flux densities of free and complexed carrier are

$$j_{\mathrm{C}} = -D_{\mathrm{C}} \frac{\mathrm{d}c_{\mathrm{C}}}{\mathrm{d}x} = (D_{\mathrm{C}}/h)[c_{\mathrm{C}}(0) - c_{\mathrm{C}}(h)], \tag{5.86}$$

$$j_{\mathrm{CA}} = -D_{\mathrm{CA}} \frac{\mathrm{d}c_{\mathrm{CA}}}{\mathrm{d}x} = (D_{\mathrm{CA}}/h)[c_{\mathrm{CA}}(0) - c_{\mathrm{CA}}(h)], \tag{5.87}$$

$$j_{\mathrm{CAB}} = -D_{\mathrm{CAB}} \frac{\mathrm{d}c_{\mathrm{CAB}}}{\mathrm{d}x} = (D_{\mathrm{CAB}}/h)[c_{\mathrm{CAB}}(0) - c_{\mathrm{CAB}}(h)], \tag{5.95}$$

and it is assumed that $j_{\mathrm{C}} + j_{\mathrm{CA}} + j_{\mathrm{CAB}} = 0$, which states that the total (i.e. free and complexed) carrier flux density must be zero because the carrier cannot partition to the aqueous phase. Furthermore, we assume that $D_{\mathrm{C}} = D_{\mathrm{CA}} = D_{\mathrm{CAB}}$, so that the above equations imply that the total carrier $c_{\mathrm{CT}} = c_{\mathrm{C}} + c_{\mathrm{CA}} + c_{\mathrm{CAB}}$ has a uniform distribution. Then, the fraction of carrier in CA and CAB forms at the membrane, solution α interface are

$$\theta_{\mathrm{CA}}(0) = \frac{c_{\mathrm{CA}}(0)}{c_{\mathrm{CT}}} = \frac{K_{\mathrm{CA}} c_{\mathrm{A}}^{\alpha}}{1 + K_{\mathrm{CA}} c_{\mathrm{A}}^{\alpha}(1 + K_{\mathrm{CAB}} c_{\mathrm{B}}^{\alpha})}, \tag{5.96}$$

$$\theta_{\mathrm{CAB}}(0) = \frac{c_{\mathrm{CAB}}(0)}{c_{\mathrm{CT}}} = \frac{K_{\mathrm{CA}} K_{\mathrm{CAB}} c_{\mathrm{A}}^{\alpha} c_{\mathrm{B}}^{\alpha}}{1 + K_{\mathrm{CA}} c_{\mathrm{A}}^{\alpha}(1 + K_{\mathrm{CAB}} c_{\mathrm{B}}^{\alpha})}, \tag{5.97}$$

and similar expressions can be written for the other interface.

Since free solute transport is not possible, the flux density of solute A across the membrane is $j_{\mathrm{A}} = j_{\mathrm{CA}} + j_{\mathrm{CAB}}$, and it can be evaluated as

$$j_{\mathrm{A}} = j_{\mathrm{C,L}} K_{\mathrm{CA}} \frac{c_{\mathrm{A}}^{\alpha}(1 + K_{\mathrm{CAB}} c_{\mathrm{B}}^{\alpha}) - c_{\mathrm{A}}^{\beta}(1 + K_{\mathrm{CAB}} c_{\mathrm{B}}^{\beta})}{[1 + K_{\mathrm{CA}} c_{\mathrm{A}}^{\alpha}(1 + K_{\mathrm{CAB}} c_{\mathrm{B}}^{\alpha})][1 + K_{\mathrm{CA}} c_{\mathrm{A}}^{\beta}(1 + K_{\mathrm{CAB}} c_{\mathrm{B}}^{\beta})]}, \tag{5.98}$$

where $j_{\mathrm{C,L}} = D_{\mathrm{C}} c_{\mathrm{CT}}/h$. As in the case of competitive binding to the carrier, it is not convenient to write this flux density in the form $j_{\mathrm{A}} = P_{\mathrm{CA}}(c_{\mathrm{A}}^{\alpha} - c_{\mathrm{A}}^{\beta})$ because the permeability P_{CA} would then be a complicated function of the concentrations c_{A}^{α}, c_{B}^{α}, c_{A}^{β}, and c_{B}^{β}, and the equilibrium constants K_{CA} and K_{CAB}. Equation (5.98) clearly shows that the sign of j_{A} is equal to the sign of $[c_{\mathrm{A}}^{\alpha}(1 + K_{\mathrm{CAB}} c_{\mathrm{B}}^{\alpha}) - c_{\mathrm{A}}^{\beta}(1 + K_{\mathrm{CAB}} c_{\mathrm{B}}^{\beta})]$ and not to the sign of $(c_{\mathrm{A}}^{\alpha} - c_{\mathrm{A}}^{\beta})$. This opens the possibility for uphill co-transport.

If $c_{\mathrm{A}}^{\alpha} > c_{\mathrm{A}}^{\beta}$ then $j_{\mathrm{A}} > 0$ and the co-transport proceeds downhill for both solutes (Fig. 5.25, left). If $c_{\mathrm{A}}^{\alpha} < c_{\mathrm{A}}^{\beta}$, we can observe that $j_{\mathrm{A}} > 0$ (i.e. solute A can flow uphill from a low concentration solution α to a high concentration solution β) provided that $(1 + K_{\mathrm{CAB}} c_{\mathrm{B}}^{\alpha})/(1 + K_{\mathrm{CAB}} c_{\mathrm{B}}^{\beta}) > c_{\mathrm{A}}^{\beta}/c_{\mathrm{A}}^{\alpha} > 1$ (Fig. 5.25, right). Obviously, this requires that $c_{\mathrm{B}}^{\alpha} > c_{\mathrm{B}}^{\beta}$ and also that a significant fraction of the carrier is in CAB form. These two conditions then imply that $j_{\mathrm{B}} > 0$ and, therefore, that solute B must be transported downhill when solute A is transported uphill. Finally, if the concentration difference

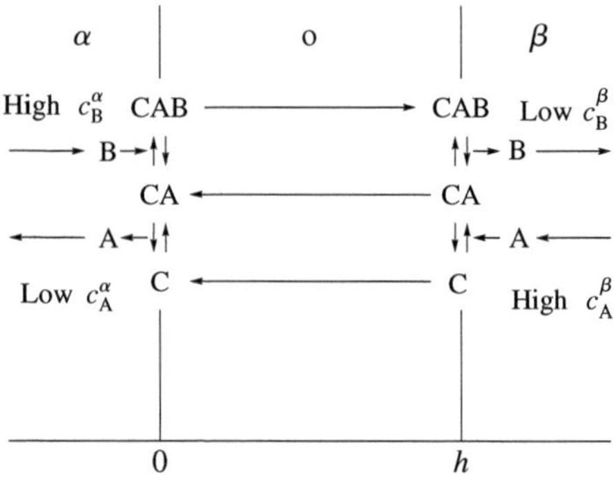

Fig. 5.26.
The carrier-mediated countertransport of two neutral solutes A and B that bind sequentially to the same carrier always proceeds downhill for both solutes.

for solute A is so high that $c_A^\beta/c_A^\alpha > (1 + K_{CAB}c_B^\alpha)/(1 + K_{CAB}c_B^\beta) > 1$, then the coupled transport of the two solutes proceeds in the opposite direction (Fig. 5.26).

In relation to the carrier-mediated co-transport illustrated in Fig. 5.25, it is interesting to note that if the complexed carrier CA were not able to diffuse across the membrane, the transport of the two solutes would satisfy a 1:1 stoichiometric relation. This stoichiometric co-transport could then be understood as the following sequence of steps:

1) at the interface with solution α, the solutes A and B bind to the carrier and form CAB,
2) the carrier complex CAB diffuses across the membrane,
3) at the opposite interface, solutes A and B are released,
4) the free carrier C diffuses back across the membrane, and the cycle is repeated.

In this simple situation the uphill co-transport of solute A can be easily understood because its transport across the membrane depends on the difference $(c_A^\alpha c_B^\alpha - c_A^\beta c_B^\beta)$, which can be positive even though $c_A^\alpha < c_A^\beta$.

5.4.4 Polyelectrolyte extraction by carrier-mediated co-transport with an acid

As an application example of the theory presented in the previous sections, we consider the extraction of an anionic polyelectrolyte PE^{m-} by carrier-mediated co-transport using supported liquid membranes. The carrier C is an amine, and the solute that is co-transported is an acid. A specific system that we have investigated is the extraction of lignosulphonate using trilaurylamine as a carrier and decanol as the membrane solvent [16]. The feed solution (solution α) contains HCl and the polyelectrolyte in sodium form Na_mPE. The strip solution (solution β) contains NaOH.

The reactions taking place at the membrane interfaces are

$$\mathrm{HCl}(\alpha) + \mathrm{C(o)} \rightleftarrows \mathrm{C : HCl(o)}\ \text{feed}, \tag{5.99}$$

$$\mathrm{Na}_m\mathrm{PE}(\alpha) + m\,\mathrm{C : HCl(o)} \rightleftarrows \mathrm{(C : H)}_m\mathrm{PE(o)} + m\,\mathrm{NaCl}(\alpha)\ \text{feed}, \tag{5.100}$$

$$\mathrm{C : HCl(o)} + \mathrm{NaOH}(\beta) \rightleftarrows \mathrm{C(o)} + \mathrm{NaCl}(\beta) + \mathrm{H_2O}(\beta)\ \text{strip}, \tag{5.101}$$

$$\mathrm{(C : H)}_m\mathrm{PE(o)} + m\,\mathrm{NaOH}(\beta) \rightleftarrows m\,\mathrm{C(o)} + \mathrm{Na}_m\mathrm{PE}(\beta) + m\,\mathrm{H_2O}(\beta)\ \text{strip}, \tag{5.102}$$

where the semicolon (e.g. in C:HCl) illustrates the electrostatic bond between the carrier and the hydrogen ion. As usual, reactions (5.99)–(5.102) are considered to be in equilibrium. Although the polyelectrolyte has several counterions, not all of them are released when it binds to the C:HCl complex in reaction (5.100). In the experiments reported in Ref. [16] it was verified that low molecular mass lignosulphonate exchanged only one counterion when binding to the amine complex, that is, $m = 1$ in reactions (5.100) and (5.102). Thus, in the following paragraphs we denote the polyelectrolyte anion with the counterions that have not been released as PE^-.

Before proceeding to the theoretical description of this transport problem, it is in order to discuss whether it corresponds to a sequential or a competitive binding mechanism. The species of interest, PE^-, cannot bind to the free carrier C unless the latter binds first to H^+, thus forming the complex C:HPE. From this point of view, it could be understood that binding is sequential, as shown in reactions (5.99) and (5.100). Nevertheless, due to the low relative permittivity of the membrane, the protonated amine $\mathrm{C:H}^+$ can only diffuse through the liquid membrane in salt (or ion pair) form, C:HCl. Thus, the theoretical modelling of this transport problem resembles one of competitive binding where the two neutral solutes that compete for using the carrier to cross the membrane are HCl and HPE, and the two carrier complexes are C:HCl and C:HPE. It should be stressed, however, that in the aqueous feed solution there is no species like HPE trying to compete with HCl for the carrier. Reaction (5.100) occurs between PE^- in the aqueous phase and C:HCl in the organic phase.

Let us then introduce the notation A = HPE, B = HCl, CA = C:HPE complex, and CB = C:HCl complex. Reaction (5.100) then reads

$$\mathrm{A}(\alpha) + \mathrm{CB(o)} \rightleftarrows \mathrm{CA(o)} + \mathrm{B}(\alpha), \tag{5.103}$$

and its equilibrium constant is[14]

$$K = \frac{c_{\mathrm{CA}} c_{\mathrm{B}}}{c_{\mathrm{CB}} c_{\mathrm{A}}}, \tag{5.104}$$

where no phase superscripts are needed because every species can only be in one phase.

[14] If there were a reaction like $\mathrm{HPE}(\alpha) + \mathrm{C(o)} \rightleftarrows \mathrm{C:HPE(o)}$, then K would be the ratio of the equilibrium constants of this reaction and (5.99), i.e. it would be $K = K_{\mathrm{CA}}/K_{\mathrm{CB}}$ in the notation of Section 5.4.3.

Under the appropriate experimental conditions, it can be assumed that:

1) the aqueous phases do not get polarized, i.e. the rate-determining step is the transport across the membrane, and hence the concentrations of the carrier complexes on the strip side are zero, $c_{CA}(h) \approx 0$ and $c_{CB}(h) \approx 0$, and
2) reaction (5.99) is so displaced towards the formation of complexed carrier C:HCl that there is no free carrier C at the feed side, $c_C(0) \approx 0$.

Figure 5.27 shows a sketch of the concentration profiles in the system, where it has been illustrated that the total carrier concentration at the interface with the feed solution, $c_{CT}(0) = c_{CA}(0) + c_{CB}(0)$, might be different from that at the interface with the strip solution, $c_{CT}(h) = c_C(h)$.

The flux densities inside the membrane are

$$j_C = -\frac{D_C}{h} c_C(h), \tag{5.105}$$

$$j_{CA} = \frac{D_{CA}}{h} c_{CA}(0), \tag{5.106}$$

$$j_{CB} = \frac{D_{CB}}{h} c_{CB}(0). \tag{5.107}$$

The condition that the carrier cannot exit the membrane phase, $j_C + j_{CA} + j_{CB} = 0$, implies that the distribution of the total carrier concentration is linear and, therefore, that

$$c_{CT}(0) + c_{CT}(h) = 2c_{CT}^{b}, \tag{5.108}$$

where c_{CT}^{b} is the average total carrier concentration. If we could use the approximation $D_C \approx D_{CA} \approx D_{CB}$, as in the previous sections, the polyelectrolyte flux density, $j_A = j_{CA}$, would be given by

$$j_A = j_{C,L} \frac{Kc_A^{\alpha}}{Kc_A^{\alpha} + c_B^{\alpha}}, \tag{5.109}$$

where $j_{C,L} \equiv D_C c_{CT}^{b}/h$. This result can be considered a particular case of eqn (5.90) corresponding to $c_A^{\beta} \approx 0$, $c_B^{\beta} \approx 0$, and $c_C(0) \approx 0$. However, the C:HPE complex is much larger that the free carrier and the C:HCl complex, and the differences in their diffusion coefficients are significant. The polyelectrolyte flux density, $j_A = j_{CA}$, can then be evaluated as

$$j_A = \frac{D_C c_{CT}^{b}}{h} \frac{2D_{CA} Kc_A^{\alpha}}{(D_C + D_{CA})Kc_A^{\alpha} + (D_C + D_{CB})c_B^{\alpha}}. \tag{5.110}$$

As expected in the case of membrane-control under study, eqn (5.110) shows that the polyelectrolyte flux density is proportional to the total carrier concentration. This is a good diagnostic criterion for the analysis of experimental results. However, the carrier concentration cannot be increased arbitrarily, as after certain limit (typically of the order of 0.1 M) the carrier begins to aggregate.

Introducing the approximation $D_{CA} \ll D_C \approx D_{CB}$, eqn (5.110) can be further simplified to

$$\frac{1}{j_A} = \frac{h}{2D_{CA}c_{CT}^{b}}\left(1 + \frac{2c_B^{\alpha}}{Kc_A^{\alpha}}\right), \tag{5.111}$$

which shows that a linear plot of measured values of $1/j_A$ against $c_B^{\alpha}/c_A^{\alpha}$ allows us to determine $D_{CA}c_{CT}^{b}/h$ and K from the intercept at the origin and slope, respectively. Note that the maximum flux density is $2D_{CA}c_{CT}^{b}/h$, where the factor 2 arises from the approximation $D_{CA} \ll D_C$. Equation (5.110) provides several diagnostic criteria for membrane control. Reducing the membrane thickness, increasing the carrier concentration, or increasing D_{CA} by reducing the solvent viscosity all increase the flux. These variations cannot, of course, be realized without limitation, as the rate-determining step may change to, for example, the stripping reaction.

5.4.5 Polyelectrolyte extraction by carrier-mediated countertransport with an acid

The experimental set-up and theoretical modelling considered in Section 5.4.4 can be used, with minor changes, to study the extraction of an anionic polyelectrolyte (PE^-) using carrier-mediated counter-transport with an acid. Most assumptions employed there, including that the transport across the membrane is the rate-determining step, are also used here. The feed solution (solution α) contains HCl and the polyelectrolyte in sodium form NaPE, and the solution receiving the polyelectrolyte (solution β) contains HCl in a concentration larger than in the feed side. Moreover, it can be assumed here that the fraction of free carriers can be neglected, $c_C \approx 0$ (Fig. 5.28). This implies that the countertransport of polyelectrolyte and acid takes place in a 1:1 stoichiometric relation, i.e. one HCl molecule is transported to the compartment for every extracted polyelectrolyte molecule.

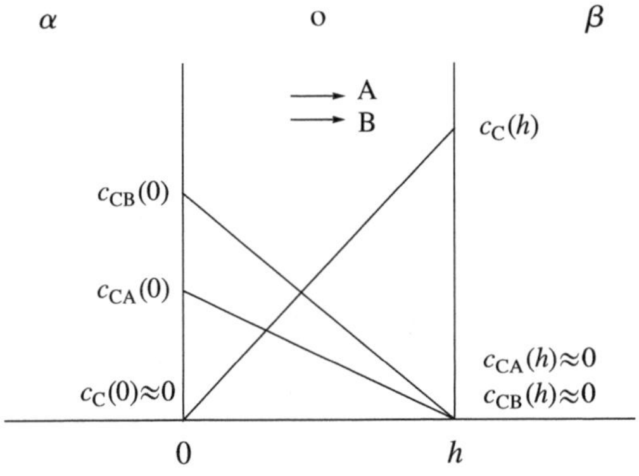

Fig. 5.27.
Schematic concentration profiles in polyelectrolyte extraction by carrier-mediated co-transport with an acid. The carrier is an amine and the complexes are CA = C:HPE and CB = C:HCl.

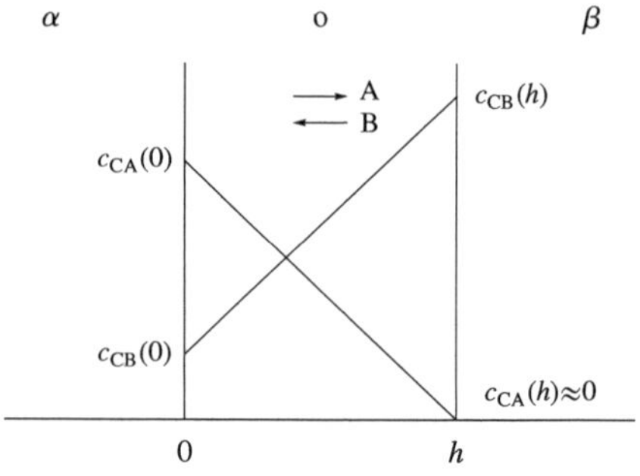

Fig. 5.28.
Schematic concentration profiles in polyelectrolyte extraction by carrier-mediated countertransport with an acid. The carrier is an amine and the complexes are CA = C:HPE and CB = C:HCl. The polyelectrolyte concentration at the receiving solution is negligible when transport through the membrane is the rate-determining step.

The flux densities inside the membrane are

$$j_{CA} = \frac{D_{CA}}{h} c_{CA}(0), \tag{5.106}$$

$$j_{CB} = \frac{D_{CB}}{h} [c_{CB}(0) - c_{CB}(h)]. \tag{5.112}$$

The condition that the carrier cannot exit the membrane phase, $j_{CA} + j_{CB} = 0$, implies that the distribution of the total carrier concentration is linear and, therefore, that

$$c_{CA}(0) + c_{CB}(0) + c_{CB}(h) = 2c_{CT}^{b}, \tag{5.113}$$

where c_{CT}^{b} is the average total carrier concentration. As already explained in Section 5.4.4, if we could use the approximation $D_C \approx D_{CA} \approx D_{CB}$, the polyelectrolyte flux density would be given by eqn (5.109). However, the C:HPE complex is much larger that the free carrier and the C:HCl complex, and the differences in their diffusion coefficients are significant. The polyelectrolyte flux density, $j_A = j_{CA}$, can then be evaluated as

$$j_A = \frac{D_{CB} c_{CT}^{b}}{h} \frac{2D_{CA} K c_A^{\alpha}}{(D_{CA} + D_{CB}) K c_A^{\alpha} + 2D_{CB} c_B^{\alpha}}. \tag{5.114}$$

Once again, since the transport is controlled by the membrane, eqn (5.114) shows that the polyelectrolyte flux density is proportional to the total carrier concentration. Moreover, introducing the approximation $D_{CA} \ll D_C \approx D_{CB}$, eqn (5.114) also simplifies to eqn (5.111). Therefore, we conclude that a linear plot of measured values of $1/j_A$ against $c_B^{\alpha}/c_A^{\alpha}$ allows us to determine $D_{CA} c_{CT}^{b}/h$ and K from the intercept at the origin and slope, respectively; that is, similar information as in co-transport would be obtained.

Contrary to the predictions from eqn (5.114), it can be experimentally observed that the polyelectrolyte flux density depends on the HCl concentration in the strip side, c_B^{β}. This interesting situation occurs when the strip reaction is kinetically controlled, as we show below. Heterogeneous reaction kinetics is often dictated by the adsorption equilibria of the reacting species. The reaction mechanism can thus be formally written as

$$\mathrm{CA(o)} + \mathrm{B(w)} \rightleftarrows \mathrm{CA}(\sigma) + \mathrm{B}(\sigma), \tag{5.115}$$

$$\mathrm{CA}(\sigma) + \mathrm{B}(\sigma) \rightarrow \mathrm{CB(o)} + \mathrm{A(w)}, \tag{5.116}$$

where σ denotes the membrane/stripping solution interface. It is reasonable to assume that reaction (5.116) is the rate-determining step. At steady state, the rate of this reaction is equal to the flux density, and we can write

$$j_{\mathrm{A}} = k\theta_{\mathrm{CA}}\theta_{\mathrm{B}}, \tag{5.117}$$

where θ_{CA} and θ_{B} are the surface coverage fractions at the strip interface and k is the kinetic rate constant of the heterogeneous reaction (5.116). The concentration $c_{\mathrm{CA}}(h)$ is probably so low that the surface coverage follows the linear form of the Langmuir adsorption isotherm

$$\theta_{\mathrm{CA}} = \kappa_{\mathrm{CA}} c_{\mathrm{CA}}(h), \tag{5.118}$$

while for the hydrochloric acid the full form of the Langmuir isotherm is needed

$$\theta_{\mathrm{B}} = \frac{\kappa_{\mathrm{B}} c_{\mathrm{B}}^{\beta}}{1 + \kappa_{\mathrm{B}} c_{\mathrm{B}}^{\beta}}. \tag{5.119}$$

In eqns (5.118) and (5.119), κ_{CA} and κ_{B} are the adsorption constants. At steady state, the governing equations are

$$\frac{c_{\mathrm{CB}}(0)}{c_{\mathrm{CA}}(0)} = \frac{c_{\mathrm{B}}^{\alpha}}{K c_{\mathrm{A}}^{\alpha}}, \tag{5.120}$$

$$j_{\mathrm{A}} = j_{\mathrm{CA}} = \frac{D_{\mathrm{CA}}[c_{\mathrm{CA}}(0) - c_{\mathrm{CA}}(h)]}{h} = \frac{D_{\mathrm{CB}}[c_{\mathrm{CB}}(h) - c_{\mathrm{CB}}(0)]}{h} = -j_{\mathrm{CB}}, \tag{5.121}$$

$$c_{\mathrm{CA}}(0) + c_{\mathrm{CB}}(0) + c_{\mathrm{CA}}(h) + c_{\mathrm{CB}}(h) = 2c_{\mathrm{CT}}^{\mathrm{b}}, \tag{5.122}$$

$$j_{\mathrm{A}} = k\theta_{\mathrm{CA}}\theta_{\mathrm{B}} = \frac{k\kappa_{\mathrm{CA}} c_{\mathrm{CA}}(h)\kappa_{\mathrm{B}} c_{\mathrm{B}}^{\beta}}{1 + \kappa_{\mathrm{B}} c_{\mathrm{B}}^{\beta}}. \tag{5.123}$$

After some algebra, the polyelectrolyte flux density can be shown to satisfy

$$\frac{c_{\mathrm{CT}}^{\mathrm{b}} K c_{\mathrm{A}}^{\alpha}}{j_{\mathrm{A}}} = \frac{h}{D_{\mathrm{CB}}} \frac{(D_{\mathrm{CA}} + D_{\mathrm{CB}}) K c_{\mathrm{A}}^{\alpha} + 2D_{\mathrm{CB}} c_{\mathrm{B}}^{\alpha}}{2D_{\mathrm{CA}}} + \frac{1}{k} \frac{K c_{\mathrm{A}}^{\alpha} + c_{\mathrm{B}}^{\alpha}}{\kappa_{\mathrm{CA}}} \left(1 + \frac{1}{\kappa_{\mathrm{B}} c_{\mathrm{B}}^{\beta}}\right), \tag{5.124}$$

which reduces to eqn (5.114) when the kinetic rate constant k is very large and describes well the experimentally observed dependence on c_B^β. Thus, although kinetic limitations have been neglected throughout this chapter because it is a good approximation in supported liquid membranes due to their relatively large thickness, this last application example serves to emphasize that the simplifying assumptions introduced in any theoretical modelling have to be supported by the experimental data.

Exercises

5.1 Assume that a weak 1:1 electrolyte is partitioning at the aqueous/organic interface, i.e. both the cation and anion, as well as the undissociated species are in equilibrium between the two phases. How does the distribution ratio c_T^o/c_T^w change as the ratio of the volumes of the phases vary?

5.2 Evaluate the facilitation factor F for the transport of a single solute A in the limit of small concentration differences $\Delta_\alpha^\beta c_A \ll \bar{c}_A$, where $\bar{c}_A \equiv (c_A^\alpha + c_A^\beta)/2$, and show that it is roughly proportional to the total carrier concentration in the membrane.

5.3 Show that the facilitation factor F for the transport of an electrolyte AB that is completely dissociated in A^+ and B^- ions in the aqueous phase and transported as ion pairs AB, either free or complexed to the carrier, in a liquid membrane is

$$F = 1 + \frac{D_C c_{CT}}{D_{AB}^o K_{AB}} \frac{K_{CAB}(c_{AB}^\alpha + c_{AB}^\beta)}{[1 + K_{CAB}(c_{AB}^\alpha)^2][1 + K_{CAB}(c_{AB}^\beta)^2]},$$

where D_{AB}^o is the diffusion coefficient of the ion pairs in the liquid membrane, $K_{AB} \equiv \sqrt{K_{A^+} K_{B^-}}$ is the partition coefficient of the electrolyte that is given by the geometric mean of the chemical partition coefficients of the ions, and K_{CAB} is the equilibrium constant of the complexation reaction $A^+(w) + B^-(w) + C(o) \rightleftarrows CAB(o)$. Assume that $D_C = D_{CAB}$ and neglect boundary-layer effects.

5.4 Show that in the case $D_C \neq D_{CA}$ the permeability of the membrane to the solute (due to the carrier-mediated transport) is

$$P_{CA} = \frac{D_C}{h} \frac{D_{CA}}{D_C(1 - \bar{\theta}) + D_{CA}\bar{\theta}} \frac{\Delta\theta\, c_{CT}^b}{\Delta c_A^w}$$

where $\bar{\theta} \equiv [\theta(0) + \theta(h)]/2$ is the average value of the fraction of complexed carrier.

References

[1] R.D. Noble and J. Douglas Way (ed.), *Liquid Membranes. Theory and Applications*, ACS Symposium Series 347, ACS, Washington, 1987; R.A. Bartsch and J. Douglas Way (ed.), *Chemical Separations with Liquid Membranes*, ACS Symposium Series 642, ACS, Washington, 1996.

[2] R.W. Barker, *Membrane Technology and Applications*, John Wiley & Sons, Chichester, 2004, Ch. 11.

[3] T. Kakiuchi, 'Equilibrium electric potential between two immiscible electrolyte solutions', in A.G. Volkov and D.W. Deamer (ed.), *Liquid-Liquid Interfaces. Theory and Methods*, CRC Press, Boca Raton, 1996, Ch. 1.

[4] L.Q. Hung, 'Electrochemical properties of the interface between two immiscible electrolyte solutions: Part I. Equilibrium situation and Galvani potential difference', *J. Electroanal. Chem.*, 115 (1980) 159–174.

[5] L.Q. Hung, 'Electrochemical properties of the interface between two immiscible electrolyte solutions: Part III. The general case of the Galvani potential difference at the interface and of the distribution of an arbitrary number of components interacting in both phases', *J. Electroanal. Chem.*, 149 (1983) 1–14.

[6] T. Kakiuchi, 'Limiting behavior in equilibrium partitioning of ionic components in liquid-liquid two-phase systems', *Anal. Chem.*, 68 (1996) 3658–3664.

[7] J.A. Manzanares, R. Lahtinen, B. Quinn, K. Kontturi, and D.J. Schiffrin, 'Determination of rate constants of ion transfer kinetics across immiscible electrolyte solutions', *Electrochim. Acta*, 44 (1998) 59–71.

[8] B.C. Pressmann, 'Induced active transport of ions in mitochondria', *PNAS*, 53 (1965) 1076–1083; B.C. Pressmann, E.J. Harris, W.S. Jagger, and J.H. Johnson, 'Antibiotic-mediated transport of alkali ions across lipid barriers', *PNAS*, 58 (1967) 1949–1956.

[9] W.J. Ward and W.L. Robb, 'Carbon dioxide-oxygen separation: facilitated transport of carbon dioxide across a liquid film', *Science*, 156 (1967) 1481–1484.

[10] E.L. Cussler, 'Membranes which pump', *AIChE J.*, 17 (1971) 1300–1303.

[11] M.S. Frant and J.W. Ross Jr., 'Potassium ion specific electrode with high selectivity for potassium over sodium', *Science*, 167 (1970) 987–988.

[12] H.C. Visser, D.N. Reinhoudt, and F. de Jong, 'Carrier-mediated transport through liquid membranes', *Chem. Soc. Rev.*, (1994) 75–81.

[13] E.L. Cussler, *Multicomponent Diffusion*, Elsevier, New York, 1975.

[14] S. Durand-Vidal, J.P. Simonin, and P. Turq, *Electrolytes at Interfaces*, Kluwer Academic Publishers, New York, 2002, Section 3.4.1.

[15] A.J. Barbero, J.A. Manzanares, and S. Mafé, 'A computational study of facilitated diffusion using the boundary element method', *J. Non-Equilib. Thermodyn.*, 20 (1995) 332–341.

[16] A.K. Kontturi, K. Kontturi, P. Niinikoski, and G. Sundholm, 'Extraction of a polyelectrolyte using a supported liquid membrane. Parts I and II', *Acta Chem. Scand.*, 44 (1990) 879–882, 883–891.

List of symbols

For those symbols that have been used with different meanings (e.g., $B, c, h, I, R, U, \ldots$) or as auxiliary variables (e.g., A, B, C, E,...) only the most frequent use is mentioned here. Other symbols (e.g., Da, Pe, Sc, l, R_i, w, ...) are used only once in the text and are not included here either. Note that the phase superscript φ is included here in many symbols, although they are used sometimes in the text without it.

Acronyms

BRF	barycentric reference frame
FRF	Fick's reference frame
HRF	Hittorf's reference frame
LRF	laboratory reference frame

Roman symbols

a	electrode radius, m
a_i^{φ}	activity of component i in phase φ, 1
A	electrode or membrane area, m^2
A_r	$\equiv -\sum_i \nu_{i,r}\mu_i$, chemical affinity of reaction r, J mol^{-1}
b	volume density of an arbitrary extensive quantity B, $[B]$ m^{-3}
B	arbitrary extensive quantity, $[B]$
$B_{i,k}$	auxiliary variable, s m^{-2}
c_i^{φ}	molar concentration of component i in phase φ, mol m^{-3}
d_{h}	hydraulic permeability of the membrane, $m^2\ s^{-1}Pa^{-1}$
$\vec{D}$	electric displacement, C m^{-2}
D_i^{φ}	diffusion coefficient of component i in phase φ, $m^2\ s^{-1}$
D_i^{γ}	$\equiv \beta_i D_i$, diffusion coefficient of component i corrected for activity, $m^2\ s^{-1}$
$D_{i,k}$	cross-diffusion coefficient, $m^2\ s^{-1}$
$\overline{D}_{i,k}$	Stefan–Maxwell diffusion coefficient, $m^2\ s^{-1}$
$D_{I,K}$	Fickian diffusion coefficient, $m^2\ s^{-1}$
e	$\equiv e_{\mathrm{k}} + u$, total energy density, J m^{-3}
e_{k}	$\equiv \rho v^2/2$, translational kinetic energy density, J m^{-3}
E	electrode potential, V
E	electrophoretic enhancement factor, 1
$\vec{E}$	electric field intensity, V m^{-1}
$\vec{E}_{\mathrm{dif}}$	$\equiv \vec{E} - \vec{E}_{\mathrm{ohm}}$, diffusion electric field, V m^{-1}
$\vec{E}_{\mathrm{ohm}}$	$\equiv \vec{I}^m/\kappa$, ohmic electric field, V m^{-1}

f	$= F/RT$, auxiliary variable, V^{-1}
F	Faraday constant, $C\ mol^{-1}$
F	facilitation factor, 1
g	volume density of Gibbs potential, $J\ m^{-3}$
G	Gibbs potential or Gibbs free energy, J
G_k	$\equiv \sum_i z_i^k j_i/D_i$, auxiliary variable, $mol\ m^{-4}$
$\Delta_r G$	rate of change of the Gibbs free energy due to a transport process, $J\ m^{-2}\ s^{-1}$
$\Delta_o^w G_i^\circ$	$\equiv \mu_i^{\circ,w} - \mu_i^{\circ,o}$, standard Gibbs free energy of transfer of component i, $J\ mol^{-1}$
h	membrane thickness, m
i	electric current, A
$\vec{I}$	$\equiv F \sum_i z_i \vec{j}_i$, conduction electric current density (in the LRF if used without superscript), $A\ m^{-2}$
$\vec{I}_d$	$\equiv \partial\vec{D}/\partial t$, displacement electric current density, $A\ m^{-2}$
$I_{L,i}$	$\equiv -nFD_i c_i^b/\nu_i\delta$, limiting diffusion current density of component i, $A\ m^{-2}$
I_L^φ	limiting current density in phase φ, $A\ m^{-2}$
$\vec{I}_T$	$\equiv \vec{I}_d + \vec{I}$, total electric current density, $A\ m^{-2}$
$\vec{j}_b$	flux density of B (in the LRF if used without superscript), $[B]\ m^{-2}\ s^{-1}$
$\vec{j}_i$	$= c_i\vec{v}_i$, molar flux density of component i in the LRF, $mol\ m^{-2}\ s^{-1}$
$\vec{j}_i^m$	molar flux density of component i in the BRF, $mol\ m^{-2}\ s^{-1}$
$\vec{j}_\upsilon$	$= \vec{v}_\upsilon$, volume flux density in the LRF, $m\ s^{-1}$
$\vec{J}_K$	$= c_K\vec{v}_K$, molar flux density of neutral component K in the LRF, $mol\ m^{-2}\ s^{-1}$
$\vec{J}_K^H$	molar flux density of neutral component K in the HRF, $mol\ m^{-2}\ s^{-1}$
k	reaction rate constant, $[k]$
K	thermodynamic equilibrium constant (of a homogeneous reaction), $[K]$
K_i	equilibrium partition constant of component i, 1
$K_{i,k}$	Stefan–Maxwell friction coefficient, $N\ s\ m^{-4}$
$l_{i,k}$	ionic phenomenological transport coefficient, $mol^2\ J^{-1}\ m^{-1}\ s^{-1}$
L_d	diffusion length, m
$L_{I,K}$	component phenomenological transport coefficient, $mol^2\ J^{-1}\ m^{-1}\ s^{-1}$
m_i	mass of component i, kg
M_i	molar mass of component i, $kg\ mol^{-1}$
n	stoichiometric number of the electron in an electrode reaction, 1
n_i	amount of matter of component i, mol
N	number of components, 1
N_A	Avogadro's constant, mol^{-1}
p	mechanical pressure, Pa

P_i^φ	permeability of component i in phase φ, m s^{-1}
Q	cumulative flux, mol m^2
r	radial position co-ordinate, m
r	(electrode) reaction rate, mol m^{-2} s^{-1}
r	auxiliary variable used for different ratios, 1
$\vec{r}$	position, m
r_i	$\equiv c_i^\alpha / c_i^\beta$, concentration ratio of component i, 1
R	universal gas constant, J K^{-1} mol^{-1}
R	electrical resistance, Ω m^2
$\vec{R}$	position along a fluid particle trajectory, m
s	volume density of entropy, J K^{-1} m^{-3}
s	variable in Laplace domain, s^{-1}
S	entropy, J K^{-1}
S	membrane permselectivity, 1
S_k	$\equiv \sum_i z_i^k c_i$, auxiliary concentration variable, mol m^{-3}
t_i	transport number of ionic species i, 1
T	temperature, K
T_i	$\equiv z_i F j_i / I$, integral transport number of ionic species i, 1
u	volume density of internal energy, J m^{-3}
u_i	mobility of component i, m mol $N^{-1}s^{-1}$
U	internal energy, J
v	solution velocity, m s^{-1}
$\vec{v}$	$\equiv \sum_i w_i \vec{v}_i$, barycentric velocity in the LRF, m s^{-1}
$\vec{v}_i$	velocity of component i in the LRF, m s^{-1}
$\vec{v}_\upsilon$	$\equiv \sum_i c_i \upsilon_i \vec{v}_i$, volume-average velocity in the LRF, m s^{-1}
V^φ	volume of phase φ, m^3
$V(x)$	electric potential component, V
$\dot{V}$	volume flow rate, m^3 s^{-1}
w_i	mass fraction of component i, 1
x	Cartesian position co-ordinate, m
X	$\equiv z_M c_M / z_2$, membrane fixed-charge concentration, mol m^{-3}
y	Cartesian position co-ordinate, m
z	Cartesian position co-ordinate, m
z_i	charge number of ionic species i, 1

Greek symbols

α	dissociation degree, 1
β_i	activity correction factor for the diffusion coefficient of ionic species i, 1
δ	diffusion boundary layer thickness, m
δ_{ik}	$= 1$ if $i = k$, $= 0$ if $i \neq k$, Kronecker delta, 1
ϕ	electric potential, V
$\Delta_\alpha^\beta \phi$	$\equiv \phi^\beta - \phi^\alpha$, potential difference between bulk phases, V
$\Delta_o^w \phi_i^\circ$	$\equiv -\Delta_o^w G_i^\circ / z_i F$, standard transfer potential of component i, V
φ	$\equiv z_2 f[\phi(x) - \phi^w]$, dimensionless electric potential, 1

$\overleftrightarrow{\gamma}'$	viscous deformation rate tensor, s^{-1}
γ_i^{φ}	molar activity coefficient of component i in phase φ, 1
Γ	auxiliary variable, 1
$\Gamma(n)$	$\equiv \int_0^\infty e^{-t}t^{n-1}dt$, gamma function of argument n, 1
$\Gamma(n,x)$	$\equiv \int_x^\infty e^{-t}t^{n-1}dt$, incomplete gamma function of argument n, 1
η	dynamic viscosity, Pa s
κ^{φ}	electrical conductivity in phase φ, Ω^{-1} m^{-1}
κ_D^{φ}	Debye parameter (or reciprocal Debye length) in phase φ, m^{-1}
κ_i	contribution of ionic species i to the electrical conductivity, Ω^{-1} m^{-1}
κ_T	thermal conductivity, W K^{-1} m^{-1}
λ_i	molar electrical conductivity of ionic species i, $m^2\Omega^{-1}$ mol^{-1}
μ_i	chemical potential of component i, J mol^{-1}
$\tilde{\mu}_i$	electrochemical potential of component i, J mol^{-1}
ν	kinematic viscosity, m^2 s^{-1}
ν_i	stoichiometric coefficient of ionic species i, 1
$\nu_{i,r}$	stoichiometric coefficient of component i in reaction r, 1
$\nu_{i,K}$	stoichiometric coefficient of component i in component K, 1
π	osmotic pressure, Pa
π_b	volume density of production rate of B, $[B]$ m^{-3} s^{-1}
π_e	volume density of production rate of total energy, Jm^{-3} s^{-1}
π_i	volume density of production rate of amount of component i, mol m^{-3} s^{-1}
π_s	volume density of production rate of entropy, J K^{-1} m^{-3} s^{-1}
$\vec{\pi}_{\vec{v}}$	volume density of external force, Pa m^{-1}
π_{w_i}	volume density of production rate of mass of component i, kg m^{-3} s^{-1}
θ	$\equiv T\pi_s$, dissipation function, J m^{-3} s^{-1}
θ	auxiliary variable (used for different concentration fractions), 1
θ_{ch}	contribution of homogeneous chemical reactions to the dissipation function, J m^{-3} s^{-1}
θ_{dif}	contribution of chemical diffusion to the dissipation function, J m^{-3} s^{-1}
θ_{ed}	$= \theta_{dif} + \theta_{ohm}$, contribution of electrodiffusion to the dissipation function, J m^{-3} s^{-1}
θ_η	contribution of viscous flow to the dissipation function, J m^{-3} s^{-1}
θ_{ohm}	contribution of electric conduction to the dissipation function, J m^{-3} s^{-1}
ρ	mass density, kg m^{-3}
ρ_i	$\equiv \rho w_i$, mass density of component i, kg m^{-3}
ρ_e	$\equiv F\sum_i z_i c_i$, electric charge density, C m^{-3}
σ	membrane surface-charge concentration, C m^{-2}
$\overleftrightarrow{\sigma}$	stress tensor, Pa
$\overleftrightarrow{\sigma}'$	viscous stress tensor, Pa
τ	transition or relaxation time, s

υ_i	partial molar volume of component i, $m^3\ mol^{-1}$
ω	angular rotation frequency, rad s^{-1}
ξ	dimensionless position variable, 1
ξ_r	molar reaction co-ordinate, mol m^{-3}
$\psi(r,x)$,	electric potential component, V
ζ	$\equiv x/(2\sqrt{D_1 t})$, Boltzmann variable, 1

Subscripts and superscripts

0	solvent
1	electroactive species (in electrode processes)
2	co-ion (in membrane processes)
3	common ion (in ternary systems)
12	electrolyte
12,d	dissociated electrolyte
12,T	total electrolyte
12,u	undissociated electrolyte
13	electrolyte
23	electrolyte
c	chemical contribution
cell	cell (potential)
ch	chemical reaction
chem dif	chemical diffusion
D	Donnan
D	Debye
e	electrostatic contribution
e	energy
ed	electrodiffusion
eff	effective value
dif	(chemical) diffusion
dif	diffusion (potential)
H	Hittorf
i	charged species
ion dif	ionic diffusion
I	neutral component
j	spatial direction
k	spatial direction
K	neutral component
L	limiting value
m	barycentric
mig	ionic migration
M	membrane
o	organic phase
ohm	ohmic or electric conduction
T	total
υ	Fick
w	aqueous phase
$\pm$	thermodynamic mean value

Mathematical symbols

$\equiv$	definition
$\langle\,\rangle$	average value (in radial direction or over the membrane volume)
$-$	average value (in axial direction or average of values in bulk solutions or over the membrane system)
$\sim$	Laplace transformed variable (except in $\tilde{\mu}$ and $\tilde{\psi}$)
$\sim$	deviation from average value
δ	change in a variable when the system undergoes an infinitesimal process
d	magnitudes referred to a volume element
Δ	increment
$\mathcal{L}$	Laplace transform operator
$\mathcal{L}^{-1}$	inverse Laplace transform operator
$\vec{\nabla}$	gradient operator, m^{-1}
$\frac{\mathrm{D}}{\mathrm{D}t}$	$\equiv \vec{v} \cdot \vec{\nabla} + \left(\frac{\partial}{\partial t}\right)_{\vec{r}}$, material or substantial time derivative, s^{-1}

Index